International Conference on Advances in Metallurgy, Materials and Manufacturing (ICAMMM-2017)

IOP Conference Series: Materials Science and Engineering
Volume 314

Salem, India
6 - 8 March 2017

Editors:

P.G. Venkatakrishnan
P. Sundararaj
V. Karthik

ISBN: 978-1-5108-5926-5
ISSN: 1757-8981

Printed from e-media with permission by:

Curran Associates, Inc.
57 Morehouse Lane
Red Hook, NY 12571

Printed by Curran Associates, Inc. (2018)

For permission requests, please contact the Institute of Physics at the address below.

Institute of Physics
Dirac House, Temple Back
Bristol BS1 6BE UK

Phone: 44 1 17 929 7481
Fax: 44 1 17 920 0979

techtracking@iop.org

Additional copies of this publication are available from:

Curran Associates, Inc.
57 Morehouse Lane
Red Hook, NY 12571 USA
Phone: 845-758-0400
Fax: 845-758-2633
Email: curran@proceedings.com
Web: www.proceedings.com

Table of contents

Volume 314

International Conference on Advances in Metallurgy, Materials and Manufacturing

6–8 March 2017, Salem, Tamilnadu, India

Accepted papers received: 7 February 2018
Published online: 1 March 2018

Preface

Papers

IOP Conf. Series: Materials Science and Engineering **314** (2018) 011001 doi:10.1088/1757-899X/314/1/011001

TEQIP II SPONSORED

INTERNATIONAL CONFERENCE
ON
ADVANCES IN METALLURGY, MATERIALS AND MANUFACTURING

March 6-8, 2017

at

CJ PALLAZZIO, SALEM

ORGANISED BY

DEPARTMENT OF METALLURGICAL ENGINEERING
GOVERNMENT COLLEGE OF ENGINEERING, SALEM

In association, with

INDIAN INSTITUTE OF METALS, SALEM CHAPTER

IOP Conf. Series: Materials Science and Engineering **314** (2018) 011001 doi:10.1088/1757-899X/314/1/011001

Editorial Board

IOP Conf. Series: Materials Science and Engineering **314** (2018) 011001 doi:10.1088/1757-899X/314/1/011001

PREFACE

The 1st International Conference on Advances in Metallurgy, Materials and Manufacturing (ICAMMM-2017) was organized by Department of Metallurgical Engineering, Government College of Engineering, Salem, India and sponsored by Technical Quality Improvement Program (TEQIP) Phase-II during 6-8th March, 2017. This conference was structured with an objective to bring the experts and growing researchers together on a common podium to confer and exchange ideas on current developments, findings and progresses on broad areas associated to materials, metallurgy and manufacturing engineering. This conference provided an excellent forum for industry institute interaction to promote collaborative research activities involving students, engineers, professionals, researchers and scientists from India and overseas. The conference papers were invited in three major themes, namely Metallurgy, Materials and Manufacturing. The received papers were subjected to peer review process. Each paper was reviewed by at least two reviewers. Out of 164 papers received, 31 papers were selected for publishing in IOP Conference Series: Materials Science and Engineering.

The organizing committee would like to thank all the reviewers and authors for their valuable contributions to this ICAMMM 2017 conference.

Organizing Committee

Convener	:	Dr. P.G. Venkatakrishnan
Organizing Secretaries	:	Prof. P. Sundararaj
		Dr. V. Karthik
Organizing Committee	:	Prof. D. Noorullah
		Prof. N. Thenammai
		Dr. K. Venkatesan
		Prof. B. Anandavel
		Prof. M. Deepak Kumar
		Prof. K.T. Thilagam
		Prof. S. Dhanalakshmi

IOP Conf. Series: Materials Science and Engineering **314** (2018) 011001 doi:10.1088/1757-899X/314/1/011001

ICAMMM 2017
Programme Schedule

DAY 1: 6th March, 2017 - Monday		
Time	**Program**	**Speaker/Title**
9:30 AM – 10:30 AM	Inaugural Function	**Chief Guest & Guests of Honour** Dr. U. Balachandran, Argonne National Laboratory, USA Shri. D. Ravichandar, CEO, JSW Steel Ltd., Salem Shri. R. Sundaram, CEO, The Salem Aeropark, Salem
Tea Break		
11: 00 AM – 11.50 AM	Plenary Lecture 1	Title: **Materials for Sustainable Energy Systems** Speaker: **Dr. U. (Balu) Balachandran**, Argonne National Laboratory, Argonne.
11:50 AM – 12:40 PM	Plenary Lecture 2	Title: **Challenges and Opportunities in Integrated Computational Materials Engineering for Metallurgical Engineers** Speaker: **Prof. Gandham Phanikumar**, Professor, Department of Metallurgical and Materials Engineering, IIT Madras.
Lunch Break		
2:00 PM – 2:50 PM	Plenary Lecture 3	Title: **Sustainability, the key to future product development** Speaker: **Prof. Satish Vasu Kailas**, Professor, Surface Interaction and Manufacturing Laboratory, Department of Mechanical Engineering, Indian Institute of Science, Bangalore.
Tea Break		
3:15 PM – 5:00 PM	Technical Session I (3 parallel sessions)	**Section A:** Keynote Lecture: **Dr. G. Balachandran**, VP (R&D), JSW Steel Ltd., Salem Title: **Challenges in Special Steel Making.**
		Section B: Keynote Lecture: **Dr. Alphons A. Antonysamy**, Principal, Materials Scientist-Additive Manufacturing, GKN Aerospace, United Kingdom. Title: **Additive Manufacturing of Ti6Al4V for Aerospace Applications.**
		Section C: Keynote Lecture: **Dr. A. Murugaiyan,** Department of Metallurgical and Materials Engineering, IIT Madras. Title: **Hot Cracking during Welding of Automotive Steels.**

Note: Technical sessions will be held under three themes namely 1. Metallurgy, 2. Materials and 3. Manufacturing

IOP Conf. Series: Materials Science and Engineering **314** (2018) 011001 doi:10.1088/1757-899X/314/1/011001

DAY 2: 7th March, 2017 - Tuesday		
Time	**Program**	**Speaker/Title**
9:30 AM – 10:10 AM	Plenary Lecture 4	Title: **Quantitative in situ deformation in the transmission electron microscope.** Speaker: **Dr. Christoph Gammer**, Scientist, Erich Schmid Institute of Materials Science, Leoben, Austria
10:10 AM – 10:45 AM	Plenary Lecture 5	Title: **Modern usage of High Strength Steels** Speaker: **Dr. V. Ramasamy**, Former Executive Director, SAIL (Salem Steel Plant & RDCIS, Ranchi).
	Tea Break	
11:00 AM – 12:45 PM	Technical Session II (3 parallel sessions)	**Section A**: Keynote Lecture: **Dr. S. Anand**, McMaster University, Canada.
		Section B: Keynote Lecture: **Dr. Ing. K.G. Prashanth,** Erich Schmid Institute of Materials Science, Leoben, Austria.
		Section C: Keynote Lecture: **Dr. N. Suresh**, Assistant Professor, Materials group, Department of Applied Mechanics, IIT Delhi
	Lunch Break	
2:00 PM – 2:30 PM	Plenary Lecture 6	Title: **Quality Assurance of Materials for Aerospace applications – Challenges.** Speaker: **Shri. M. Arumugam,** Scientist- G & Group Head, Quality Control and Inspection Group, Liquid Propulsion Systems Centre, ISRO, Trivandrum.
2:30 PM – 3:00 PM	Plenary Lecture 7	Title:**Characterization of novel materials - their synthesis and manufacturing** Speaker: **Dr.P.Parameswaran,** Head, XRDSES, Physical metallurgy Division, , IGCAR, Kalpakkam.
	Tea Break	
3:15 PM – 5:00 PM	Technical Session III (3 parallel sessions)	**Section A:** Keynote Lecture: **Dr. M. Ravi**, Senior Principal Scientist, NIIST-CSIR, Trivandrum.
		Section B: Keynote Lecture: **Dr. D. Nagarajan,** Assistant Professor, Dept. of Metallurgical and Materials Engineering, NIT Rourkela.
		Section C Keynote Lecture: **Dr. I. Balasundar**, Scientist, Aeronautical Materials Division, DMRL, Hyderabad.

Note: Technical sessions will be held under three themes namely 1. Metallurgy, 2. Materials and 3. Manufacturing

IOP Conf. Series: Materials Science and Engineering **314** (2018) 011001 doi:10.1088/1757-899X/314/1/011001

DAY 3: 8th March, 2017 - Wednesday		
9:30 AM – 10:15 AM	Plenary Lecture 8	Title: **In-situ synchrotron experiments as a tool to probe defect mediated plasticity in nanostructured materials.** Speaker: **Mag. Dr. Florian Spieckermann,** Senior Scientist, Montanuniversitat Leoben, Austria.
10:15 AM – 11:00 AM	Plenary Lecture 9	Title: **Degradable magnesium based Implants: Opportunities and Challenges.** Speaker: **Dr. T.S.N. Sankara Narayanan,** Research Professor, School of Dentistry, Chonbuk National University, Jeonju 561-756, South Korea.
Tea Break		
11:30 AM – 12: 45 PM	Conference Workshops & Poster Session	Workshop 1: **Computational Thermodynamics.** **Dr. K. R. Ravi**, Associate Professor, PSG Institute of Advanced Studies, Coimbatore. Workshop 2: **Advances in Welding Processes and Power Sources.** **Shri. P. Sasi Anand**, Zonal Manager South, Fronius India Private Limited, Pune.
Lunch Break		
2:00 PM – 3:30 PM	Technical Session IV (3 parallel sessions)	**Section A**: Keynote Lecture:**Shri. T. Muthukumar,** DGM (Quality), SAIL, Salem Steel Plant, Salem.
		Section B: Keynote Lecture: **Dr. Ajoy Kumar Pandey**, Assistant Professor, Department of Metallurgical and Materials Engineering, NIT Warangal.
		Section C: Keynote Lecture: **Shri. Prasan Swaminathan,** Management Consultant, Chennai.
Tea Break		
4:00 PM – 4:30 PM	Valedictory Function	

Note: Technical sessions will be held under three themes namely 1. Metallurgy, 2. Materials and 3. Manufacturing

IOP Conf. Series: Materials Science and Engineering **314** (2018) 011001 doi:10.1088/1757-899X/314/1/011001

Peer review statement

All papers published in this volume of *IOP Conference Series: Materials Science and Engineering* have been peer reviewed through processes administered by the proceedings Editors. Reviews were conducted by expert referees to the professional and scientific standards expected of a proceedings journal published by IOP Publishing.

IOP Conf. Series: Materials Science and Engineering **314** (2018) 012001 doi:10.1088/1757-899X/314/1/012001

Distortion control in 20MnCr5 bevel gears after liquid nitriding process to maintain precision dimensions

M Mahendiran[1] and M Kavitha[2]

[1]Eppinger Tooling Asia, Coimbatore, Tamilnadu 642 109, India
[2]Department of Metallurgical Engineering, PSG College of Technology, Coimbatore, Tamilnadu 641 004, India

ABSTRACT. Robotic and automotive gears are generally very high precision components with limitations in tolerances. Bevel gears are more widely used and dimensionally very close tolerance components that need stability without any backlash or distortion for smooth and trouble free functions. Nitriding is carried out to enhance wear resistance of the surface. The aim of this paper is to reduce the distortion in liquid nitriding process, though plasma nitriding is preferred for high precision components. Various trials were conducted to optimize the process parameters, considering pre dimensional setting for nominal nitriding layer growth. Surface cleaning, suitable fixtures and stress relieving operations were also done to optimize the process. Micro structural analysis and Vickers hardness testing were carried out for analyzing the phase changes, variation in surface hardness and case depth. CNC gear testing machine was used for determining the distortion level. The presence of white layer was found for about 10-15μm in the case depth of 250± 3.5μm showing an average surface hardness of 670 HV. Hence the economical liquid nitriding process was successfully used for producing high hardness and wear resistant coating over 20MnCr5 material with less distortion and reduced secondary grinding process for dimensional control.

1. Introduction

Bevel Gears are used in the Robotic and automotive gear boxes to transmit drive from a drive shaft to a driven shaft which requires a very close dimensional accuracy for high precision assembly. Secondary machining process is involved to maintain dimensional accuracy for a perfect assembly [1-5]. During secondary processes, grinding will also be employed, which makes production delay by consuming huge time and does not satisfy the requirement of close tolerances. To avoid the above mentioned problem, a new process to be developed for high precision gear part without further grinding in order to reduce the rejections and loss. Since bevel gears are exposed to high torque and high speed operations hardening process is inevitable, a strategic structure was formed to capitalize this process to maintain the required tolerance levels [6]. The durability of the gears is important for high torque operations hence alloy steels are most preferred [3,7]. 20MnCr5 alloy steels are preferred for the production of bevel gears [6]. Case-hardening or surface hardening process is carried for all types of gears. This process of hardening the surface of a metal object while allowing the metal deeper underneath to remain soft, thus forming a thin layer of harder metal at the surface. There are various types of case hardening process available to improve the hardness of the material they are flame and induction hardening, carburizing, cyaniding, carbo nitriding, ferritic nitro carburizing and nitriding.

Nitriding is a heat treating process that diffuses nitrogen into the surface of a metal to create a case-hardened surface this process involves with ferritic region in low temperature (550°C) which is below the transformation temperature [1,8,9]. These processes are commonly used on low-carbon, low-alloy steels. Nitriding is used in gear manufacturing units widely because of high dimension stability offered to the gears and no further secondary process requirement. Nitriding can be done as liquid, gas and Plasma method. 20MnCr5 is a case hardening steel having low carbon and low alloying elements which is widely used for high precision gears for the purpose of core toughness and surface wear resistivity through case hardening process. For gears lead, profile and run out parameters are important and hence examination of these is important. Surface hardness and case depth should be ensured for gears for reliable operation.

2. Material and Methods

The selected material 20MnCr5 alloy steel with the chemical composition given in Table 1, is exposed to liquid nitriding process to determine the distortion free feasible nitriding process parameter combinations. The results will lead to successful solution to avoid further grinding process after nitriding in bevel gears.

Table 1. Chemical Composition of 20MnCr5 steel

Elements	C	Si	Mn	P	S	Cr	Fe
Wt.%	0.17-0.22	0.40 max	1.10-1.40	0.025 max	0.035 max	1.00-1.30	Balance

Bevel gears are properly processed in CNC gear manufacturing machine maintained with proper pre dimensions determined towards nominal nitriding growth. Surface cleaning reagents are used to remove the surface impurities like oil, grease, corrosion products and dust particles. Pre heating process carried out at 350 °C for two hrs. Liquid nitriding process was conducted in several trials to optimize the process parameters like bath strength, temperature and soaking time. Post washing is done with green solutions Bio cleaning for the surfaces is followed by nitriding process. Hardness testing, microstructural analysis and case depth measurements were done by metallurgical microscopes. Future-tech image analyser (FM800) is used for dimensional measurements and GMS Gleason-350 CNC gear testing machine are used to observe the cumulative dimensional parameters like lead, profile and run out.

2.1. Nitriding

Bevel gears are processed as per the required dimensions (7th quality) and tolerances (±10μm) with 20MnCr5 steel. The dimensions of the final component has been inspected, ensured and were exposed to liquid nitriding with various trial combinations of temperatures from 540°C to 560°C for 2 to 4 hrs duration under the potentiality of 18 to 20 percent strength of liquid bath to obtain feasible results. 20MnCr5 material bevel gears are liquid nitrided under the three range of process combinations are given as, (i) constant temperature at 560°C and 18 percent potentiality with variable soaking time from 2 to 4 hrs durations, (ii) constant soaking time 3 hrs and bath strength 18 percent under variable temperature 540 °C to 560 °C, and (iii) the other trials done with variable potentiality from 18 to 20 percent strength with constant temperature 560 °C for 3 hrs constant soaking time. The liquid nitriding process was carried out in pit type salt bath nitriding furnace at Eppinger tooling Asia Private Limited with components proper loading and jigging in to the Furnace. The nitrided samples as shown in Figure 1 had been cleaned by bio cleaning system to remove the surface residues. The dimensional inspection has been carried out in computerized gear tester (Gleason 350 GMS) to identify the distortion due to nitriding. The observed dimensions and gear teeth parameters has been noted and compared with initial readings.

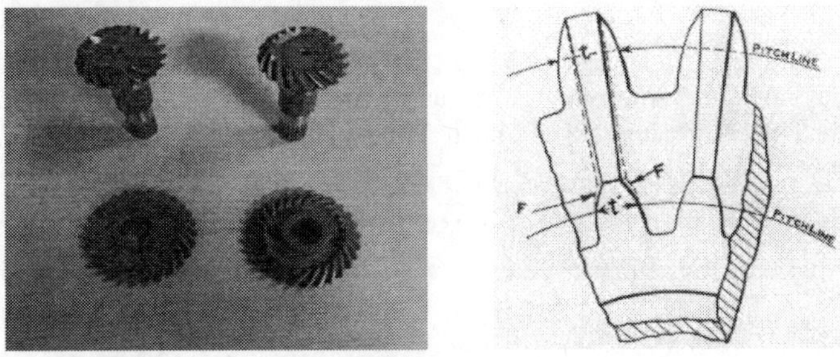

Figure 1. (a) 20MnCr5 bevel gear component after nitriding; (b) Bevel gear tooth profile

2.2. Testing and characterization

Future tech FM-800 model Micro Hardness tester was used to determine the hardness and case depth. Micro Hardness was determined using diamond indentor with a load of 500 gm with dwell time for 20 seconds. Case depth, white layer thickness, surface hardness and distortion level are characterized to ensure nitriding parameters. Nitrided gears with all the three process parameter combinations have been cut into small pieces and polished to get mirror like finish. The polished sample has been etched in 3% Nital for 15 seconds. Microscopic examination has been carried out in optical microscope to study the micro structure of all the nitrided samples. Case depth obtained in various nitriding processes have been analyzed recorded. The white layer measurement has been carried out since the white layer is the major reason for the dimensional distortion. Longitudinal deviation in dimension of gear tooth is called Lead error. Flank error - both left and right flank deviations are called profile error. These errors including profile errors are measured using Gleason 350 GMS gear testing machine.

3. Results and Discussion

Table 2 shows that the measured dimensions and distortion values observed in the nitride samples with varying combinations of critical nitriding parameters.

Table 2. Consolidated dimensional/distortion and metallurgical results of the nitride samples.

Process	Based on Temperature (2 hrs with 18% potentiality)			Based on Time (540°C with 18% potentiality)			Based on Bath Strength (560°C for 2hrs)		
	540°C	550°C	560°C	2 hrs	3 hrs	4 hrs	18%	19%	20%
Avg. Surf. Hardness ±10 (HV @ 0.5kgf)	530	595	783	535	770	789	770	786	850
Case Depth(0.25 +/- 0.10 mm)	0.05	0.14	0.20	0.08	0.22	0.29	0.19	0.23	0.32
White Layer (µm)	3.5	6	11.5	3.5	11	21	105	11.5	21
Distortion Level(mm)	0.0075	0.0075	0.0085	0.0030	0.0075	0.0480	0.0090	0.0090	0.0480
Lead (mm)	0.008	0.007	0.011	0.001	0.005	0.025	0.008	0.008	0.029
Profile (mm)	0.005	0.006	0.009	0.003	0.009	0.040	0.011	0.012	0.043
Run out (mm)	0.010	0.011	0.012	0.001	0.004	0.054	0.013	0.013	0.057

From the results shown in Figure 2, it is observed that the bevel gear is having less distortion in combination of temperature 560°C in 3 hrs soaking time under 18 percent bath strength over the other process parameters. The low temperature and lower soaking time nitriding is having moderate distortion than the above combination.

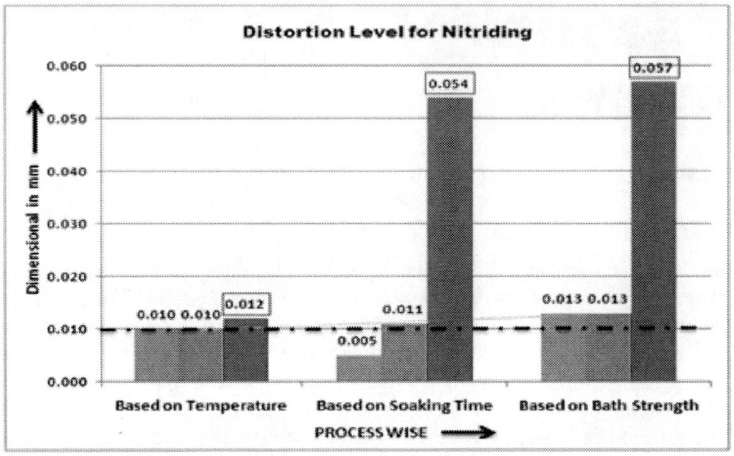

Figure 2. Comparison of Distortion level in different processes

From the results obtained as shown in Figure 3, it was observed that very high surface hardness and case depth was obtained in long soaking period than other nitriding processes, specified hardness was obtained in the combination of 560°C , 3 hrs soaking and 18 percent bath strength.. The low hardness observed in less soaking time.

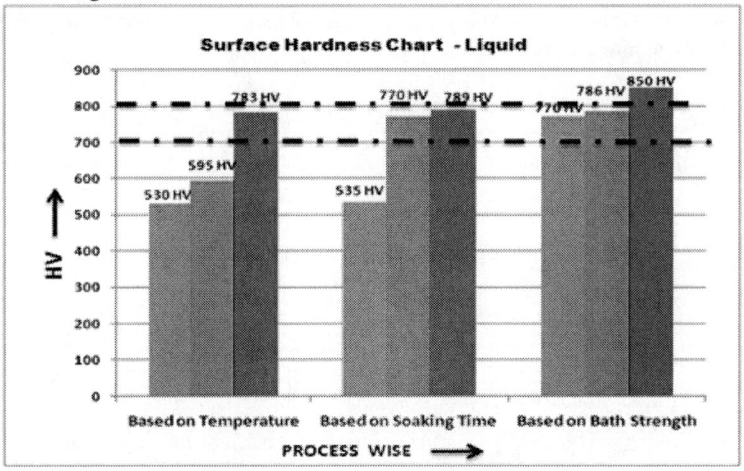

Figure 3. Surface hardness analysis

Microstructural analysis of the samples shown in Figure 4 reveals the martensite formation with retained austenite in all the samples. It is found that homogeneous distribution of martensite needles were observed in all the conditions/processes.

IOP Conf. Series: Materials Science and Engineering **314** (2018) 012001 doi:10.1088/1757-899X/314/1/012001

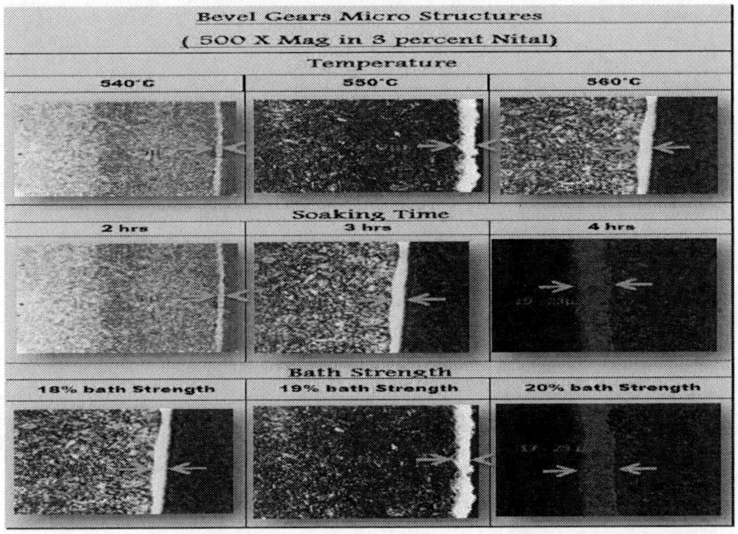

Figure 4. Microstructure of nitrided components with varying parmeters.

White layer is varied in the combination of temperature, soaking time and bath strength variations as shown in the figures 4, 5 and table 2.

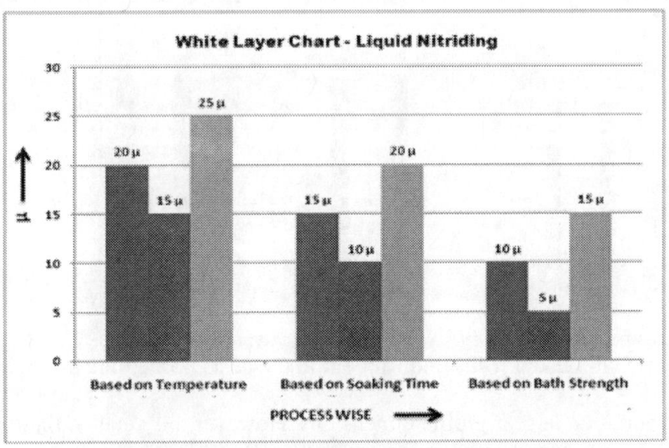

Figure 5. Analysis of white layer formation

The case depth, hardness and distortion in gear profiles are found very high in higher soaking time as well as higher bath strength combinations than the moderate nitriding process. Specified case depth obtained 560°C , 3 hrs soaking and 18 percent bath strength refer Figures 6 and 7.

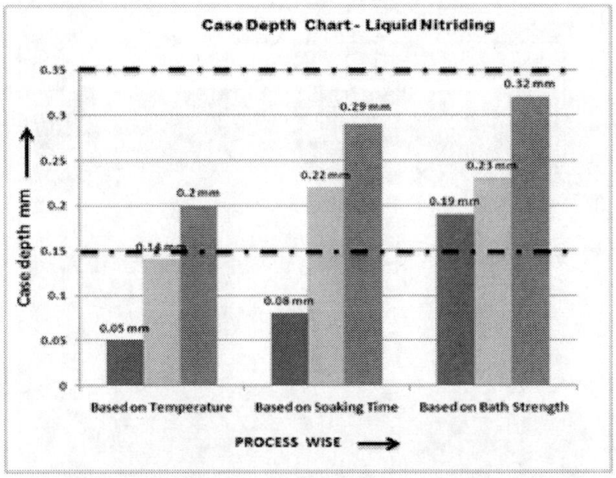

Figure 6. Case depths in various trials

As per AGMA Standard, the tooth profile errors like runout, Lead and profile error are to be measured. The eccentricity of the datum circle relative to the datum axis, wobble of the gear blank relative to the datum axis of rotation, tooth alignment variation are measured.

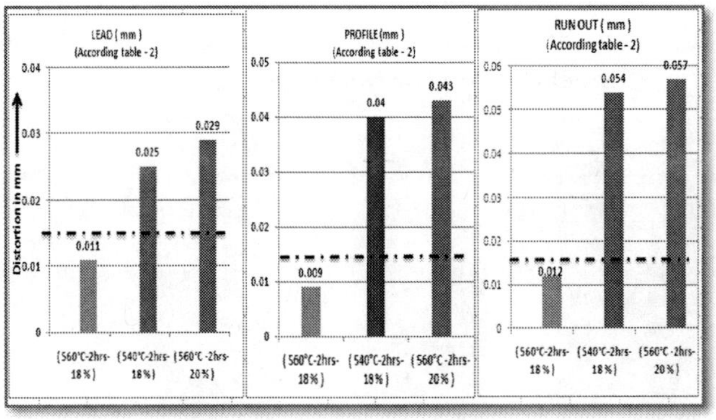

Figure 7. Variation in Lead, Profile and Run out in the gear profile dimensions

Above results show a small variation in gear profile dimensions. However this small variation could led to more wear and tear in the components during sliding each other. Hence it is necessary to analyze the effect of different process parameters. The surface nitrogen concentration and increase in time of contact increased diffusion efficiency and hardness along with nitrogen. Though the % carbon is relatively less than the required amount of 0.30% for complete transformation to martensite, presence of Mn and Cr promote the formation of matensite along with diffused Nitrogen.

4. Conclusions

Surface hardening of high precision automotive bevel gears made of 20MnCr5 alloy steel are carried out using liquid nitriding process with the combinations process parameters. Three different combinations of liquid nitriding process with temperature (540-560°C), soaking time (2-4hrs)and liquid bath strength from 18-20% potentiality was success fully done. Characterization of the microstructure and variation in the level of distortion are studied to eliminate further grinding process after nitriding. The 4 hrs soaking time and 20% bath strength are inducing the high distortion up to 57 microns than requirement of below 10 microns. Higher surface hardness of about 790 – 850 HV is observed in the above conditions. Lower soaking time and the bath strength have resulted in the surface hardness of about 480 – 535 HV as well as minimum distortion level. Formation of white layer (FeN compound layer) on the surfaces is observed too high in 4 hrs of soaking time at 560°C under 18 percent bath strength. Hence, it is observed that the liquid nitriding at 560°C under 18 percent liquid bath strength for 3 hrs soaking time is reliable process parameter combination to get better metallurgical properties with limited distortion level. The selected nitrided samples were assembled in the main assembly without further grinding process. Additionally, the pre-dimensional concept introduced before nitriding process considering the nitride layer growth of about 5–10 microns. It was also successful to control and maintain the bevel gear dimensional tolerances.

5. References

[1] Pye D, 2003 *Handbook on Practical Nitriding and Ferritic Nitrocarburizing* (Ohio: ASM International)

[2] Fukuoka T, Kikuchi S, Komotori J, Fukazawa K, Misaka Y, Kawasaki K 2012 *J. Japan Inst. of Metals* 7 **76** pp 422-428

[3] Haracic N, Catovic F 2011 *Proc. Int. Conf. on Trends in the Development of Machinery and Associates Technology* (Czech Republic) pp 196-200

[4] Hiraoka Y, Watanabe Y, Umezawa O 2016 *J. Japan Inst. of Metals* 4 **80** pp 253-258

[5] Somers MAJ, Mittemeijer EJ 2000 Proc. Int. Conf. on *Modeling the Kinetics of the Nitriding and Nitro carburizing of Iron* (ASM International) pp 15-18

[6] Lotze TH 2003 *J. Application Bulletin*, (USA: Super Systems Inc.), pp 1-4

[7] Widi KA, Wardana ING, Sujana W 2013 *J. Materials, Mechanics and Manufacturing* 2 **1** pp 114-116

[8] Pye D 2013 Nitriding and Nitrocarburizing *Encyclopedia of Tribology* pp 2421-2428 (*Preprint* 10.1007/978-0-387-92897-5_1191)

[9] Funatani K 2004 *J. Metal Science and Heat Treatment* 7 **46** pp 277-281

Surface buffing and its effect on chloride induced SCC of 304L austenitic stainless steel

Pandu Sunil kumar[1], Swati Ghosh Acharyya[1*], S.V. Ramana Rao[2], Komal Kapoor[2]

[1] *School of Engineering Sciences and Technology, University of Hyderabad, Hyderabad-500046, India*

[2] *Nuclear Fuel Complex, Department of Atomic Energy, Government of India, Hyderabad-500062, India*

*Email: swati364@gmail.com

Abstract: The study focuses on the impact of buffing operation on the stress corrosion cracking (SCC) susceptibility of 304L austenitic stainless steel (SS). The SCC susceptibility of the buffed surfaces were determined by testing in boiling magnesium chloride ($MgCl_2$) environment as per ASTM G 36. Test was conducted for 3hr, 9hr and 72hr to study the SCC susceptibility. Buffed surfaces were resistant to SCC even after 72hr of exposure to boiling $MgCl_2$. The surface and cross section of the samples were examined for both before and after exposure to boiling $MgCl_2$ and was characterized using optical microscopy. The study revealed that buffing operation induces compressive residual stresses on the surface, which helps in protecting the surface from SCC.

1. Introduction.

Austenitic stainless steels (SS) are used in various industrial applications, owing to several advantages like excellent mechanical properties and good corrosion resistance **[1]**. AISI 304L stainless steels being metastable in nature are highly susceptible to martensitic transformation by the application of plastic deformation or by cooling to a temperature below 0°C **[2]**. The transformation of austenite to martensite phase during mechanical and thermo-mechanical treatment of AISI 304L SS has been studied **[3–5]**. The formation of strain-induced martensite not only results in high hardness near the surface but also results in increased dissolution of ions in presence of chloride environment resulting in early initiation of SCC [**6**]. Surface working has been shown to result in more damage when compared to bulk deformation of 304L SS **[6-9]**. However, surface working operations are indispensible in the industrial fabrication of components. The literature establishes the deleterious effects of surface working operations on the SCC susceptibility of 304L SS. However, a solution to the problem of SCC is absent in the literature. The present study aims to use one of the commonly used surface working operations, namely buffing for modifying the surface residual stress distribution of SS. Surface buffing is a conventional finishing process of a metal surface to give the desired finish. The purpose of buffing is to improve the surface appearance of a metal and produce a smooth tight surface **[10]**. The present study has investigated the effect of surface buffing on SCC susceptibility of 304L austenitic SS.

IOP Conf. Series: Materials Science and Engineering **314** (2018) 012002 doi:10.1088/1757-899X/314/1/012002

2. Experimental techniques

2.1. Material and methods

AISI grade 304L austenitic SS was used in this study with a chemical composition of (wt. %): 0.03 C, 18 Cr, 8 Ni, 1.6 Mn, 0.04 P, 0.4 Si, 0.013 S and balance Fe. The material was sectioned into 100 mm x 25 mm x 5 mm. These samples were solution annealed at 1025°C for 15 min. Subsequently, the material was subjected to buffing operation using 240, 400, and 600 grit paper at a fixed rotating speed of 3600 rpm which resulted in thickness reduction of 50 μm. The surface roughness was measured using surface profilometer (contact mode) and Vickers hardness measurements were taken along cross-section for surface buffed and As-received samples. X-ray diffraction studies were conducted to confirm the phases present in the material. Residual stresses distribution of the surface was determined using X ray diffraction technique.

2.2. Stress corrosion cracking tests

The stress corrosion cracking (SCC) susceptibility of 304L SS was determined by boiling $Mgcl_2$ testing as per ASTM G 36 [**11**]. Strips of the surface buffed samples (dimensions 10mm x 10mm x 5mm) were exposed to boiling $MgCl_2$ solution maintained at 155°C ±1°C in a Erlenmeyer flask with a water cooled condenser and a provision of a thermometer. The condenser prevents the loss of vapor during boiling. Care has been taken to maintain the temperature inside the flask. List of apparatus required for the SCC setup are listed in Figure1 are 1) Erlenmeyer flask.2) Water-cooled condenser. 3) water Inlet/outlet pipe. 4) Thermometer (250°c). 5) Hot plate. The test has been conducted for 3hr, 9hr and 72hr. Followed by detailed characterization of the nature and morphology of the cracks.

Figure1: Stress corrosion cracking (SCC) test setup [ASTM G36-94]

3. Results:

3.1 Surface roughness and Hardness measurements

The surface roughness of the As-received and surface buffed samples were measured using surface profilometer (contact mode) and micro-vickers hardness was measured across the thickness of the samples using 100gf load and dwell time of 10 s, for each case two sample were taken and three readings were taken. Hardness profile is plotted in Figure 2 and Results are tabulated in table 1.

Table 1: Hardness measurements and surface roughness readings

SAMPLE 304L stainless steel	Average Vickers Hardness (HV)	Average Surface roughness (Ra)
As-received 304L SS	170±14	0.5μm
Surface buffed (SB)	197±18	0.05μm

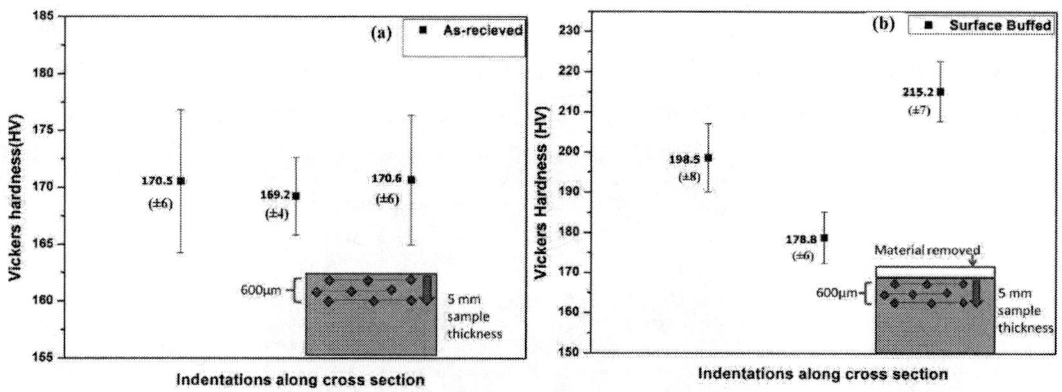

Figure 2: Hardness profile of SS 304L SS a) As-received and b) Surface buffed condition.

3.2. X-ray diffraction studies

XRD technique was used to determine the phases present by using Cu-Kα radiation, wavelength of 1.54Å. Accelerating voltage 40 V, diffraction angle from 40°-100° and step size of 0.02° was used. Both Figure 3(a) and Figure 3(b) showed the typical austenite peaks and additional peaks of strain induced martensite was evident in buffed samples.

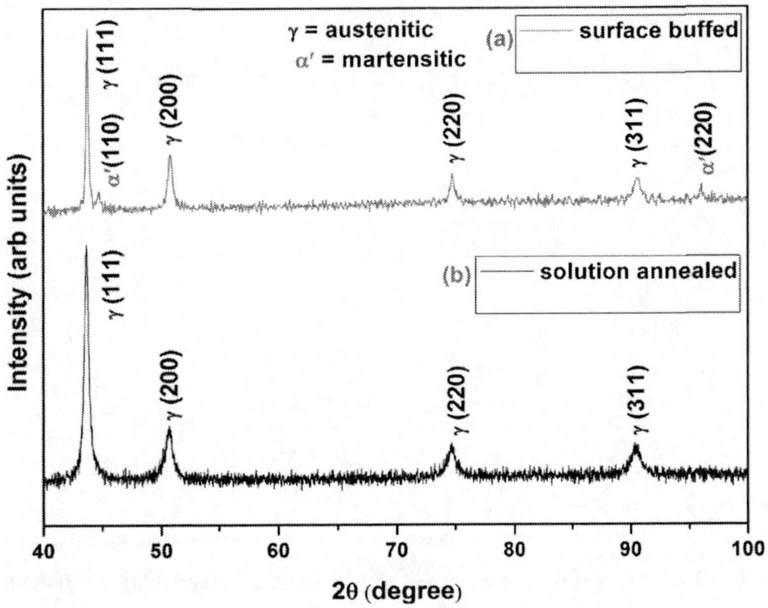

Figure 3: X-ray diffraction patterns for 304L SS in a) Surface buffed and b) Solution annealed condition

3.3. Residual stress measurements

Residual stresses present on the surface for as-received stainless steel 304L samples and surface buffed samples was determined using X-ray diffraction technique [**12**] with Cr-Kα as X-ray radiation source with an applied voltage of 27 kV and current of 70 mA. The wavelength used was 2.28 Å. The austenitic peak (311) plane was considered for diffraction with a step size of 0.1°, Bragg angle (2θ = 147.6°) was kept constant. The collimeter diameter was set to 4mm to measure the surface residual stresses with an exposure time of 20 S. X-ray penetration was around 10μm. From Figure 4, residual stress values are calculated and tabulated in Table 2.

Table 2: Residual stress values of as-received 304L SS and surface buffed operations

SAMPLE ID 304L stainless steel	Longitudinal direction (0 degree) MPa	Transverse direction (90 degree) MPa
As-received 304L SS	-298±26	-152±10
Surface buffed (SB)	-518±28	-481±14

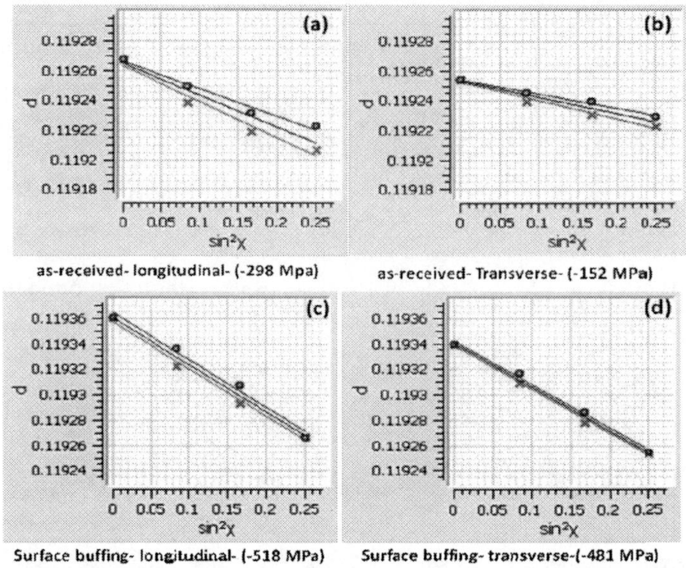

Figure 4: Residual stresses plot of d vs. $\sin^2\chi$ of (a-b) as-received 304L (c-d) surface buffed.

3.4 Stress corrosion cracking (SCC)

Figure: 5 (a-c) shows the optical micrographs of surface buffed samples after exposure in boiling $MgCl_2$ solution for 3hr, 9hr and 72hr respectively. Figure: 6(a-c) shows the cross-sectional micrographs along the thickness of buffed samples for 3hr, 9hr and 72hr respectively. Buffed samples did not show any signs of cracking. However on prolonged exposure to chloride environment these samples showed intergranular corrosion which is expected. However, no SCC occurred in the case of buffed samples even after 72hr exposure to boiling $MgCl_2$ in cross-sectional images. The SCC behavior of the machined surfaces could be correlated to the residual stress distribution on the surfaces as discussed in Section 3.3. Buffed samples had compressive stresses on the surface which protected these from SCC. The increased SCC resistance of the 304L SS samples subjected to buffing operation would have extensive application for major industrial components facing severe corrosive environments. If the surface after machining is subjected to buffing operation prior to its industrial application, the nature of residual stress distribution on the surface can be modified and the surface would be made more resistant to SCC. Hence buffing operation which would directly lead to an increase in the service life of the component.

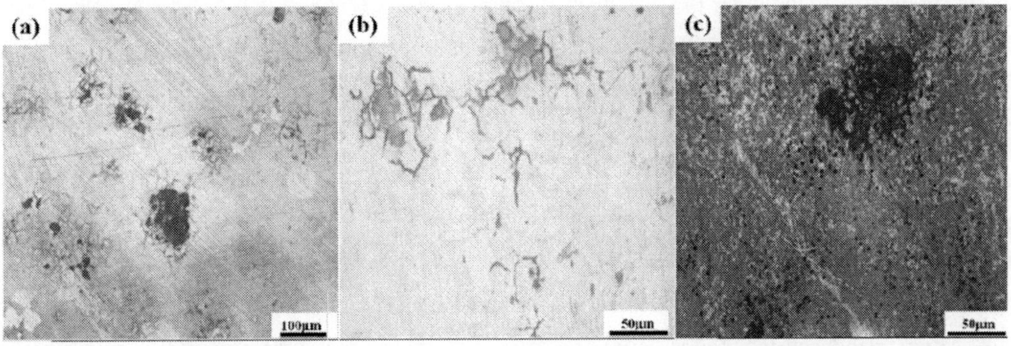

Figure 5: optical micrographs of the surface of 304L SS in buffed condition after exposure to boiling $MgCl_2$ a) 3 hr, b) 9hr, and c) 72hr.

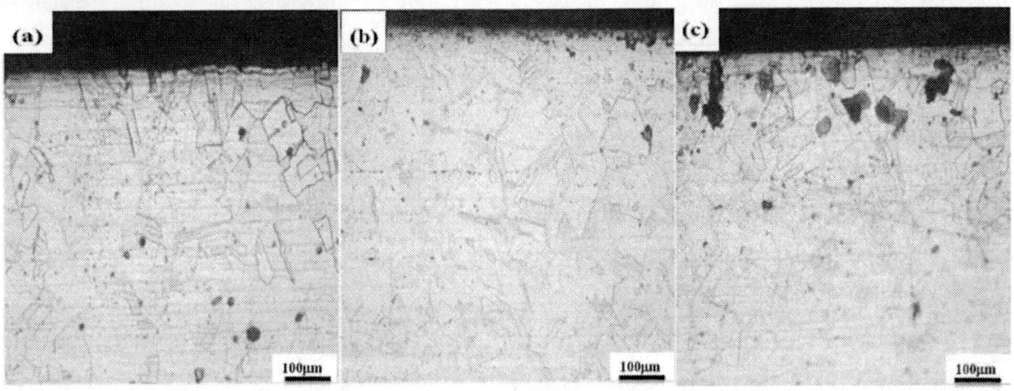

Figure 6: cross sectional micrographs of buffed 304L SS after exposure to boiling $MgCl_2$ a) 3 hr, b) 9hr, and c) 72hr.

Conclusions:

The effect of surface buffing on chloride-induced SCC of austenitic SS 304L has been investigated. The results obtained from the study are summarized as follows:

(a) Buffing induced high compressive residual stresses on the surface of the 304 L SS
(b) Buffed samples did not undergo SCC even after 72hr of exposure to boiling $MgCl_2$.
(c) Buffing operation made 304L SS more resistant to SCC in an aggressive chloride environment.

References:

[1] Lo KH, Shek CH, Lai JKL. Recent developments in stainless steels. Mater. Sci. Eng. R 2009; 65:39–104.

[2] Bayerlein M, Mughrabi H, Kesten, Meier B. Improvement of the strength of a metastable austeniti stainless steel by cyclic deformation-induced martensitic transformation at 103 K. Mater. Sci.Eng. A 1992; 159:35–41.

[3] Maxwell PC, Goldberg A, Shyne JC. Stress-assisted and strain-induced martensites in FE–NI–C alloys. Metall Trans 1974; 5:1305–24.

[4] Manganon Jr PL, Thomas G. Structure and properties of thermal-mechanically treated 304 stainless steel. Metall Trans 1970; 1:1587–94.

[5] Hecker SS, Stout MG, Staudhammer KP, Smith JL. Effect of strain state and strain rate on deformation induced transformation in 304 stainless steel: magnetic measurements and mechanical behavior. Metall Trans 1982; 13A:619– 26.

[6] S. Ghosh, V.Kain, Microstructural changes in AISI 304L stainless steel due to surface machining: Effect on its susceptibility to chloride stress corrosion cracking, J.Nucl.Mater.403 (2010) 62–67.

[7] T. Shoji, Progress in mechanistic understanding of BWR SCC and its implication to prediction of SCC growth behavior in plants, in 11th International Symposium on Environmental Degradation of Materials in Nuclear Power Systems-Water Reactors, Stevenson, WA, 2003.

[8] G.G. Scatigno, M.P.Ryan, F.Giuliani, M.R.Wenman, The effect of prior cold work on the chloride stress corrosion cracking of 304L austenitic stainless steel under atmospheric conditions, J.Mater.Sci.Eng. A 668(2016)20–29.

[9] D.T. Spencer, M.R. Edwards, M.R. Wenman, C. Tsitsios, G.G. Scatigno, P.R. Chard-Tuckey, The initiation and propagation of chloride-induced transgranular stress-corrosion cracking (TGSCC) of 304L austenitic stainless steel under atmospheric conditions, Corros. Sci. 88 (2014) 76–88.

[10] Al Dickman, Mechanical surface preparation- polishing and buffing (1999).

[11] ASTM, G36-94, Standard Practice for Evaluating Stress-Corrosion-Cracking Resistance of Metals And Alloys in a Boiling Magnesium Chloride Solution, Reapproved 2006.

[12] P. Withers, H. Bhadeshia, Residual stress. Part 1 – measurement techniques, Mater. Sci. Tech.17 (2001) pp.355-365.

IOP Conf. Series: Materials Science and Engineering **314** (2018) 012003 doi:10.1088/1757-899X/314/1/012003

Failure Analysis of Gold Dust defect in 430 grade Stainless Steel

S Mithra[1*], P G Venkatakrishnan[1], D Noorullah[1], V Karthik[1]

[1]Department of Metallurgical Engineering, Government College of Engineering Salem.

Abstract— Gold dust is a surface defect that usually observed on the surface of 430 grade stainless steel. Gold dusting is characterized by sparkling appearance. They occur as small flakes of metal on the cold rolled surface elongated in the rolling direction. As this defect occurs on the surface of the component, it affects the intended application of the steel for which they have been designed. Further loss of material in the powder form can affect the mechanical properties of the component. The defect is analyzed through optical and Scanning electron microscope with EDAX. From the results it is observed that the defect is due to sensitization. The paper is based on the work to identify the sources of the defect and solutions to overcome this defect.

Keywords: 430 grade stainless steel, Gold Dust, Sensitization, Surface Defect.

1. Introduction

Surface investigation of gold dust in 430 grade stainless steel involves the failure analysis of the cold rolled surface. Gold dusting occurs on the surface after the cold rolling process. The defect appears as the metal powders deposited on the on the rolled plates as shown in figure 1 that are visible to naked eye and they easily stick on the hands when toughed. These powders are usually identified by placing a transparent tape. Once the tape is removed most of the powders stick in the tape. Hence they can be easily wiped off and a simple grinding process can remove the powders on the surface. The grinding of the surface layer will not only cost a lot but also result in continuous precipitation on further ageing after treatment.

The surface grinding of rolled surface is not possible all the time. The defect is also not continuous throughout the surface of the cold rolled plate. They occur randomly on the rolled surface to certain distance. They also do not occur in all the coils of the same grade. So these are the peculiar defects that occur under certain conditions. These precipitates have sparkling appearance and the deposit appears similar to the dust on the surface and hence the name Gold dusts.

This defect is commonly called as Metal powder defect as the precipitates come from inside of the metal component. The defect usually occurs in the cold rolled surface after annealing and pickling cycle. Analysing this defect mainly requires the information regarding the condition under which the defect occurs which enables a route to find the solution.

The reason for precipitation of gold dust can be found if and only if the component which is precipitating out is identified. Thorough understanding of the cause of the failure is required to rectify this defect. The analysis of the defect is by comparison between the cold rolled plate with the gold dust defect and without the defect

IOP Conf. Series: Materials Science and Engineering **314** (2018) 012003 doi:10.1088/1757-899X/314/1/012003

Figure 1: Gold Dust defect in 430 grade Stainless Steel

. The investigation is made through the optical microscope, Scanning Electron Microscope and Energy Dispersive Spectroscopy to compare the gold dust sample with the non-defective sample

2. Problems due to the defect

Stainless steel grade 430 is non-hardenable steel containing straight chromium, and belongs to the ferritic group of steels. This steel is known for its good corrosion resistance and formability, coupled with practical mechanical properties. It can be used in certain chemical applications due to its resistance to nitric acid. The 430 grade stainless steels have their application in Linings for dish washers, Refrigerator cabinet panels, Automotive trim, Lashing Wire, Element Supports, Stove trim rings, Fasteners, Chimney Liners where they require intact contact. The precipitation of the gold dust will disturb the lining surface. Hence the application of 430 grade stainless steel is affected. The grinding of the surface layer will not only cost a lot but also result in continuous precipitation on further ageing after treatment. The surface grinding of rolled surface is not possible all the time. Through the precipitation process one of the components of the 430 grade stainless steel is depleted. So this may affect the properties of the material under operation due to the loss of the material.

3. Experimental works:

3.1 Chemical composition of coils under examination:

A number of coils are taken of examination and the defect does not occur in the entire coil and occurs randomly at certain regions. The coils are sampled with and without defect are taken for analysis. The coil number mentioned here are for our experimental identification.The chemical composition of the coils compared is listed in the table 3.1. The samples with and without gold dust are collected from the same coil to make an effective comparison.

Table 1 Chemical composition of the coils taken for analysis

COIL NO	C %	Si%	Mn%	P%	S%	Cr%	Ni%	N %

136329	0.027	0.31	0.62	0.31	0.002	16.29	0.14	0.056
139821	0.055	0.23	0.70	0.40	0.007	16.20	0.12	0.12
139667	0.045	0.24	0.65	0.37	0.001	16.12	0.12	0.043

Several other coils are also examined to study the variation of the microstructure from case to core. The chemical compositions of the coils listed in the table are the values before rolling operation.

3.2 Microstructural analysis

The microscopic analysis is made to fine the difference between the samples with and without the gold dust at the microscopic level. As mentioned earlier once the metal powders on the surface are removed it again precipitates. This precipitation will result in the depletion of the metal which will cause the microstructure variation which can be depicted through comparison. The samples of size 4mm are cut from the rolled plate. The procedures for metallographic inspection are similar in other cases. As they are stain less steels electrolytic etching is done

Electrolytic etching: Specimen should be connected to terminal of the unit and stainless

steel sheet of approximately 50 X 30 mm size to be attached to negative terminal. For 430 grade Ferritic stainless steels, Etchant is 5%HCl +95% Methyl alcohol, Current density: 3.8 Amp/cm^2, Voltage: 5 V, Time of Etching: 30 sec.

The micrographs with and without gold dust defect for two different coils are shown in the figure 2 and 3.

Coil no.136329

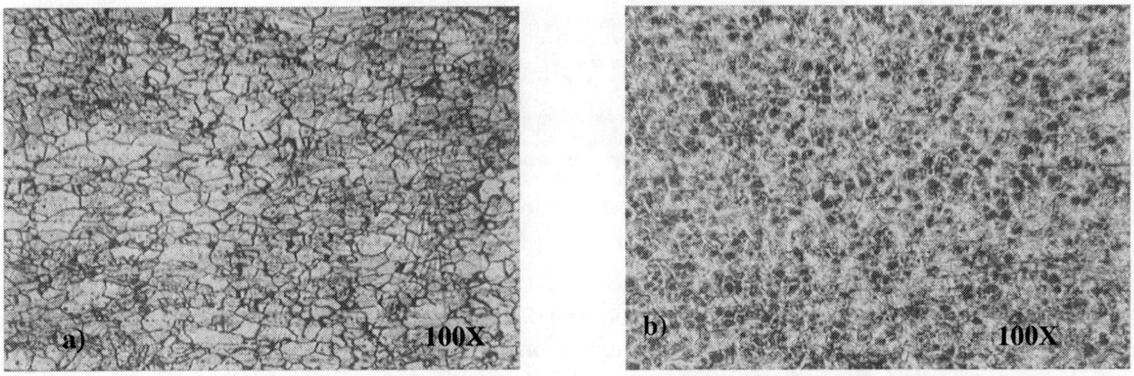

Figure 2: a)Fine recrystallized structure with distributed carbides (Micrograph of defect free sample) b)Partially recrystallized structure with carbide precipitates at grain boundaries (Micrograph of defect sample).

Coil no.139821

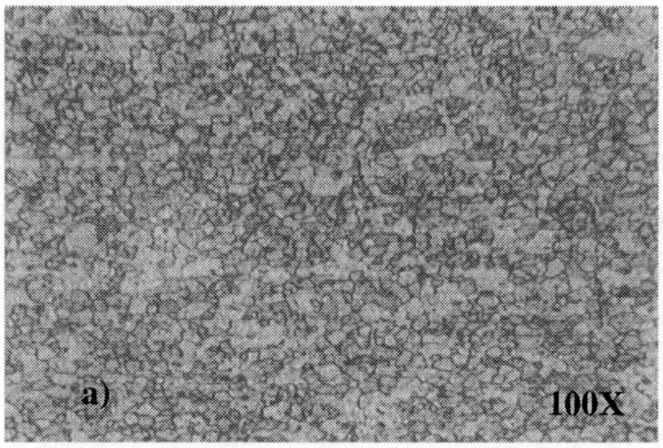

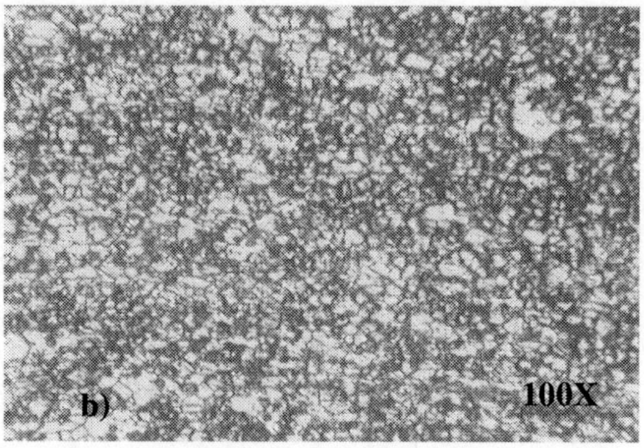

Figure 3: a) Fine recrystallized ferritic grains of size 8 throughout the structure (Micrograph of defect free sample). b) Fine recrystallized ferritic grains of size 8 throughout with profuse carbide precipitation. (Micrograph of defect sample).

The sample with the defect shows profuse precipitation when compared with the defect free sample. The comparison of the microstructure from case to core is also made which is shown in the figure 4. The grain boundaries are broken in the case whereas towards the core they are complete.

Coil No: 139667

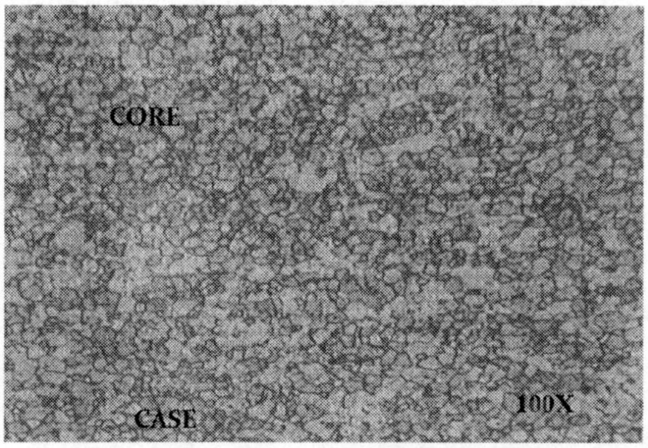

Figure 4 : Comparison of Case and Core.

3.3 SEM Analysis

To view the gold dust at higher magnification and for the better topographical analysis Scanning Electron Microscope is used. The specimen preparation is required before SEM analysis. The samples of thin strips of rolled plate of 25 X 25 mm and 0.8 mm thickness were taken for analysis. SEM images were taken till 1000X for comparison.

The SEM image of the defect free sample at 1500X magnification is shown in the figure 5(a). The SEM image of gold dust sample at 150X is shown in figure 5 (b). The SEM image of gold dust sample at 1000X is shown in figure 6.

IOP Conf. Series: Materials Science and Engineering **314** (2018) 012003 doi:10.1088/1757-899X/314/1/012003

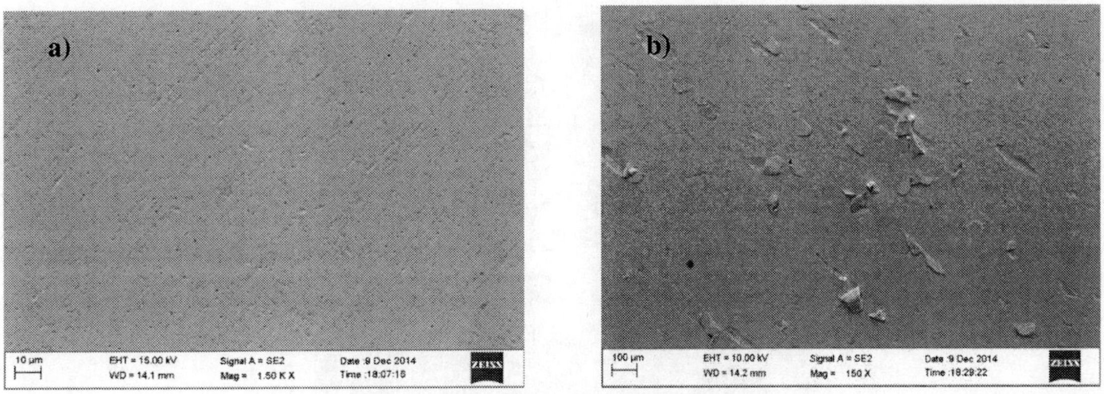

Figure 5: a) Scanning elecron photomicrograph of cold rolled surface of 430 grade stainless steel at 1500X Coil no.139821 (without gold dust defect). b) Scanning elecron photomicrograph of cold rolled surface of 430 grade stainless steel at 150X Coil no.136329 (with gold dust defect).

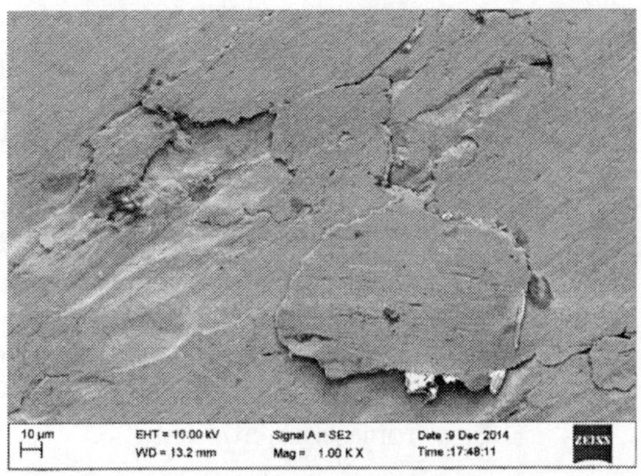

Figure 6: Scanning elecron photomicrograph of cold rolled surface of 430 grade stainless steel at 1000X , Coil no.136329 (with gold dust defect).

3.4 EDAX Analysis

Energy dispersive X-ray spectroscopy (EDS) is a technique to detect characteristic X-rays on the basis of their energy to obtain chemical composition of a specimen bombarded by a focused beam of electrons. The EDAX is employed to know the chemical composition of the metal powder. Thus the material which precipitates out can be analyzed through the EDAX image.

IOP Conf. Series: Materials Science and Engineering **314** (2018) 012003 doi:10.1088/1757-899X/314/1/012003

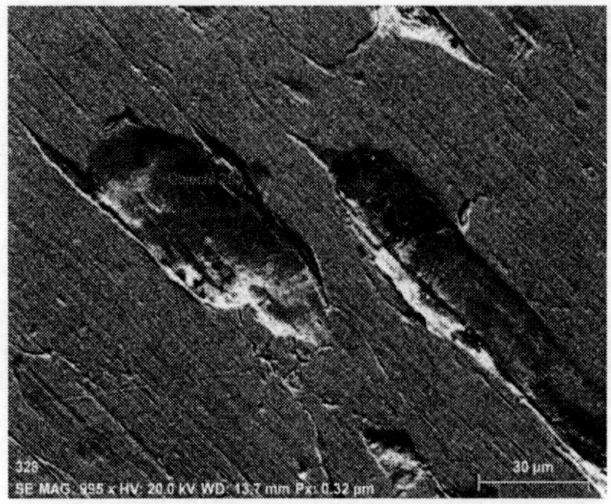

Figure 7: Micrograph from energy dispersive spectroscopy.Analysis is done at two points: one at the defect area and another on the suface near the defect.

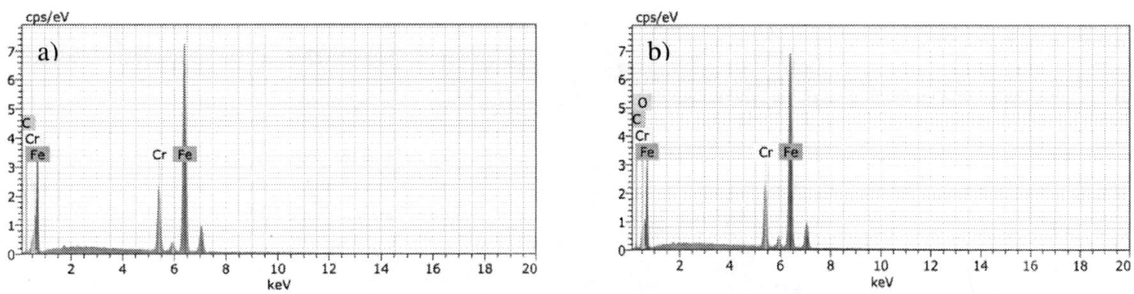

Figure 8: a) Graph obtained at the area of the defect at point 254 in Energy Dispersive Spectrometer micrograph b) Graph obtained at the area near the defect at point 253 in Energy Dispersive Spectrometer micrograph

Table 2 The Spectrum obtained from objects 254 shows

El	AN	Series	unn. C [wt.%]	norm.C [wt.%]	Atom.C [wt.%]	Error (1 sigma)
Fe	26	K series	80.25	82.68	73.59	2.17
Cr	24	K series	13.81	14.22	13.60	0.41
C	6	K series	3.01	3.10	12.81	0.76
		Total	97.07	100.00	100.00	

Table 3: The Spectrum obtained from objects 253 shows

El	AN	Series	unn. C [wt.%]	norm.C [wt.%]	Atom.C [wt.%]	Error (1 sigma)
Fe	26	K series	78.93	78.87	61.48	2.13
Cr	24	K series	12.87	12.86	10.77	0.38
C	6	K series	5.83	5.82	21.11	1.19
O	8	K series	2.44	2.44	6.64	0.49
		Total	100.08	100.00	100.00	

The chemical composition of the specimen at two points are taken one on the surface of the gold dust and the other on the matrix near the defect. The variation in the chemical composition is shown in the graph and the spectrum at two points such as spectrum 254 and 253 are listed below the graph in figure 8 a) ad b) respectively.

The EDAX results show that there is a significant decrease in the chromium percentage and increase in the carbon percent and oxidation has occurred at or near the region of the defect when compared with the matrix far away from the defect.

4. Observation and Inference

From metallurgical micrograph (Figure 2(a) and Figure 3(a)), there is fine recrystallized grain grains whereas in in Gold dust sample shown in figure 2(b) and 3(b), the areas near the grain boundaries appear black indicating that there is a profuse carbide precipitation at the grain boundaries as a result of sensitization. Further from the figure 4 the grain boundaries are distorted towards the surface due to the precipitation.

In Scanning Electron Microscope and Energy Dispersive Spectroscopy, Gold dust samples show a significant decrease in Cr% and increase in C% clearly indicating that the area affected by defect suffers Chromium depletion and the related Chromium Carbide precipitation. There is also increase in O_2% which is due to the fact that there is no sufficient Cr for passivation and hence an oxide formation has occurred at the surface.

5. Conclusion

From the results of chemical analysis, microstructure and SEM analysis we find that this defect is accompanied by a Chromium Carbide precipitation. Visual observation shows that this defect is characterized by a sparkling appearance, which results from small flakes of metal on the cold rolled surface. The apparent cause of such defect is nothing but "The sensitization of ferritic stainless steel which occurs during the annealing process".

Use of a low carbon alloy will completely eradicate the above problem but while considering the compatibility of such steel with respect to the application it is required, there are a number of practical ways to prevent or lower the sensitization which include:

1. Limiting of the annealing temperature to 840°C (which is still sufficient for recrystallization).

2. Lowering of the cooling rate after annealing (by the use of batch annealing, for example), and a limiting of grain growth.

3. Stabilization using Titanium or Niobium-which readily forms carbides than chromium and hence lowering the chances for Chromium Carbide precipitation.

6. References

[1] Hodges R J 1971 *Corrosion,* **1** 164-167.
[2] Frankenthal R P and Pickeringh W 1973 *J Electrochem Soc* **120** 23-26.
[3] Bond A P 1969 *T Metall Soc AIME* **245** 2127-34.
[4] Nishimura R 1992 *Corrosion* **48** 882–890.
[5] Semchysen M, Bond A P, and Dunda H J 1972 Effects of composition on ductility and toughness of ferritic stainless steel *The Metallurgical Evolution of Stainless Steel* ed F B Pickering ASM pp 260–273
[6] Pistorius P C and Coetzee M 1995 *J South Afr Inst Min Metall* **96** 119-125

Experimental investigation on hard turning of AISI 4340 steel using cemented coated carbide insert

J Pradeep Kumar[1,2], K P Kishore[1], M Ranjith Kumar[1], K R Saran Karthick[1] and S Vishnu Gowtham[1]

[1]Department of Production Engineering, PSG college of Technology, Coimbatore 641 004, India.

Abstract. Hard turning is a developing technology that offers many potential advantages compared to grinding, which remains the standard finishing process for critical hardened surfaces. In this work, an attempt has been made to experimentally investigate hard turning of AISI 4340 steel under wet and dry condition using cemented coated carbide insert. Hardness of the workpiece material is tested using Brinell and Rockwell hardness testers. CNC LATHE and cemented coated carbide inserts of designation CNMG 120408 are used for conducting experimental trials. Significant cutting parameters like cutting speed, feed rate and depth of cut are considered as controllable input parameters and surface roughness (R_a), tool wear are considered as output response parameters. Design of experiments is carried out with the help of Taguchi's L9 orthogonal array. Results of response parameters like surface roughness and tool wear under wet and dry condition are analysed. It is found that surface roughness and tool wear are higher under dry machining condition when compared to wet machining condition. Feed rate significantly influences the surface roughness followed by cutting speed. Depth of cut significantly influences the tool wear followed by cutting speed.

1. Introduction

Manufacturers around the world constantly strive for lower cost solutions in order to maintain their competitiveness, on machined components and manufactured goods. Globally, part quality has been found to be at acceptable levels and it continues to improve, while the pressure for part piece cost is enormous and is constantly being influenced downward by competition and buyer strategies. The trend is toward higher quality, lower cost and smaller batch sizes. In current industrial scenario CNC machine tools, which operate with mature technology and provide both consistency and reliability, have now become the biggest contributor to part quality and cost. A rapid adoption of these newer and more cost effective manufacturing techniques will be constantly required if manufacturing operations are to remain competitive. The technology evolution for addressing machining of hard materials results in a newer solution called hard turning.

Hard turning is the process of machining hardened ferrous material with a hardness value more than 45HRC in order to obtain finished work pieces directly from hardened parts. The cutting tools

[2] Corresponding author. Tel.: +91 – 9443872965.
E – mail address : jp.psgtech@gmail.com

of choice are typically Cubic Boron Nitride (CBN), ceramic, carbide and sometimes cermet. Hard turning does provide an alternative for those applications that do not require the high end processing capability of a grinder, or for pregrind roughing operations. Clearly, for many applications involving very close tolerance work, grinding will still remain the process of choice. More specifically, hard turning does not eliminate the need for grinding but can relieve the production burden on the more expensive grinders for the properly chosen application. The tooling choice will need to be matched to the application, desired production rates and the operating cost goals. The current tooling technology allows the user to be able to choose between wet or dry operations. For parts, such as gears, that have interrupted cuts, dry condition is preferred. That is because the insert would experience thermal shock while entering and exiting cuts would likely cause too or part breakage. Cutting without coolant provides obvious cost benefits as well. Wet operations refer to processes under flood or high-pressure with a water-soluble coolant. Coolant can be helpful for continuous cutting applications in terms of providing longer tool life and better surface finishes. Machine requirements for hard turning include machine rigidity, part rigidity, rigid tools, solid work holding and vibration dampening.
The range of hard turned applications will vary based upon the part requirements, tolerance levels, surface finish and very importantly the machine tool. On a daily basis, parts are being hard turned in the following industry segments; automotive, bearing, marine, punch and die, mold, hydraulics and pneumatics, machine tool and aerospace. Commonly processed hard turned materials would include steel alloys such as bearing steels, hot and cold-work tool steels, high speed steels, die steels, case hardened steels. Advantages of hard turning are process flexibility, higher metal removal rates, improved surface finish, reduced time consumption, easier chip control. Limitations of hard turning include white layer formation, tooling and machine process capability.

2. Literature survey

Hamza Bensouilah [1] et al. (2016) concluded that surface roughness is minimal at higher cutting speed and lower feed rate. Coated ceramic inserts performed significantly better than uncoated ceramic inserts. According to ANOVA it was observed that feed rate significantly influenced surface roughness. The optimum process parameters for better surface finish were found to be cutting speed = 150 m/min, feed rate = 0.04 mm/rev and depth of cut = 0.1 mm. D'Addona D M [2] et al. (2016) investigated the analysis of surface roughness in hard turning using wiper insert geometry and observed that wiper insert geometry gives superior surface finish as compared to conventional inserts and it can give comparable surface finish with grinding operation. Feed is found to be most significant parameter for surface roughness. After feed, depth of cut and type of insert are found to have statistically significant effect on surface roughness. The cutting conditions to achieve good surface roughness are wiper geometry with 1.2 mm nose radius, 1200 RPM, 0.08 mm/rev feed rate and 0.1 mm depth of cut. Dipti Kanta Das[3] et al. (2014) investigated on Investigations on hard turning using coated carbide insert and reported the good surface quality of roughness about 0.42 microns during hard machining of EN 24 steel with the help of coated carbide insert. Using grey-based Taguchi approach, the optimal parametric combination for surface quality characteristics (Ra and Rz) have been obtained to be depth of cut: 0.4 mm, feed: 0.04 mm/rev and cutting speed: 130 m/min respectively. Feed was considered to be the most dominant parameter for both surface roughness parameters Ra and Rz. Sathish Chinchanikar [4] et al. (2014) studied on comparative evaluations on surface roughness during dry and wet condition and concluded that hard turning with cutting fluids has not significantly improved the surface finish in comparison to dry cutting. However coconut based cutting fluid produced lower values of surface roughness especially at higher values of cutting parameters in comparison to hard turning with dry and with water based cutting fluid. Surface roughness decreased initially with increase in speed but when it exceeded 150-160 m/min, Ra value increased. Ra values of about 0.5 – 0.6 μm was obtained. However it has been observed that by limiting cutting parameters hard turning under dry condition was the better option in comparison to wet condition. Varaprasad.Bh [5] et al. (2014) investigated the effect of machining parameters on tool wear in hard turning of AISI D3 steel. The authors observed that significant parameter for tool flank wear was depth of cut. The speed and feed have little influence on the total variation The RSM based DOE was found to be an effective way in determining the optimal cutting parameters to be speed of 165m/min, feed rate of 0.05mm/rev and depth of cut of 0.3mm to achieve a low tool wear of 0.148mm. A.Srithar [6] et al. (2014) reported on experimental investigation and surface roughness analysis on hard turning of AISI D2 steel using coated carbide insert. Investigations were carried out on conventional lathe using the prefixed cutting conditions. The results specify that the increase of cutting speed decreases the surface roughness. The results shows that feed rate is highly control the parameter, which influence the surface roughness parameters in machining of AISI D2 steel. Gaurav Bartarya [7] et al. (2012) reported on state of art in hard turning and observed that radial force seem to be dominant in hard turning. White layer formation was due to phase transformation to martensite. Coolant may help in reducing white layer thickness and tensile residual stress. For finish hard turning depth of cut was less than the nose radius of the tool. Guddat J [8] et al. (2011) reported on hard turning of AISI 52100 using PCBN wiper geometry inserts and the resulting surface integrity. The assessment of surface and subsurface integrity reveals that, when compared to conventional geometry inserts, the application of PCBN wiper inserts leads to significantly improved surface roughness and higher compressive residual stresses. The author summarized that using wiper inserts leads to superior surface integrity compared to conventional inserts. At the same time, higher productivity can be achieved. Vikram Kumar R [9] et al. (2007) investigated on performance of coated tools during hard turning under minimal fluid application and reported that overall performance of cutting tools during minimal cutting fluid application was found to be superior to that of dry turning. Cutting performance mainly depends on fluid application parameters such as cutting force, number of pulses and amount of cutting fluid. Grzesik W[10] et al. (2006) studied on the hard turning of quenched alloy parts using conventional and wiper inserts and concluded that hard turning with wiper provide higher surface finish.

From the extensive literature survey, it was found that hard machining has the advantages of cost, time and coolant savings compared to grinding. The biggest challenge for hard machining was the surface quality whether that was close to grinding or not. Therefore optimized process parameters are highly essential for its successful implementation in the industry. Therefore, the experimental investigation on hard turning was becoming increasingly important. As seen from the literature review, the hard machining operation was usually done by costly ceramic and CBN insert. The attempt has been made in the present work utilizing low cost coated carbide insert in hard turning under dry and wet condition to justify its usability compared to costlier CBN and ceramic insert. Thus the objective is to investigate surface quality characteristics (Ra), tool wear in hard turning of AISI 4340 steel using cemented coated carbide insert. The methodology adopted in the present work is shown in 'figure 1'.

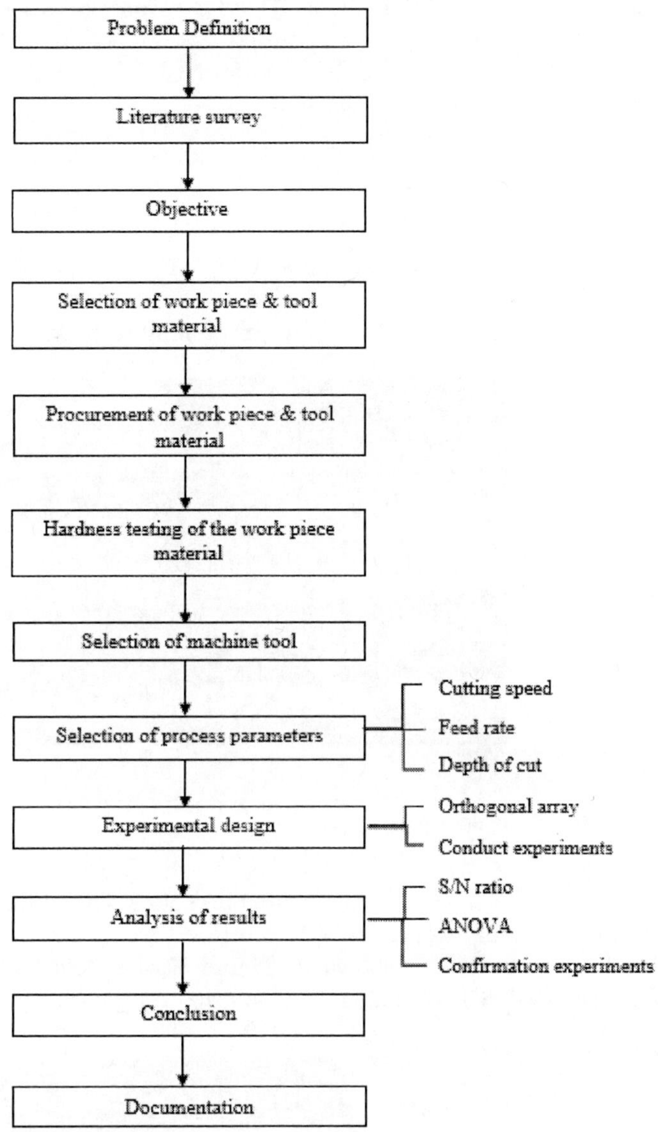

Figure 1. Graphical abstract of the work.

3. Experimental details

3.1. Workpiece material

The work material is AISI 4340 steel. It is a general-purpose steel having a wide range of application in automobile and allied industries by virtue of its good hardenability enabling it to be used in fairly large sections. Bars of 36 mm diameter and 100 mm length are used in the present investigation. The composition of the work material is shown in table 1.

IOP Conf. Series: Materials Science and Engineering **314** (2018) 012004 doi:10.1088/1757-899X/314/1/012004

Table 1. Chemical composition of AISI 4340 steel.

%C	%Mn	%Ni	%Cr	%Mo
0.40	0.70	1.80	0.80	0.25

3.2. Machine tool
The CNC LATHE (Galaxy Midas 6) used for machining is shown in 'figure 2'.

Figure 2. CNC LATHE.

3.3. Cutting tool
Cemented carbide inserts (commercially available) comprising of thick TiCN and thick Al_2O_3 coatings are selected for the present work. The selected insert with a general specification of CNMG 120408 and chip breaker of type DM is shown in 'figure 3'. The tool holder of hole type clamping with a general specification of PCLNR 2525M12 is selected for the present investigation.

Figure 3. CNMG 120408 insert.

3.4. Brinell hardness testing
The hardness of the material is measured by applying load of 3000kg in different locations on the surface of the material. Ball indenter of 10 mm diameter is used to measure the hardness. The indentation time is 15 seconds. Three trials are conducted as shown in table 2, for evaluation of work material Brinell hardness. The Brinell hardness is calculated using equation (1).

$$\text{Brinell hardness} = \frac{2P}{\pi D\left(D-\sqrt{D^2-d^2}\right)}\ \text{BHN} \quad (1)$$

where, P is the load (kg) applied, D is the diameter of indenter (mm) and d is the diameter of indentation (mm).

Table 2. Hardness (BHN) of AISI 4340 steel.

S.No	Diameter of indenter 'D' (mm)	Load 'P' (Kg)	Diameter of indentation 'd' (mm)	BHN
1.	10	3000	2.85	461.512
2.	10	3000	2.84	464.503
3.	10	3000	2.83	467.816

3.5. Rockwell hardness testing

The hardness of the material is also measured by applying load of 150kg in different locations on the surface of the material. Diamond indenter of 120° is used to measure the hardness. The indentation time is 15 seconds. Three trials are conducted as shown in table 3, for evaluation of work material Rockwell hardness.

Table 3. Hardness (HRC) of AISI 4340 steel.

S.No	Load 'P' (Kg)	HRC
1.	150	48
2.	150	51
3.	150	49

3.6. Selection of parameters

Taguchi method is used for execution of the plan of experiments, L9 orthogonal array is used for experimentation. The three dominant process parameters such as cutting speed, feed rate and depth of cut each varying at three levels are selected as controllable parameters for carrying out experimental trials. The range of these parameters are obtained based on initial experimental trials, literature survey, insert catalogue and machine specifications is shown in table 4.

Table 4. Design factors along with levels.

Cutting Speed (m/min)	150	250	350
Feed rate (mm/rev)	0.04	0.08	0.12
Depth of cut (mm)	0.2	0.4	0.6

The experimental treatments of combination of these parameters are shown in table 5.

Table 5. L9 orthogonal array with design factors.

Trial No	Speed (m/min)	Feed rate (mm/rev)	Depth of cut (mm)
1	150	0.04	0.2
2	150	0.08	0.4
3	150	0.12	0.6
4	250	0.04	0.4
5	250	0.08	0.6
6	250	0.12	0.2
7	350	0.04	0.6
8	350	0.08	0.2
9	350	0.12	0.4

3.7. Experimental trials

Each experimental trial is conducted twice based on Taguchi L9 orthogonal array and the average values are taken for the response variables such as surface roughness value (R_a) and tool wear. Taguchi method is an experimental design technique, which is useful in reducing the number of experiments, decrease experimental time, reduce the cost and find out significant factors in a shorter time period. Taguchi's smaller the better criteria is used for evaluation of surface roughness value and tool wear. For smaller the better, the S/N ratio is calculated using equation (2).

$$\frac{S}{N} = -10\log\frac{1}{n}\left(\sum_{i=1}^{n} y_i^2\right) \qquad \text{.................................... (2)}$$

where, n is the number of trials and y_i is the response value. In order to establish the statistical significance of the cutting parameters on the output parameters analysis of variance (ANOVA) is used. The statistical significances of linear model are evaluated by P – values of ANOVA. When P – values are less than 0.05 (95% confidnce) the obtained models are considered to be statistically significant. Analysis of the experimental data obtained through Taguchi experimental design is carried out using MINITAB 17 software. The response variables such as surface roughness and tool wear are measured during dry turning and wet turning. The specimen after hard turning is shown in 'figure 4'.

Figure 4. Specimen after hard turning.

3.8. Surface roughness measurement test

The surface roughness of hard turned AISI 4340 steel under dry and wet conditions is measured using a Mitutoyo surface roughness tester SJ 201 with cut-off length of 0.8 mm. The apparatus used for surface roughness measurement is shown in 'figure 5'.

Figure 5. Surface roughness tester.

3.9. Tool wear measurement

During the course of experimentation under dry and wet conditions, the tool wear of worn out inserts are measured with the help of a Tool maker's microscope and digital weighing machine The apparatus used for tool wear measurement is shown in 'figure 6'.

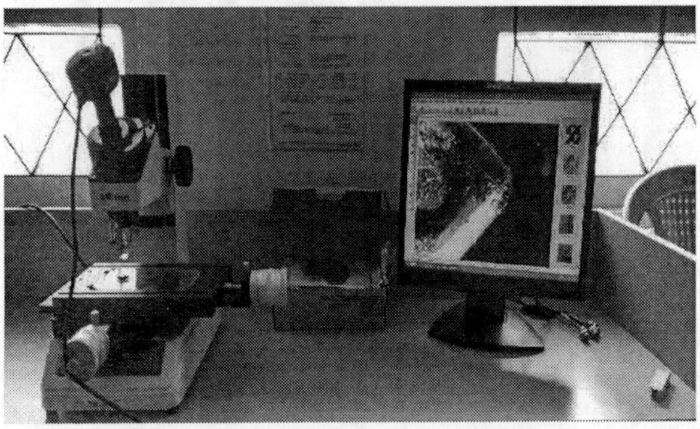

Figure 6. Tool maker's microscope.

4. Results and Discussion

4.1. Surface roughness test

The surface roughness values thus obtained as an average of two experimental trials for each combination of process parameters based on L9 orthogonal array in wet condition and dry condition is shown in table 6.

Table 6. R_a under dry and wet condition.

Trial No.	Wet condition R_a (μm)	Dry condition R_a (μm)
1	0.44	0.84
2	0.52	0.92
3	0.59	0.98
4	0.39	0.79
5	0.45	0.85
6	0.50	0.90
7	0.37	0.77
8	0.40	0.80
9	0.51	0.91
Average	0.46	0.86

The 'figure 7', depicts the comparison of surface roughness values during hard turning under wet condition and dry condition. Heat is carried away from tool and work by means of cutting fluid which at same time reduce friction between tool and chip and between tool and work and also facilitates chip formation. So, surface roughness (R_a) is minimum in wet machining when compared to dry machining.

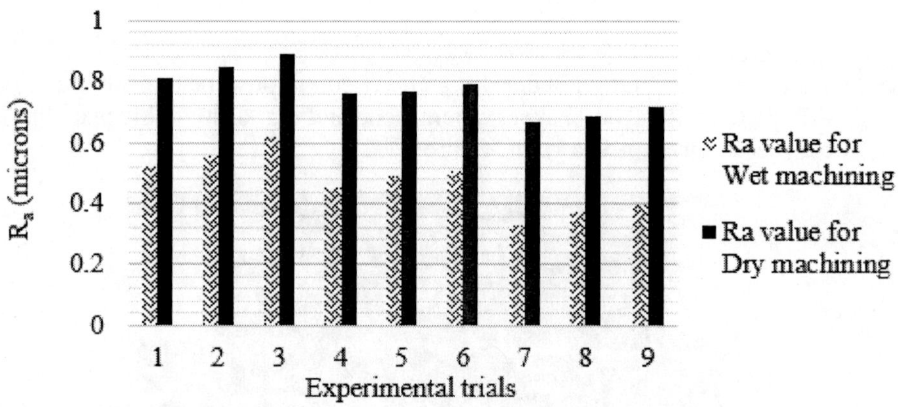

Figure 7. Comparison of R_a values.

4.2. Influence of process parameters on R_a – wet condition

From the table 7, it can be inferred that the response variable R_a is influenced more by feed rate followed by cutting speed and depth of cut respectively under wet condition.

Table 7. Response table for S/N ratio of R_a – wet condition.

Level	Cutting speed (m/min) A	Feed rate (mm/rev) B	Depth of cut (mm) C
1	5.798	7.982	7.037
2	7.045	6.858	6.569
3	7.481	5.484	6.718
Delta	1.683	2.498	0.468
Rank	2	1	3

Table 8, shows ANOVA results for S/N ratio of Ra under wet condition. It can be seen that contribution of feed rate is 65.121%, cutting speed is 31.75% and depth of cut is 2.37% in influencing the surface roughness under wet condition.

Table 8. ANOVA for S/N ratio of R_a – wet condition.

Source	DF	SS	MS	F	P	Contribution(%)	Remarks
Cutting speed (m/min)	2	4.5785	2.28923	42.34	0.023	31.75	Significant
Feed rate (mm/rev)	2	9.390	4.69501	86.84	0.011	65.121	Most Significant
Depth of cut (mm)	2	0.3425	0.1712	3.17	0.240	2.37	Insignificant
Error	2	0.1081	0.05406	-	-	-	-
Total	8	14.4191	-	-	-	-	-

4.3. Optimum combination of process parameters for R_a – wet condition

It can be inferred from 'figure 8', that cutting speed at high level (350m/min), feed rate at low level (0.04mm/rev) and depth of cut at low level (0.2mm) provides optimum combination i.e. A3B1C1 for achieving the surface roughness value under wet condition.

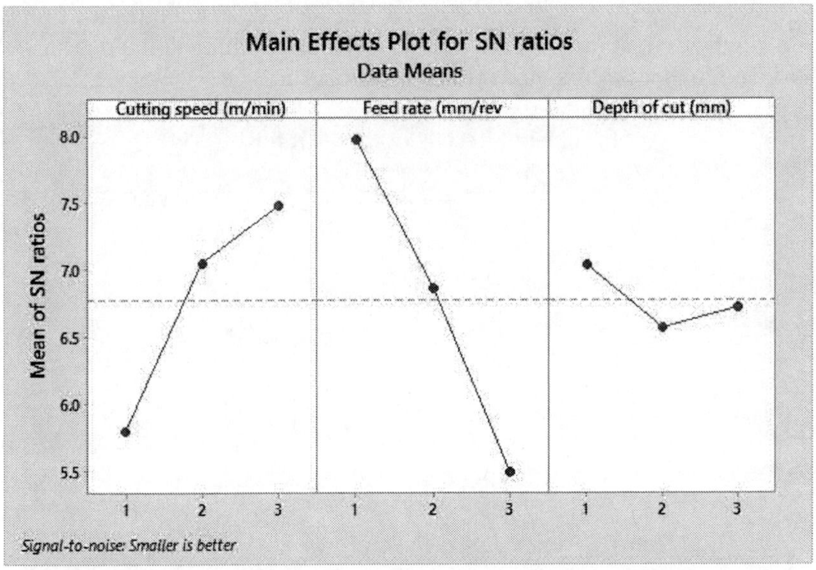

Figure 8. Main effects plot for S/N ratio of R_a - wet condition.

4.4. Influence of process parameters on R_a – dry condition

From the table 9, it can be inferred that the response variable R_a is influenced more by feed rate followed by cutting speed and depth of cut respectively under dry condition.

Table 9. Response table for S/N ratio of R_a – dry condition.

Level	Cutting speed (m/min) A	Feed rate (mm/rev) B	Depth of cut (mm) C
1	0.8047	1.9440	1.4559
2	1.4581	1.3580	1.1970
3	1.6759	0.6366	1.2858
Delta	0.8711	1.3074	0.2590
Rank	2	1	3

Table 10, shows ANOVA results for S/N ratio of Ra under dry condition. It can be seen that contribution of feed rate is 65.31%, cutting speed is 31.30% and depth of cut is 2.63% in influencing the surface roughness under dry condition.

Table 10. ANOVA for S/N ratio of R_a – dry condition.

Source	DF	SS	MS	F	P	Contribution (%)	Remarks
Cutting speed (m/min)	2	1.2332	0.61660	42.15	0.023	31.30	Significant
Feed rate (mm/rev)	2	2.5731	1.2865	87.96	0.011	65.31	Most significant
Depth of cut (mm)	2	0.10390	0.05195	3.55	0.220	2.63	Insignificant
Error	2	0.02926	0.01463	-	-	-	-
Total	8	3.93954	-	-	-	-	-

4.5. Optimum combination of process parameters for R_a – dry condition

It can be inferred from 'figure 9', that cutting speed at high level (350m/min), feed rate at low level (0.04mm/rev) and depth of cut at low level (0.2mm) provides optimum combination i.e. A3B1C1 for achieving the surface roughness value under dry condition.

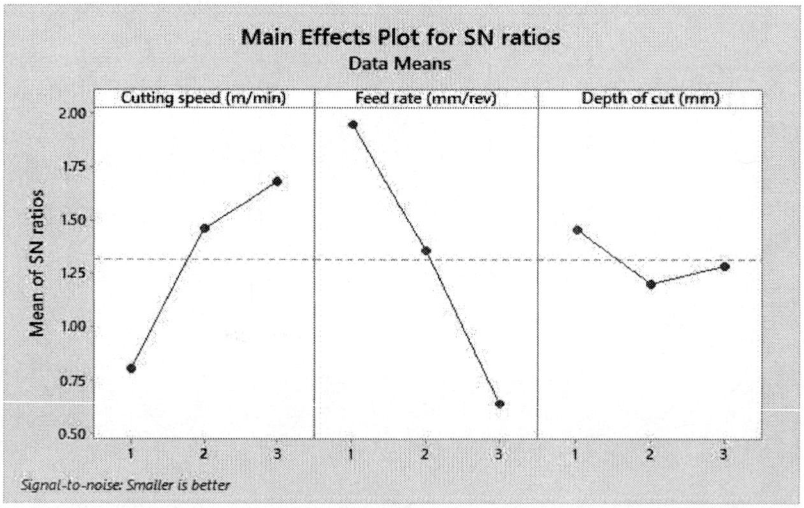

Figure 9. Main effects plot for S/N ratio of R_a – dry condition.

4.6. Tool wear test

Table 11, shows the results of tool wear of inserts under dry and wet condition for various combinations of cutting conditions (cutting speed, feed rate and depth of cut) as per the design matrix.

Table 11. Tool wear – dry and wet condition.

Trial No.	Wet condition Tool wear (mg)	Dry condition Tool wear (mg)
1	0.44	1.0
2	0.55	1.2
3	0.64	1.5
4	0.56	1.4
5	0.5	1.7
6	0.47	1.2
7	0.71	1.8
8	0.59	1.3
9	0.68	1.6
Average	0.58	1.41

The 'figure 10', depicts the comparison of tool wear during hard turning under wet condition and dry condition.

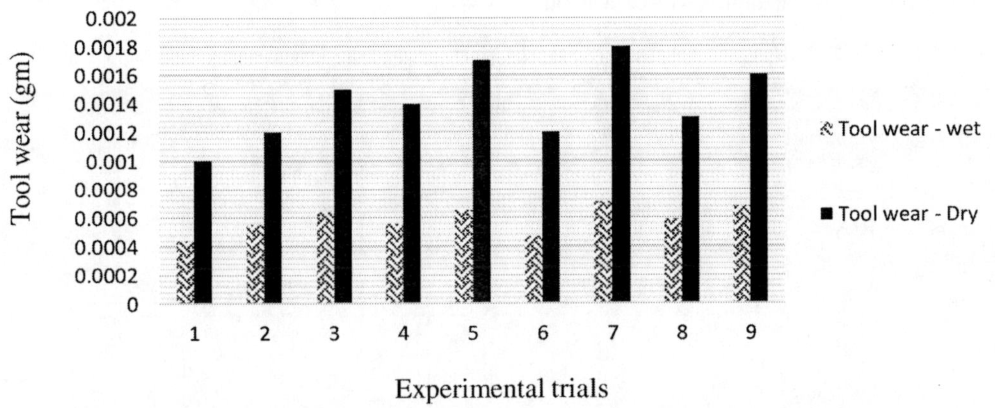

Figure 10. Comparison of tool wear.

4.7. Influence of process parameters on tool wear – wet condition

From the table 12, it can be inferred that the response variable tool wear is influenced more by depth of cut followed by cutting speed and feed rate respectively under wet condition.

Table 12. Response table for S/N ratio of tool wear – wet condition.

Level	Cutting speed (m/min) A	Feed rate (mm/rev) B	Depth of cut (mm) C
1	5.40	5.04	6.09
2	5.11	4.50	4.53
3	3.63	4.59	3.53
Delta	1.76	0.54	2.56
Rank	2	3	1

Table 13, shows ANOVA results for S/N ratio of tool wear under wet condition. It can be seen that contribution of depth of cut is 61.95%, cutting speed is 33.32% and feed rate is 3.13% in influencing the tool wear under wet condition.

Table 13. ANOVA for S/N ratio of tool wear – wet condition.

Source	DF	SS	MS	F	P	Contribution (%)	Remarks
Cutting speed (m/min)	2	5.3742	2.6871	21.08	0.045	33.32	Significant
Feed rate (mm/rev)	2	0.5060	0.2530	1.98	0.335	3.13	Insignificant
Depth of cut (mm)	2	9.9897	4.9949	39.18	0.025	61.95	Most significant
Error	2	0.2549	0.1275	-	-	-	-
Total	8	16.1249	-	-	-	-	-

4.8. Optimum combination of process parameters for tool wear – wet condition

It can be inferred from 'figure 11', that cutting speed at low level (150m/min), feed rate at low level (0.04mm/rev) and depth of cut at low level (0.2mm) provides optimum combination i.e. A1B1C1 for minimizing the tool wear under wet condition.

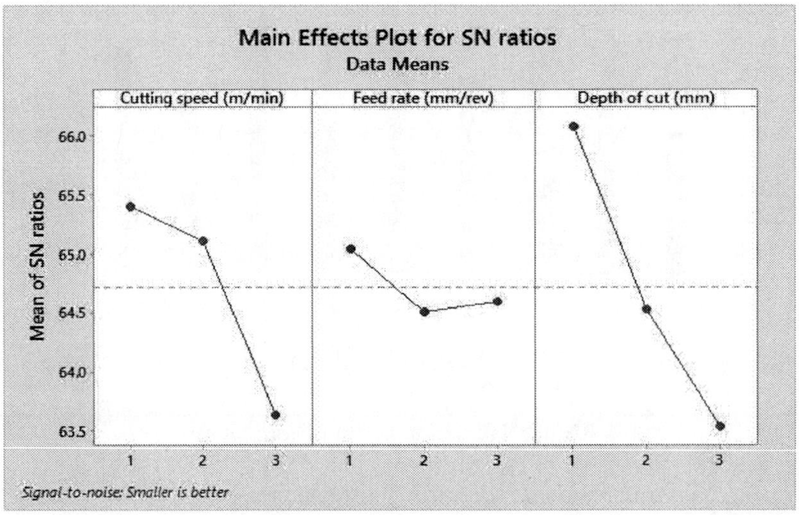

Figure 11. Main effects plot for S/N ratio of tool wear – wet condition.

4.9. Influence of process parameters on tool wear – dry condition

From the table 14, it can be inferred that the response variable tool wear is influenced more by depth of cut followed by cutting speed and feed rate respectively under dry condition.

Table 14. Response table for S/N ratio of tool wear – dry condition.

Level	Cutting speed (m/min) A	Feed rate (mm/rev) B	Depth of cut (mm) C
1	-1.702	-2.676	-1.287
2	-3.038	-2.824	-2.863
3	-3.822	-3.063	-4.412
Delta	2.12	0.39	3.12
Rank	2	3	1

Table 15, shows ANOVA results for S/N ratio of tool wear under dry condition. It can be seen that contribution of depth of cut is 67.14%, cutting speed is 31.61% and feed rate is 0.010% in influencing the tool wear under dry condition.

Table 15. ANOVA for S/N ratio of tool wear – dry condition.

Source	DF	SS	MS	F	P	Contribution (%)	Remarks
Cutting speed (m/min)	2	6.8970	3.44852	163.65	0.006	31.61	Significant
Feed rate (mm/rev)	2	0.2283	0.11417	5.42	0.156	0.010	Insignificant
Depth of cut (mm)	2	14.6449	7.32245	347.41	0.003	67.14	Most Significant
Error	2	0.0421	0.02107	-	-	-	-
Total	8	21.8124	-	-	-	-	-

4.10. Optimum combination of process parameters for tool wear – dry condition

It can be inferred from 'figure 12', that cutting speed at low level (150m/min), feed rate at low level (0.04mm/rev) and depth of cut at low level (0.2mm) provides optimum combination i.e. A1B1C1 for minimizing the tool wear under dry condition.

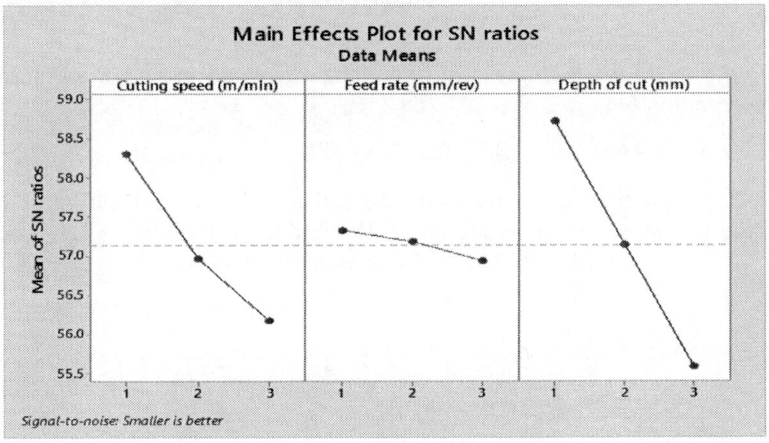

Figure 12. Main effects plot for S/N ratio of tool wear – dry condition.

IOP Conf. Series: Materials Science and Engineering **314** (2018) 012004 doi:10.1088/1757-899X/314/1/012004

4.11. Estimating the mean

Once an experiment is conducted and the optimum treatment condition within the experiment is determined, the most direct way to estimate the mean for that treatment condition is to average all the results for the trials which are set at those particular levels.

4.11.1. Estimating the mean for R_a - wet condition. The estimated mean for surface roughness (R_a) is obtained using the equation (3).

$$\mu = T + (A_3 - T) + (B_1 - T) + (C_1 - T) \quad (3)$$

where, T is the overall grand mean of the experimental results and has the value of 0.463 microns. Table 16, shows the means of surface roughness (R_a) under wet condition. The mean for a selected trial condition for parameters at ($A_3B_1C_1$) is 0.348 microns.

Table 16. Response table for means of R_a – wet condition.

Level	Cutting speed (m/min) A	Feed rate (mm/rev) B	Depth of cut (mm) C
1	0.5167	0.4000	0.4467
2	0.4467	0.4567	0.4733
3	0.4267	0.5333	0.4700
Delta	0.0900	0.1333	0.0267
Rank	2	1	3

Table 17, shows ANOVA results for means of surface roughness (R_a) under wet condition.

Table 17. ANOVA for means of R_a – wet condition.

Source	DF	SS	MS	F	P
Cutting speed (m/min)	2	0.013400	0.006700	201.00	0.005
Feed rate (mm/rev)	2	0.026867	0.013433	403.00	0.002
Depth of cut (mm)	2	0.001267	0.000633	19.00	0.050
Error	2	0.000067	0.000033	-	-
Total	8	0.041600	-	-	-

4.11.2. Estimating the mean for R_a - dry condition. The estimated mean for surface roughness (R_a) is obtained using the equation (4).

$$\mu = T + (A_3 - T) + (B_1 - T) + (C_1 - T) \quad (4)$$

where, T is the overall grand mean of the experimental results and has the value of 0.862 microns. Table 18, shows the means of surface roughness (R_a) under dry condition. The mean for a selected trial condition for parameters at ($A_3B_1C_1$) is 0.75 microns.

Table 18. Response table for means of R_a – dry condition.

Level	Cutting speed (m/min) A	Feed rate (mm/rev) B	Depth of cut (mm) C
1	0.9133	0.8000	0.8467
2	0.8467	0.8567	0.8733
3	0.8267	0.9300	0.8667
Delta	0.0867	0.1300	0.0267
Rank	2	1	3

Table 19, shows ANOVA results for means of surface roughness (R_a) under dry condition.

Table 19. ANOVA for means of R_a – dry condition.

Source	DF	SS	MS	F	P
Cutting speed (m/min)	2	0.012356	0.006178	79.43	0.012
Feed rate (mm/rev)	2	0.025489	0.012744	163.86	0.006
Depth of cut (mm)	2	0.001156	0.000578	7.43	0.119
Error	2	0.000156	0.000078	-	-
Total	8	0.039156	-	-	-

4.11.3. Estimating the mean for tool wear - wet condition. The estimated mean for tool wear is obtained using the equation (5).

$$\mu = T + (A_1 - T) + (B_1 - T) + (C_1 - T) \quad (5)$$

where, T is the overall grand mean of the experimental results and has the value of 0.58 milligram. Table 20, shows the means of tool wear under wet condition. The mean for a selected trial condition for parameters at ($A_1B_1C_1$) is 0.45 milligram.

Table 20. Response table for means of tool wear – wet condition.

Level	Cutting speed (m/min) A	Feed rate (mm/rev) B	Depth of cut (mm) C
1	0.543	0.570	0.500
2	0.560	0.597	0.597
3	0.660	0.597	0.667
Delta	0.117	0.27	0.167
Rank	2	3	1

Table 21, shows ANOVA results for means of tool wear under wet condition.

Table 21. ANOVA for means of tool wear – wet condition.

Source	DF	SS	MS	F	P
Cutting speed (m/min)	2	0.0238	0.0119	38.39	0.025
Feed rate (mm/rev)	2	0.0014	0.0007	2.29	0.304
Depth of cut (mm)	2	0.0420	0.0210	67.54	0.015
Error	2	0.0006	0.00031	-	-
Total	8	0.0679	-	-	-

4.11.4. Estimating the mean for tool wear - dry condition. The estimated mean for tool wear is obtained using the equation (6).

$$\mu = T + (A_1 - T) + (B_1 - T) + (C_1 - T) \qquad (6)$$

where, T is the overall grand mean of the experimental results and has the value of 1.4 milligram. Table 22, shows the means of tool wear under dry condition. The mean for a selected trial condition for parameters at ($A_1B_1C_1$) is 1 milligram.

Table 22. Response table for means of tool wear – dry condition.

Level	Cutting speed (m/min) A	Feed rate (mm/rev) B	Depth of cut (mm) C
1	1.23	1.40	1.16
2	1.43	1.40	1.40
3	1.56	1.43	1.66
Delta	0.033	0.03	0.5
Rank	2	3	1

Table 23, shows ANOVA results for means of tool wear under dry condition.

Table 23. ANOVA for means of tool wear – dry condition.

Source	DF	SS	MS	F	P
Cutting speed (m/min)	2	0.1688	0.0844	76.0	0.013
Feed rate (mm/rev)	2	0.0022	0.0011	1.0	0.5
Depth of cut (mm)	2	0.3755	0.1877	169.0	0.006
Error	2	0.0022	0.0011	-	-
Total	8	0.5488	-	-	-

4.12. Confidence interval around the mean

The estimate of the mean obtained is only a point estimate based on average results of the experiment. This means there is a 50% chance of actual mean being greater than estimated mean and another 50% chance that the actual mean is less than the estimated mean. Thus a confidence interval is to be established within which actual mean lies at some stated level of confidence. There are three different types of confidence intervals proposed by Taguchi depending on purpose of the estimate. For the present work, the formula for calculating the confidence interval (CI) around the estimated mean in the confirmation experiment is given in equation (7).

$$CI = [F(\alpha, 1, v_e)\, V_e\, [1/\eta_{eff} + 1/r]]^{1/2} \quad (7)$$

where, $F(\alpha, 1, v_e)$ is the value of 'F' from F – tables for confidence level of $(1-\alpha)$, α is the level of risk, V_e is the error variance, v_e is the degrees of freedom for the error, η_{eff} is the effective number of replications and r is the number of trials.

4.12.1. Confidence interval around the mean for R_a – wet condition. The confidence interval (CI) is calculated as follows. η_{eff} is calculated using equation (8).

$$\eta_{eff} = \frac{N}{1 + U_{\text{items in the estimate}}} \quad (8)$$

N = Total number of experiments conducted = 18

$U_{\text{items in the estimate}} = 4$

Substituting in the above formula, $\eta_{eff} = 3.6$
$\alpha = 1$ – confidence limits (95%) = 0.05.
$F_{ratio}(0.05, 1, 2) = 18.51$ (from F tables)
$CI = +/- [18.51 \times 0.000033[1/3.6 + 1/5]]^{1/2}$
CI = +/- 0.017
The 95% confidence level of the predicted optimum of the surface roughness (R_a) under wet condition is given by:
$[\mu - CI] < \mu < [\mu - CI]$, $0.331 < 0.348 < 0.365$

4.12.2. Confidence interval around the mean for R_a – dry condition. The confidence interval (CI) is calculated as follows. η_{eff} is calculated using equation (8).
N = Total number of experiments conducted = 18
$U_{\text{items in the estimate}} = 4$
Substituting in the equation (8), $\eta_{eff} = 3.6$
$\alpha = 1$ – confidence limits (95%) = 0.05.
$F_{ratio}(0.05, 1, 2) = 18.51$ (from F tables)
$CI = +/- [18.51 \times 0.000078[1/3.6 + 1/5]]^{1/2}$
CI = +/- 0.026
The 95% confidence level of the predicted optimum of the surface roughness (R_a) under dry condition is given by:
$[\mu - CI] < \mu < [\mu - CI]$, $0.724 < 0.75 < 0.776$

4.12.3. Confidence interval around the mean for tool wear – wet condition. The confidence interval (CI) is calculated as follows. η_{eff} is calculated using equation (8).
N = Total number of experiments conducted = 18
$U_{\text{items in the estimate}} = 4$
Substituting in the equation (8), $\eta_{eff} = 3.6$
$\alpha = 1$ – confidence limits (95%) = 0.05.
$F_{ratio}(0.05, 1, 2) = 18.51$ (from F tables)
$CI = +/- [18.51 \times 0.000311[1/3.6 + 1/5]]^{1/2}$

CI = +/- 0.052
The 95% confidence level of the predicted optimum of the tool wear under wet condition is given by:
[μ - CI] < μ < [μ - CI], 0.398 < 0.45 < 0.502

4.12.4. Confidence interval around the mean for tool wear – dry condition. The confidence interval (CI) is calculated as follows. η_{eff} is calculated using equation (8).
N = Total number of experiments conducted = 18
$U_{\text{items in the estimate}} = 4$
Substituting in the equation (8), $\eta_{eff} = 3.6$
α = 1 – confidence limits (95%) = 0.05.
F_{ratio} (0.05, 1, 2) = 18.51 (from F tables)
CI = +/- $[18.51 \times 0.001111[1/3.6 + 1/5]]^{1/2}$
CI = +/- 0.099
The 95% confidence level of the predicted optimum of the tool wear under dry condition is given by:
[μ - CI] < μ < [μ - CI], 0.901 < 1.0 < 1.099

4.13. Confirmation experiments
A successful confirmation experiment is defined as one where the average of the samples falls within the predicted confidence interval of the true mean. When the average of the results from the confirmation experiment falls within the confidence interval, it providence evidence that the significant factors as well as their levels are properly chosen. If the average of the results of the confirmation experiment does not fall within the confidence interval, then there has been some form of misinterpretation of the significant factors. Also, the repeatability and reproducibility of the measurement system should be verified.

Five confirmation experiments have been conducted at the optimum settings of the process parameters obtained from the experiment. The results of the confirmation run are given in table 24. The average of the response variables such as surface roughness (R_a) and tool wear are found to be within the confidence interval around the mean. Therefore, the selected factors and their levels are properly chosen and are significant.

Table 24. Results of the confirmation test.

Trial No.	Wet condition R_a (μm)	Dry condition R_a (μm)	Wet condition Tool wear (mg)	Dry condition Tool wear (mg)
1	0.352	0.742	0.48	1.04
2	0.334	0.764	0.40	0.98
3	0.342	0.727	0.5	1.07
4	0.361	0.770	0.44	0.97
5	0.337	0.754	0.41	1.02
Average	0.345	0.751	0.44	1.01

4.14. Microscopic images of tool wear

From the above microscopic images 'figure 13 – figure 16', it is observed that the tool wear during hard turning under dry condition is comparatively higher to tool wear under wet condition.

Figure 13. Before machining – wet.

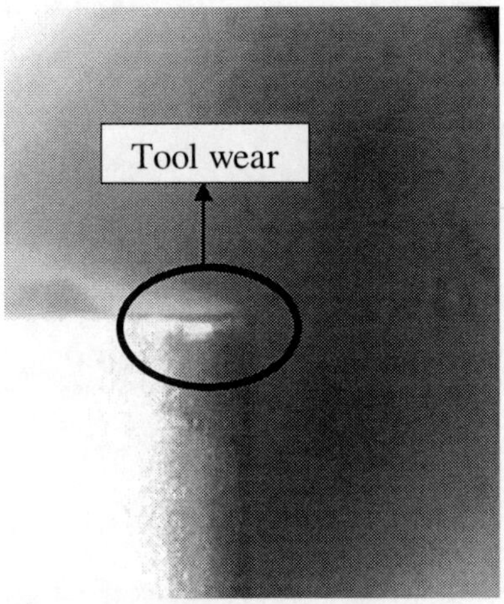

Figure 14. After machining – wet.

Figure 15. Before machining – dry.

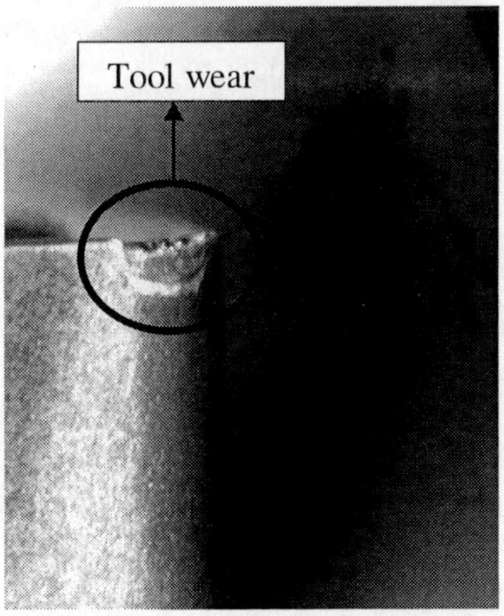

Figure 16. After machining – dry.

4.15. Study on chips

'Figure 17' and 'figure 18', shows comparison of chip forms during hard turning of AISI 4340 steel under dry and wet condition respectively. It is observed that tightly coiled chips are formed during wet turning that could be handled easily where as long snarled chips are prevalent during dry turning.

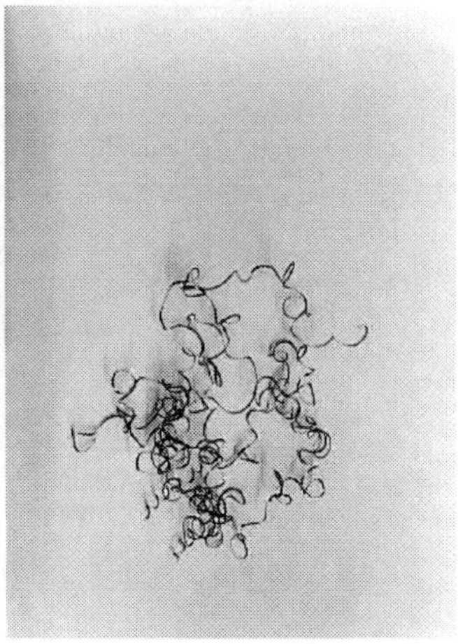

Figure 17. Snarled chips - dry.

Figure 18. Coiled chips – wet.

5. Conclusion

Analysis of surface roughness, tool wear and chip forms in hard turning of AISI 4340 steel using insert CNMG 120408 - DM under wet and dry conditions are presented. Within the range of the parameters under investigation following conclusions can be drawn.

- Surface roughness of AISI 4340 steel during hard turning under dry condition is found to be 46.51% higher to surface roughness generated during wet machining.
- Feed rate is found to be most significant parameter influencing surface roughness with 65.121% contribution under wet condition and 65.31% contribution under dry condition.
- Tool wear of insert during hard turning under dry condition is found to be 58.37% higher to tool wear under wet condition.
- Depth of cut is found to be most significant parameter influencing tool wear with 61.95% contribution under wet condition and 67.14% contribution under dry condition.
- Optimal parameters combination for better surface finish under wet and dry conditions are determined as A3B1C1 i.e. cutting speed at 350 m/min, feed rate at 0.04 mm/rev and depth of cut at 0.2mm.
- Optimal cutting parameters for minimizing tool wear under wet and dry machining are found to be A1B1C1 i.e. cutting speed at 150 m/min, feed rate at 0.04 mm/rev and depth of cut at 0.2 mm.
- Basic study on comparison of chips under wet and dry conditions are carried out. It is clear that snarled chips formed during dry condition may damage the tool or workpiece.

6. References

[1] Hamza Bensouilah, Hamdi Aouici, Mohamed Athmane Yallese, Tarek Mabrouki and Francois Girardin 2016 *Measurement* **82** 1-18

[2] D'Addona D M and Sunil J Raykar 2016 *48th CIRP Conf. on manufacturing systems (Italy)* vol 41 (Amsterdam: Elsevier) pp 841 - 846

[3] Dipti Kanta Das, Ashok Kumar Sahoo, Ratnakar Das and Routara B C 2014 *3rd Int. conf. on materials processing and characterization (Hyderabad)* vol 6 (Amsterdam: Elsevier) pp 1351 - 58

[4] Sathish Chinchanikar, Salve A V, Netake P, More A, Kendre S and Kumar R 2014 *Int. conf. on advances in manufacturing and materials engineering (Karnataka)* vol 5 (Amsterdam: Elsevier) pp 1966 - 75

[5] Varaprasad Bh, Srinivasa Rao ChB and Vinay P V 2014 *12th Global congress on manufacturing and management (Vellore)* vol 97 (Amsterdam: Elsevier) pp 338 - 345

[6] Srithar A, Palanikumar K and Durgaprasad B 2014 *12th Global congress on manufacturing and management (Vellore)* vol 97 pp 72 - 77

[7] Gaurav Bartarya and Choudhury R K 2012 *Int. J. Machine Tools & Manufacture* **53** 1 - 14

[8] Guddat J, M'Saoubia R, Alm P and Meyer D 2011 *1st CIRP conf. on surface integrity (Nottingham)* vol 19 (Amsterdam: Elsevier) pp 118 - 124

[9] Vikram Kumar and CH R Ramamoorty B 2007 *J. Materials Processing Technology* **185** 210 - 216

[10] Grzesik W and Wanat T 2006 *Int. J. Machine Tools & Manufacture* **46** 1988 – 95

[11] Ross PJ 2005 *Taguchi techniques for quality engineering* (New Delhi: Tata McGraw – Hill)

Acknowledgments

Authors wish to sincerely thank the management and principal of PSG college of Technology, Coimbatore for providing infrastructure and other facilities to carry out this research work.

Effect of Crusher Type and Crusher Discharge Setting On Washability Characteristics of Coal

Ahila P, S. Battacharya
Department of Fuel and Mineral Engineering, Indian school of mines, Dhanbad-826004, Jharkhand, India

Abstract. Natural resources have been serving the life of many civilizations, among these coals are of prime importance. Coal is the most important and abundant fossil fuel in India. It accounts for 55% of the country's energy need. Coal will continue as the mainstay fuel for power generation. Previous researches has been made about the coal feed size and coal type had great influence on the crushing performance of the same jaw crusher and amount of fines generated from a particular coal depends not only upon coal friability but also on crusher type. Therefore, it necessitates crushing and grinding the coal for downstream process. In this paper the effect of crusher type and crusher discharge setting on washability characteristics of same crushed non-coking coal has been studied. Thus four different crushers were investigated at variable parameters like discharge settings, different capacities and feed openings. The experimental work conducted for all crushers with same feed size and HGI (Hardgrove Grindability Index). Based on the investigation the results indicate that the four crushers which has been involved for the experimental work shows that the variation in not only the product size distribution and also reduction ratio. Maximum breakage has been occurred at coarsest size fraction of irrespective of crusher type and discharge setting.

Keywords: *Coal, discharge setting, breakage efficiency, washability characteristics, crusher*

1. Introduction

Coal is a sedimentary, combustible rock derived from vegetable debris which has undergone many physical and chemical changes during the very long course of millions of years. It consists primarily of elemental carbon [1]. About 67% of electricity produced in India is by combustion of coal. The total estimated reserves of coal in world are estimated to be909.064 BT and for India the same is estimated to be 301.24 BT [2]. And such Coal is a heterogeneous substance, which is consisted of combustible (organic matter) and non-combustible (moisture and mineral matter) materials. Comminution behavior or grindability of coal, which is a measure of its resistance to crushing and grinding, is related to its physical properties as well as its rank, chemical, and petrographical compositions. Examining the grinding behavior of coal is significant, because the comminution of coal is essential for any kind of its utilization such as combustion, carbonization, gasification, and beneficiation [3-4].
Coal requires a certain treatment before it is marketed as an end product. The treatment may be simple size preparation involving crushing and screening or elaborate beneficiation involving physical cleaning of the coal and/or chemical processing. Coal is crushed in many stages. There are many crusher manufactures and there are many varieties of machines made for crushing of minerals and coals such as jaw crusher, roll crushers, impact crusher and hammer mill [5].

According to Deniz [6] coal feed size and coal type had great influence on the crushing performance of the same jaw crusher used by him in a laboratory investigation and indicated significant variation in crushed coal size distribution. An investigation of this nature could assist plant manager in the selection of appropriate coal crushing equipment. Mikhal et al [7] have shown that the amount of fines generated from a particular coal depends not only upon coal friability but also on crusher type. Similar dependency was observed for crusher product size distribution also. Four crushers, jaw crusher, hammer mill, coalpactor and Bradford breaker, were used to reduce the top size of five coals from different operating mines. The Bradford breaker generated the least fines and the jaw crusher generated the greatest amount of fines.

Similarly the Bradford breaker produced the largest average particle size while the hammer mill and the coalpactor produced the smallest. This article presents a study effect of crusher and crusher discharge setting on washbility characteristics on coal.

2. Materials and methods

Coal samples with top size of 63.5x3mm were collected from Calcutta Electric Supply Corporation power used for experimental studies. Proximate analysis [8] of the sample which is showed in Table 1 has high ash content, low volatile matter and low fixed carbon. Therefore it is non-coking coal. Hargrove Grindability Index (HGI) test were made for coal sample and it was found that average HGI of the coal 61[9]. Crushers used in the experimental work listed in Table 2. These crushers with varied capacities, feed openings and discharge settings were used. Coal was crushed by all the four crushers in parallel followed by size by size float – sink [10] test. Plots of the size distribution curve of feed material were given in Fig 1.

Table 1. Proximate Analysis of Coal Sample Using in Experiments

Parameter	Ash	Volatile matter	Moisture content	Fixed carbon
Wt. (%)	46.86	19.75	3.82	29.57

Table 2. Crushers used in the experimental work

Reduction Ratio	Crusher I	Crusher II	Crusher III	Crusher IV
d 100	1.27	1.22	1.27	1.27
d 90	1.66	1.09	2.04	2.14
d 80	1.63	1.12	1.8	2.11
d 50	1.26	1.05	1.35	1.9
d 100/Discharge setting	2.63	1.59	2.94	3.85

* Calculated as a ratio of feed opening and discharge setting

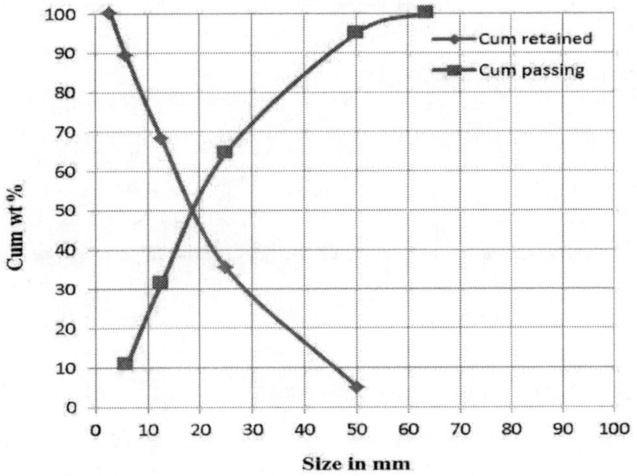

Fig 1 Feed size distribution curve

3. Result and discussion

3.1 Feed and all four crusher product size distribution curves

Comparing feed PSD curve with products of different crushers shown in Figure 2. Tables 3 and 4 indicate that d100 product sizes are not showing significant difference as compared to the feed size, which means notable crushing, occurs at d100. But when it comes to d90 and d80 product size improved crushing is observed. D50 values of both the jaw crusher are quite close, 19mm and 15mm. So, in case with the roll crusher 14mm and 10mm. Such closeness in value however does not get translated into a similarity in reduction ratios. Fig 3 shows how crusher having different capacities and different discharge settings contribute to different average particle sizes for the same coal. Coal has same HGI and which produce different average particle size for the crushed coal.

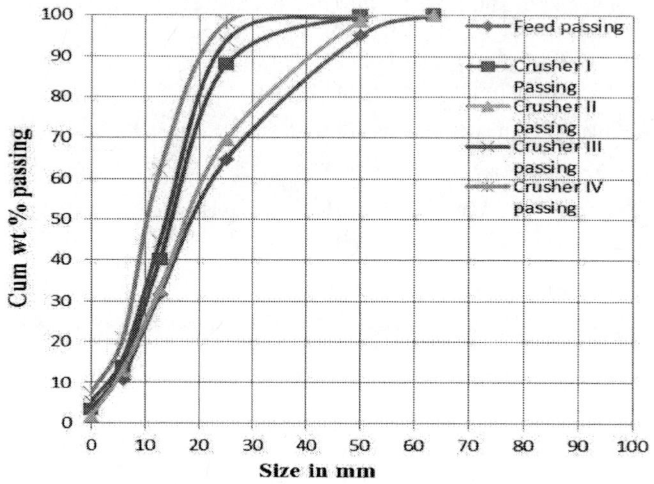

Fig 2 Feed and all crusher product size distribution

Table 3. Feed size and all crusher product size for the same feed size coal

	Feed size, mm	Crusher I product, size mm	Crusher II product, size mm	Crusher III product size, mm	Crusher IV Product size, mm
d100	63.5	50	52	50	50
d 90	45	27	41	22	21
d 80	36	22	32	20	17
d 50	19	15	18	14	10

Table 4. Reduction ratio of all crushers for the same feed size coal

Reduction Ratio	Crusher I	Crusher II	Crusher III	Crusher IV
d 100	1.27	1.22	1.27	1.27
d 90	1.66	1.09	2.04	2.14
d 80	1.63	1.12	1.8	2.11
d 50	1.26	1.05	1.35	1.9
d 100/Discharge setting	2.63	1.59	2.94	3.85

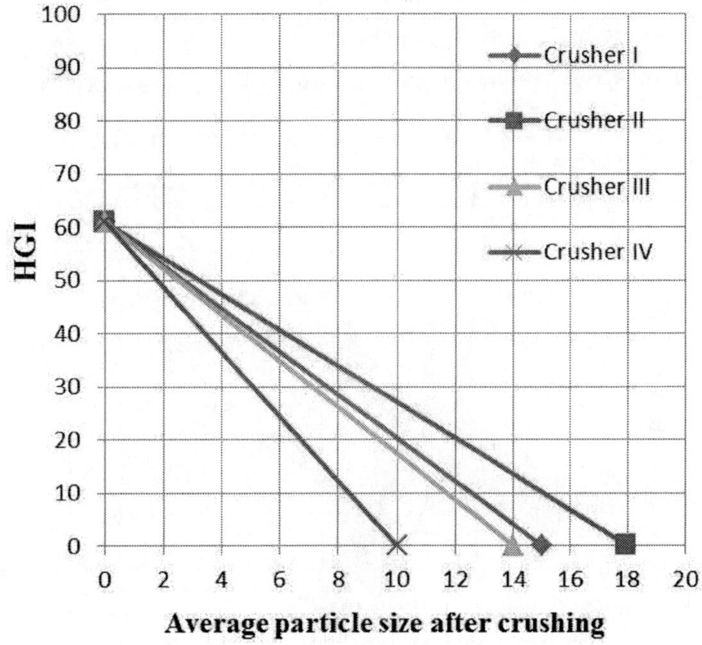

Fig 3 Relationship between HGI and average particle size after crushing

Table 5 shows that irrespective of crusher type and discharge setting, maximum breakage occurs for the coarsest size fraction. Table 3 shows that the ratio between d100 i.e. Maximum product size and discharge setting is highest for roll crusher.

Table 5 Size by size breakage efficiency for crusher products

Size Fraction, mm	Breakage Efficiency (%)			
	Crusher I	Crusher II	Crusher III	Crusher IV
-63.5+50	96.18	72.48	100	100
-50+25	60.46	4.18	79.09	94.56
-25+12.7	-	-	-	-
-12.7+6	-	4.82	-	-
-6+3	5.94	-	2.04	-
-3				

3.2 Comparison of washability curves for all crushed products

Fig 4 shows significant differences between the float curves of the coarsest size fraction obtained in the crusher products, same is the case with yield gravity curves. Only the float curves obtained for the roll crusher products have certain similarly. These differences however decrease as the product size becomes smaller (Fig. 5-8). The differences however increase somewhat for the finest size fraction (Fig.9) due to discharge setting of the crusher and size of the particles

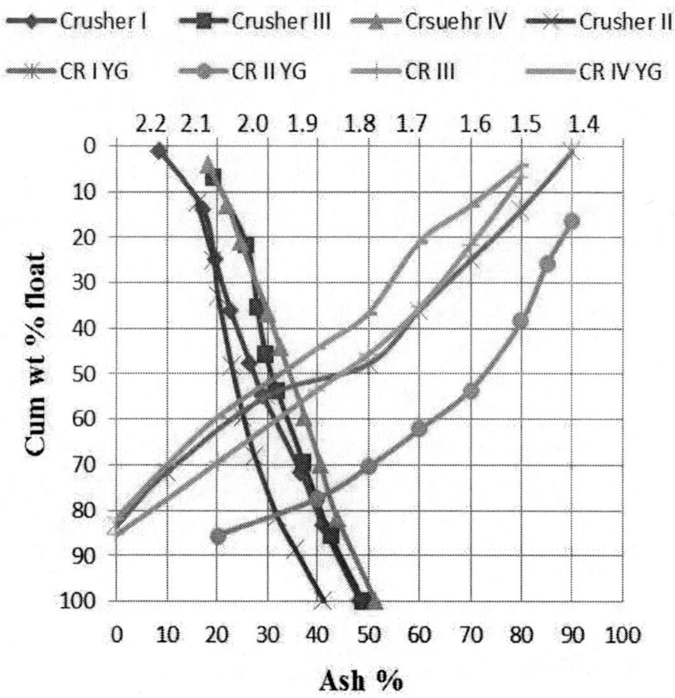

Fig 4. Float and yield curves for all crusher product at -50+25mm

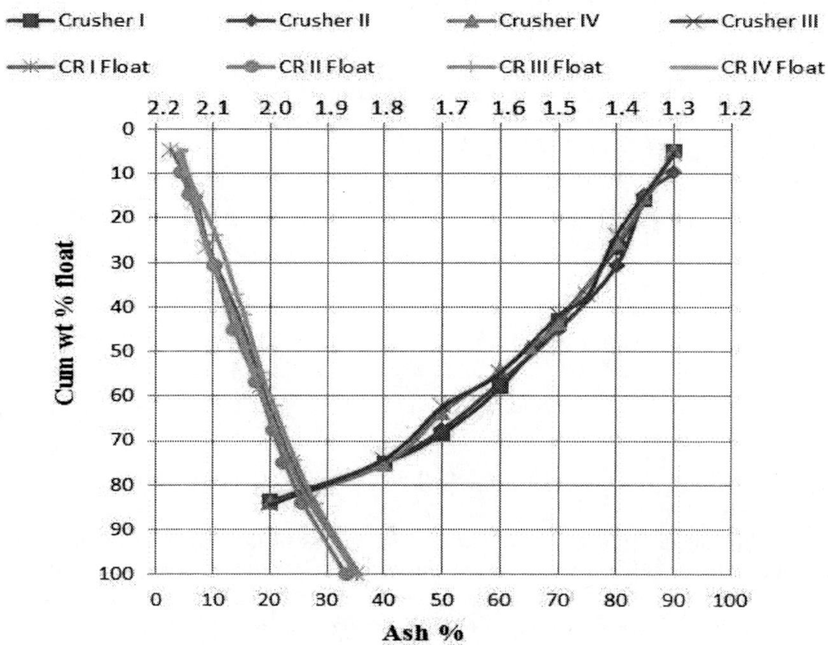

Fig 5 Float and yield curves for all crusher product at -25+12.7mm

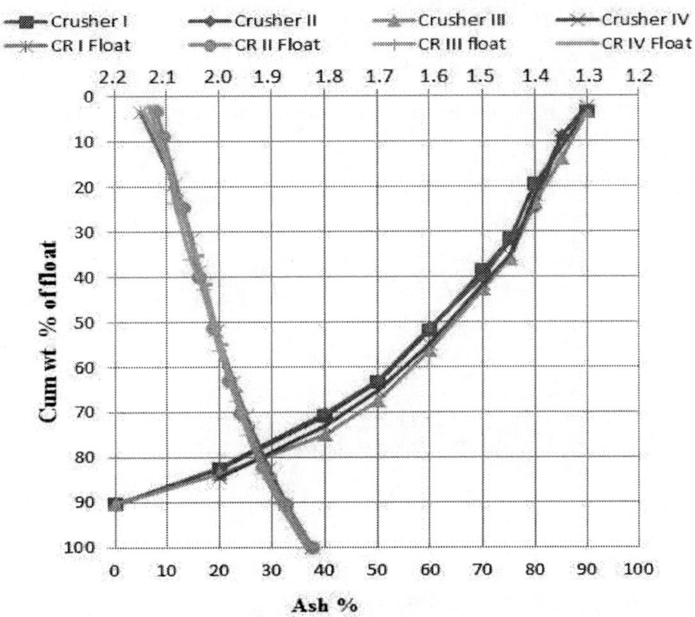

Fig 6 Float and yield curves of all crusher products at -25+12.7mm

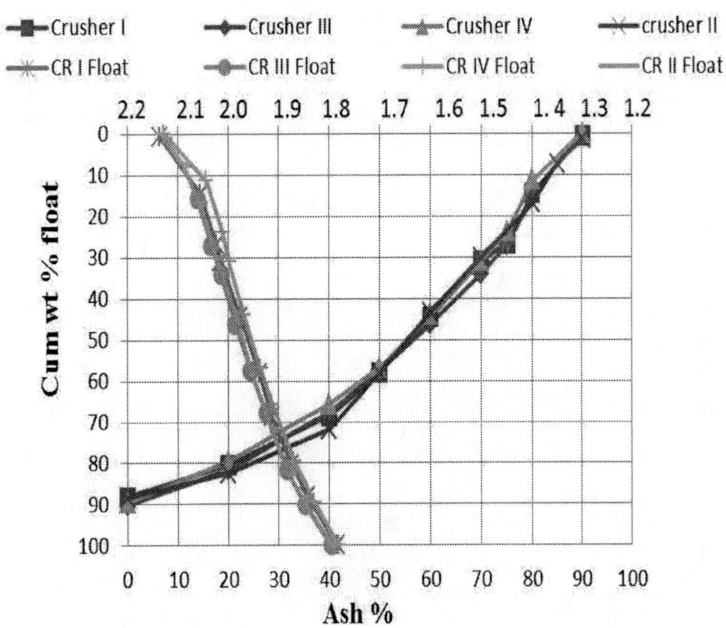

Fig 7 Float and yield curves for all crusher product at -12.7+6mm

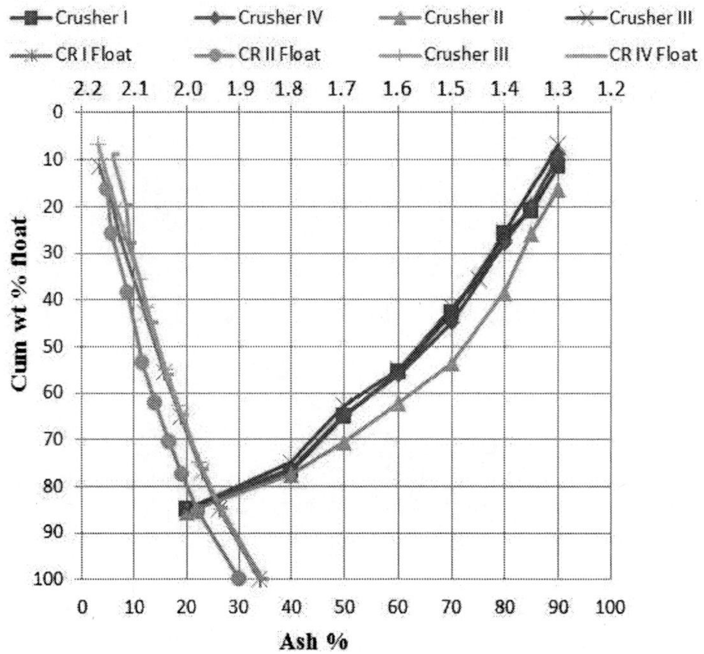

Fig 8 Float and yield curves for all crusher product at -6+3mm

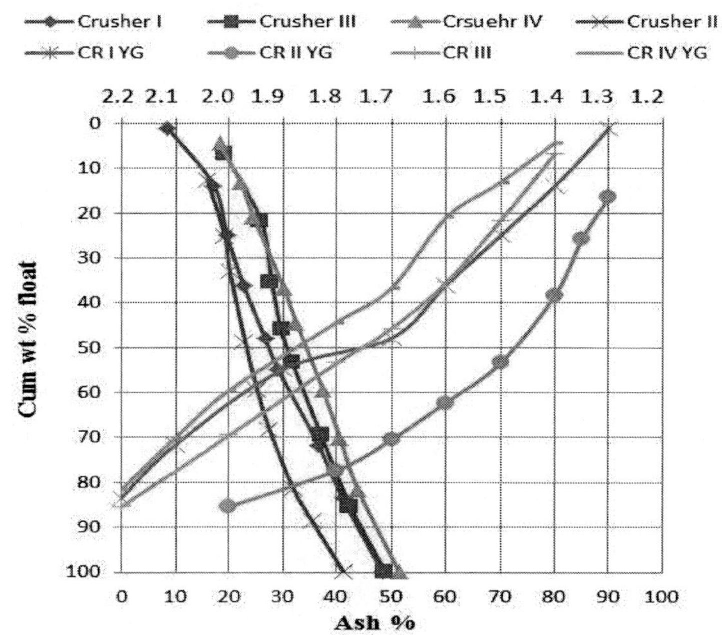

Fig 9 Float and yield curves for all crusher product at -3mm

Table 6 shows typical cleaning parameters as obtained from the washability curves. For every size fraction clean coal yield differ between the crusher products. Since it is a high ash non coking coal target ash for the clean coal has been set at 30%. Cut density however varies only for the two coarsest size fractions because these are having bigger size ranges. Yield variations is however nominal for the finest size fraction indicating similar level of liberation irrespective of the crusher discharge setting and crusher type. Reject ash variation is however similar between the size fractions except the coarsest one. NGM variation is rather significant for a given size fraction between the crusher type and discharge setting. Table 6 illustrates how far a given feed, crusher type and discharge setting can affect the washability results and thereby actual performance of the wash plant designed and built by using the set of data generated through this work.

Table 6 Typical washability parameters for the different crusher products

		Acc = 30%, cc-clean coal			**R- Reject**	
Size mm	Crusher	Y_{CC}	CD	NGM	A_R	Y_R
At-50+25mm	Crusher I	57	1.80	14	72.7	43
	Crusher II	74.5	1.86	13	74.7	25.5
	Crusher III	49	1.78	17	67.5	50.5
	Crusher IV	36	1.70	23.5	64.0	64
At-25+12.7 mm	Crusher I	74	1.89	23	73.3	26
	Crusher II	78	1.88	14	77.4	22
	Crusher III	76	1.94	12	76.8	24
	Crusher IV	72	1.87	13	72.0	28
At-12.7+6 mm	Crusher I	85	1.90	10	80.6	15
	Crusher II	85	1.90	11.5	80.4	15
	Crusher III	88	1.90	8	84.2	12
	Crusher IV	96	1.90	13.5	81.5	4
At -6+3 mm	Crusher I	87	1.90	11	80.0	11
	Crusher II	94	1.90	10	84.2	6
	Crusher III	89	1.90	16	82.1	11
	Crusher IV	90	1.90	13	80.7	10
At-3 mm	Crusher I	93	1.90	13	84.8	7
	Crusher II	100	NA**	NA	NA	NA
	Crusher III	93	1.90	18	88.5	7
	Crusher IV	92	1.90	13	80.7	8

**Not applicable

4. Conclusion

The following conclusion can be made from this work.

The results indicate that the four crushers which have been used in the experimental work shows variation not only in the product size distribution but also in reduction ratios. Maximum breakage has occurred for the coarsest size fractions irrespective of crusher type and discharge setting. Each crusher though crushing the same coal produced a different average size for the crushed coal.

It has been shown that there is significant difference between the float curves and yield curves of the same coarsest size fractions in the four crusher products. The difference however decreases with the smaller size fractions though it again increases for the finest size fractions. Washability analysis of all the crusher products have shown that for every size fraction clean coal yield varies between the crushers and cut density variation occurred at coarser size fractions, but not in finer size fractions. For a given feed coal, crusher type and discharge setting can therefore affect the crushing efficiency and breakage product size distribution and washability results and thereby the actual performance of the wash plant designed on the basis of this set of laboratory data.

IOP Conf. Series: Materials Science and Engineering **314** (2018) 012005 doi:10.1088/1757-899X/314/1/012005

5. References

[1] Joseph W. (Editor) Leonard (1979), Coal Preparation Technology, 4th edition AIME , New York p:1-11

[2] www.wikipediacoal reserves.com and www.coal.nic.in ,last accessed 27-04-2016.

[3] Vuthaluru, H. B et al. 2003. Effects of Moisture and Coal Blending on Hardgrove Grindability Index of Western Australian Coal. Fuel Processing Technology 81: 67–76.

[4] Chelgani, S. C et al 2008. Prediction of Coal Grindability Based on Petrography, Proximate and Ultimate Analysis using Multiple Regression and Artificial Neural Network Models. Fuel Processing Technology 89: 13–20.

[5] Deniz, V., and Y. Umucu. Forthcoming Interrelationships between the Bond Grindability with Physicomechanical and Chemical Properties of Coals. Energy Sources (Part A), doi: 10.1080=15567036.2010.504942.

[6] Vedat Deniz (2014), A study on the effects of coal feed size and coal type on the performance of a laboratory jaw crusher Energy source, Part A, 36:1249-1255

[7] Mikhal M.W et al. The effect of different crusher of fines generation from western Canadian friable coals, XVI International Mineral Processing Congress, Stockholm pp:1573-1583.

[8] Indian Standard: 1350 Part 1, 1984, Methods of Test for Coal and Coke: Proximate analysis (Second Revision), Bureau of Indian Standards, New Delhi.

[9] Indian Standard: 4433, 1979, Determination of Hargrove Grindability Index of Coal, Bureau of Indian Standards, New Delhi.

[10] Indian Standard: 13810, 1993, Code of Practice for Float and Sink analysis of Coal, Bureau of Indian Standards; New Delhi.

IOP Conf. Series: Materials Science and Engineering **314** (2018) 012006 doi:10.1088/1757-899X/314/1/012006

Denitrogenation model for vacuum tank degasser

R Gobinath[1] and R Vetrivel Murugan[1]
[1]JSW Steel Limited, Salem Works, Tamilnadu, India

Email: gobinath.r@jsw.in

Abstract. Nitrogen in steel is both beneficial and detrimental depending on grade of steel and its application. To get desired low nitrogen during vacuum degassing process, VD parameters namely vacuum level, argon flow rate and holding time has to optimized depending upon initial nitrogen level. In this work a mathematical model to simulate nitrogen removal in tank degasser is developed and how various VD parameters affects nitrogen removal is studied. Ladle water model studies with bottom purging have shown two distinct flow regions, namely the plume region and the outside plume region. The two regions are treated as two separate reactors exchanging mass between them and complete mixing is assumed in both the reactors. In the plume region, transfer of nitrogen to single bubble is simulated. At the gas-liquid metal interface (bubble interface) thermodynamic equilibrium is assumed and the transfer of nitrogen from bulk liquid metal in the plume region to the gas-metal interface is obtained using mass transport principles. The model predicts variation of Nitrogen content in both the reactors with time. The model is validated with industrial process and the predicted results were found to have fair agreement with the measured results.

1.Introduction

With technological advances the current steel quality has improved drastically with comparison with quality a decade ago. With emphasis on clean steel technology inclusion content and gas content are controlled to greater extent. Nitrogen acts as a detrimental and beneficial element in steel based on its nature of post processing and applications. Grades of steel which undergoes deep drawing requires lower nitrogen content. For such grades control of nitrogen starts from primary steelmaking to till caster. To get desired nitrogen one has to control incoming nitrogen through use of low nitrogen ferroalloys and controlling nitrogen pickup during the entire process. The other way is to remove it through degassing operation from the liquid steel by purging argon from bottom of the ladle through the porous plug system. As the argon bubbles rise through the plume, it picks up nitrogen gas and leaves to gases maintained at low pressure at the top. In the present work a mathematical model to predict temporal evolution of nitrogen in liquid steel during vacuum degassing for different operating parameters namely argon flow rate, vacuum level and holding time is developed.

2.Model conceptualization

Water model studies [1] [2] [3] on ladle with bottom purging have shown presence of two distinct flow regions inside the ladle, namely the plume region (mixture of liquid steel and air bubbles) and outside plume region (the recirculating region) and this is illustrated in figure 1 (a). In the model these two regions are considered as separate reactors for model formulation. In the plume region, say reactor 1, the argon bubbles are formed and they rise up in the bath. During this ascent of a bubble, transfer of nitrogen from liquid steel to bubble takes place. The liquid from the plume region flows to the reactor 2

IOP Conf. Series: Materials Science and Engineering **314** (2018) 012006 doi:10.1088/1757-899X/314/1/012006

at the top of the ladle as shown in the figure 1 (b), and vice versa at the bottom. This mass transfer of liquids results in mixing.

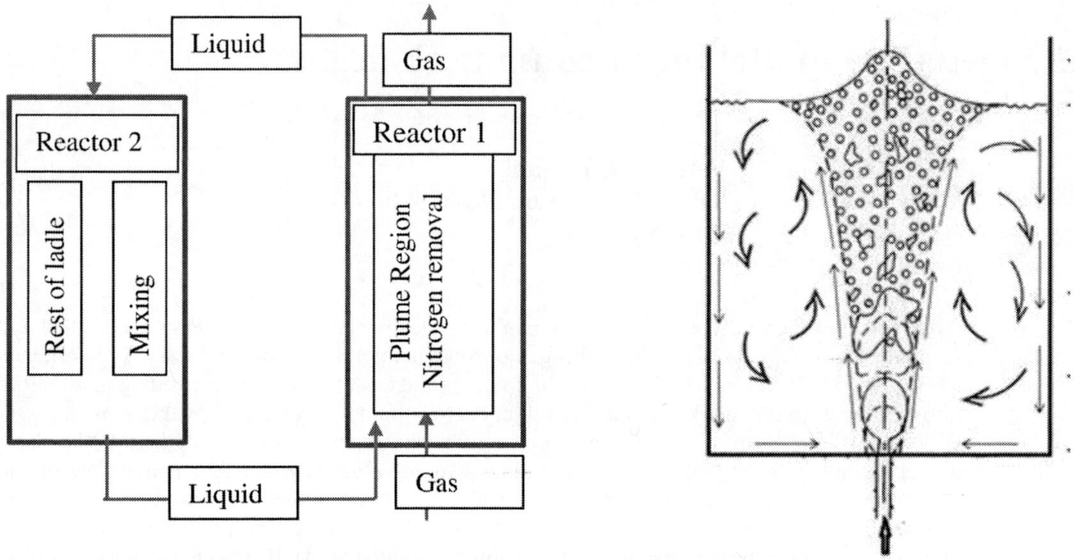

(a) Model conceptualization showing two reactors (b) Situation prevailing in a gas stirred ladle [1]

Figure 1. Model conceptualization

3.Transfer of nitrogen into a Single Bubble

Modelling transfer of species across a single bubble is taken from works of Haijuan Wang et.al. [4]. Nitrogen removal from liquid steel takes place in the following three steps.

1. Liquid phase mass transfer of nitrogen from liquid steel to the gas-liquid interface.
2. Conversion of N into N_2 takes place by interfacial reaction given by equation (1).

$$\underline{N} = \frac{1}{2} N_2 \tag{1}$$

3. Transfer of N_2 into the argon bubble.

When argon is purged into liquid steel argon bubbles are produced. Argon bubble at the bottom of the ladle has zero partial pressure of nitrogen and as it travels up, the amount of nitrogen and partial pressure inside the bubble keeps on increasing. The nitrogen transfer from liquid steel to the gas-liquid steel interface is governed by mass transfer. At this interface dissolved N in liquid steel is assumed to be in equilibrium with N_2 in the gas phase and this assumption is fairly valid under such high steel making temperatures. The N_2 from the interface is transferred to the bulk gas in the bubble by convective mass transport. When the partial pressure of nitrogen inside the bubble attains equilibrium, then the transfer of nitrogen into the bubble ceases. During the travel of any particular bubble, no compositional change occurs in plume region because bubble travel time is assumed much shorter than the refining time. The rate of nitrogen transfer (moles/sec) into argon bubble is given mathematically by equation (2) and equation (3) gives equilibrium nitrogen in the bath

$$\dot{n}_{N_2} = \frac{A_b k_N^m \rho}{100 M_{N_2}} (wt\%N^B - wt\%N^{eq}) \tag{2}$$

$$wt\%N^{eq} = P_{N_2}^{\frac{1}{2}} \,/\, (K_{eq} * f_N) \tag{3}$$

K_{eq} is the equilibrium constant for the reaction [1].

At any height of the ladle the moles of nitrogen (n_{N_2}) inside the bubble can be traced by doing a mass balance of nitrogen as given by equation (4).The molar balance for the bubble travelling a small distance '*dh*'is

$$\frac{d(U_b n_{N_2})}{dh} = \dot{n}_{N_2} \tag{4}$$

U_b is the velocity of bubble at any height '*h*' in the plume region.
The next step in the model is predicting rate of nitrogen removal in the plume region, which can be calculated from equation (5)

$$\frac{d(wt\%N)_p}{dt} = \frac{\dot{N}_b * n_{N_2 top} * 100 * M_{N_2}}{m_p} \tag{5}$$

The term '$n_{N_2 top}$' is the number of moles of nitrogen inside a bubble when it escapes at the top of melt surface, 'm_p' is the plume liquid mass and equation (6) gives number of bubbles produced per unit time $\dot{N}_b$

$$\dot{N}_b = \dot{n}_{Ar} * R * \frac{T}{V_b\left(P_v + \rho g Z + \left(\frac{4\sigma}{d_b}\right)\right)} \tag{6}$$

The term $4\sigma/d_b$ in the above equation is the excess pressure due to surface tension of liquid steel ($\sigma = 1.8$ N/m) which restricts the increase in bubble number.

4.Bubble parameters and mass transfer

Numerous cold model studies were done to study bubble dispersion modes with modified parameters. At higher gas flow rates through porous plug, a single large bubble is formed at plug exit [1] [2] [3]. The rising bubble disintegrates into number of small bubbles during its upward movement. Sano and Mori [5] [6] found that the large bubble disintegrates into small bubbles within 12cm from the orifice and estimated average bubble diameter after disintegration and empirical relation for determining bubble size is given by equation (7).

$$d_b = 0.091\left(\frac{\sigma}{\rho_l}\right)^{0.5} V_s^{0.44} \tag{7}$$

In this model the number of bubbles at any time is assumed to be constant and change in bubble size during a bubbles upward movement happens due to change in pressure inside the ladle with height. The increase in bubble size with respect to entrant of nitrogen in to bubble is negligible. When the bubbles move upward, the bubble expands due to decrease in ferrostatic pressure and it breaks into number of bubbles. Subsequently, the ascending bubbles also coalescence by which the number of bubbles decreases [1][2]. Due to simultaneous breakage and coalescence, constant bubble population is assumed throughout the bath. Spherical capped bubbles are formed during ladle refining operation [7] and equation (8) gives the empirical relation for velocity of spherical capped bubble is [7].

$$U_b = 1.02\left[\frac{g d_b}{2}\right]^{0.5} \tag{8}$$

The mass transfer rate depends on the mass transfer coefficient k_N^m. Mass transfer coefficient of nitrogen, equation (9) is given by Richardson and Bradshaw is used in the model [8].

$$k_N^m = 1.08\, d_b^{-\frac{1}{4}} D^{\frac{1}{2}} g^{\frac{1}{4}} \tag{9}$$

5.Plume Parameters

As nitrogen removal takes place in plume region, plume dynamics plays crucial role in rate of denitrogenation. Empirical relations for plume parameters for a conical plume is available in works of Krishnapisharody et.al. [9] [10]. The plume cone angle varies from 15 to 20 degrees. Plume parameters namely gas fraction (α) equation (10) and liquid velocity (U_l) equation (11) was derived with plume angle as 17.5 degrees in their work, which is used in the present model also

$$\alpha = 1.84Q^{0.63}Z^{-1.57} \tag{10}$$

$$U_l = 2.57Q^{0.31}Z^{-0.28} \tag{11}$$

Mass of steel in the plume region is calculated by equation (12)

$$m_p = \frac{\rho.steel\,fraction.V_P}{1000} \tag{12}$$

From the gas fraction (ratio of volume of gas in the plume region to volume of plume) in plume region, steel fraction can be calculated.

6.Mixing inside the ladle

Porous plug is used to assist mixing in the ladle for homogenising composition and temperature. This mixing is favoured by rising bubbles in the plume region. The rising bubble creates mass flow inside the ladle which makes the mixing possible. Tanks in series model approach is employed for simulating mixing in an agitated bath [11] [12]. In modelling mixing in VOD process, the vessel is divided into two well stirred tanks between which there is a constant exchange of the liquid [12]. As discussed earlier the current model was formulated by dividing the entire ladle into two well stirred reactors or tanks for simulating the mixing behaviour. Both the reactors are treated as continuously stirred reactors. This mixing leads to change in nitrogen concentration in both the reactors with time and nitrogen evolution inside plume and outside plume can be evaluated by performing a mass balance. The difference equations are given by equation (13) and equation (14).

$$N_{OP_{t+dt}} = \frac{m_{OP}N_{OP_t} + \dot{m}_l N_{P_{t+dt}}\Delta t}{m_{OP} + \dot{m}_l \Delta t} \tag{13}$$

$$N_{P_{t+dt_{mixing}}} = \frac{m_P N_{P_{t+dt}} + \dot{m}_l N_{OP_{t+dt}}\Delta t}{m_p + \dot{m}_l \Delta t} \tag{14}$$

where $\dot{m}_l$ is mass flow rate of liquid exchanged between the two reactors.

Water model study [1] shows the flow of the liquid is predominantly vertical in the plume region and the movement becomes horizontal at the top of plume region. Liquid transfer from plume region to the rest of the ladle and vice-versa happens as shown in figure 1. The mass flow rate $\dot{m}_l$ between the two reactors is given by using the equation (15)

$$\dot{m}_l = (1 - \alpha)A_c\rho U_l \tag{15}$$

7.Plant process and trial

The process route at JSW Steel Limited Salem works is through Energy Optimizing Furnace-Ladle Furnace-Vacuum Degassing-Caster route. The plant produces low alloy special steel long products for automobile application. The model is developed for a 45-ton ladle. After the refining process the ladle is taken to vacuum degassing station at average temperature of 1620° C. Vacuum is obtained through steam ejectors. The liquid steel temperature is assumed to be 1600 ° C and no change in temperature occurs during the degassing process. LECO oxygen nitrogen analyser is used to measure nitrogen content before and after the degassing process. In the model denitrogenation at surface is ignored. Slag is not considered in the simulation. Table 1 shows process parameters for few heats for which model validation was done.

Table 1. Vacuum parameters for few heats taken for model validation

Case no	Holding time (mins)	Argon flow rate (Nm3/hr)	mbar	BVD Nppm	Measured value AVD Nppm
1	8	7	0.30	43.0	36.6
2	9	5	0.40	58.0	42.9
3	9	10	0.40	82.0	48.1
4	8	8	0.48	58.3	44.1
5	8	8	0.50	44.3	38.7
6	8	8	0.50	44.6	34.4

8.Model assumptions

- The model is assumed to be isothermal and temperature of liquid steel is 1600° C. The effect of temperature drop during vacuum degassing is ignored.
- When the bubbles move upward, there is a volume expansion due to decrease in ferrostatic pressure. The volume expansion results in breakage of bubbles and more number of bubbles are generated, but simultaneously coalescence of bubbles also takes place. Due to simultaneous breakage and coalescence, the bubble population is assumed to be same throughout the bath. Similarly, for the above reason an average bubble size has been assumed.
- Composition of liquid steel inside the plume remains the same during the travel of any particular bubble.
- Slag is neglected in the simulation.
- No CO bubbles are formed due to decarburization as oxygen ppm before degassing is in the range of 10 -15 ppm.
- In JSW process the average oxygen and sulphur level before degassing is 15 ppm and 120 ppm, their effect as surface effective element is ignored. This assumption is valid as after degassing process the sulphur level is 10 ppm.

9.Effect of Steel composition

The presence of other alloying elements also affects the removal of nitrogen. The activity coefficient of nitrogen is influenced by the other solutes such as C, Mn, Si, Al, P, S, Cr, Ti, V and others. When the activity coefficient of nitrogen is raised then the removal of nitrogen will be more because of the decrease in equilibrium nitrogen level. Sulphur, phosphorus and aluminium are generally present in steel in the range 0.005-0.015, 0.005-0.015 and 0.020-0.030 respectively. Their influence on activity coefficient is insignificant. For a low alloy steel, the activity coefficient of nitrogen is approximately around 1. Since the second order interaction parameter for nitrogen on nitrogen is zero, the change in nitrogen while denitrogenation makes no change in the activity. So thermodynamically it is the partial pressure which determines the effectiveness of denitrogenation.

10.Model validation

Without validation the model cannot be accepted. The model was validated with the plant data as given in table 1. The model was developed and the governing equations are solved in MATLAB software. For the simulation diffusion coefficient for nitrogen, the equilibrium constant K_{eq} is calculated from standard Gibbs energy data from MSTS [13]. Interaction parameters for nitrogen in liquid iron is taken from Sigworth and Elliot [14]. The model validation covers for different flow rates and different initial nitrogen levels to ascertain the applicability of the model for various process conditions. The inputs for the model namely, liquid steel weight, chemical composition of the liquid steel and VD parameters are taken for simulation. The evolution of nitrogen concentration is calculated using a small time step Δt for a given mass of liquid steel. The model predicts nitrogen concentration in both the reactors (plume region and outside plume region) with time, which is shown in figure 2 for case 4. From figure 2 it can be seen that due to the mixing phenomena, the nitrogen concentration in both the reactors converges with progress in degassing time and finally almost same value is achieved. The measured and model predicted value (44.1 ppm) is same for this case. Figure 3 shows comparison between predicted and measured values for 6 heats and close agreement is seen, which shows the model is reliable in predicting the final nitrogen for a vacuum tank degassing process. The effect of vacuum parameters is also analysed in the current work which shows lower vacuum level and increased argon flow rate favours better denitrogenation, as in works of other researchers. As the current work is focussed on nitrogen prediction for given parameters the results on effect of VD parameters is ignored here.

The limitations of the model are denitrogenation on the surface layer is ignored and the model cannot be used if oxygen and sulphur contents are high enough to affect denitrogenation.

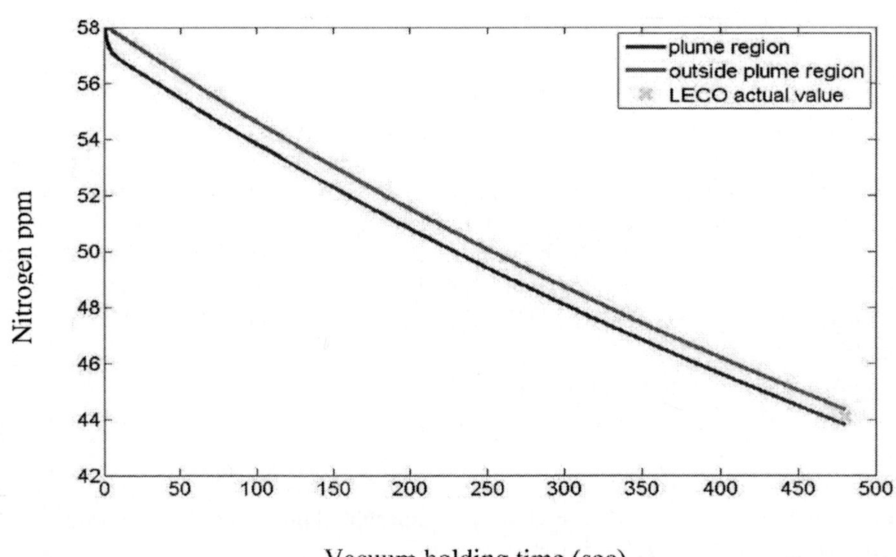

Figure 2. Model output for case 4

IOP Conf. Series: Materials Science and Engineering **314** (2018) 012006 doi:10.1088/1757-899X/314/1/012006

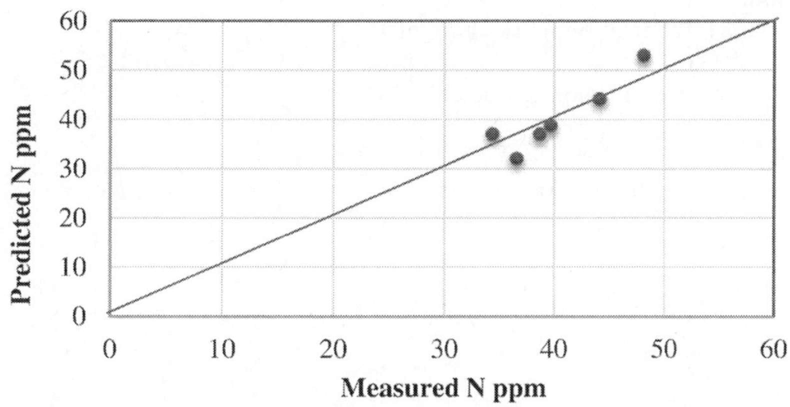

Figure 3. Comparison between model predicted value and plant measured value

11.Conclusion

Based on thermodynamics, mass transfer coupled with bubble dynamics and mixing principles a mathematical model to simulate nitrogen removal in vacuum tank degassing process was developed. The model features nitrogen removal through single argon bubble and nitrogen composition in the plume region with time. Nitrogen evolution in rest of the ladle is captured by incorporating the mixing phenomena. Model validation with plant trials showed close agreement between model value and measured value. Lower vacuum levels and increased argon flow rate shows better denitrogenation. As the results are comparable with actual values the model can be used for predicting final nitrogen for a tank vacuum degassing process.

12. List of symbols

Variables

A_b	Surface area of bubble (m^2)
A_c	Cross sectional area of plume (m^2)
d_b	Diameter of bubble (m)
dh	Small distance (m)
dt	Small time (sec)
D_i	Diffusion coefficient of species 'i'(m^2/s)
f_N	Henrian activity coefficient
g	Gravitational constant (m/s^2)
N	Dissolved nitrogen in liquid steel
k_i^m	Mass transfer coefficient of species 'i'(m/s)
M	Molar mass (kg/ mol^{-1})
m	Mass of steel (kg)
n	Moles
$\dot{n}$	Moles/sec
P	Pressure (bar)
P_{N_2}	Partial Pressure of Nitrogen (atmosphere)
Q	Argon flow rate (m^3/s)
R	Gas constant (m^3 atm K^{-1} mol^{-1})
T	Temperature (Kelvin)
t	Time (sec)
V	Volume (m^3)
V_s	Superficial velocity (m/sec)
Z	Axial height of liquid steel in ladle (m)

Greek Symbols

ρ	Density of liquid steel (kg/m^3)
α	Gas fraction
σ	Surface tension of liquid steel (N/m)

Superscript

B	Bulk nitrogen concentration
eq	Equilibrium

Subscript

Ar	Argon
h	Height
P	Plume
OP	Outside plume
v	Vacuum pressure

13.References

[1] Anagbo P E, Brimacombe J K and Castillejos A H 1989 *Canadian Metallurgical Quarterly.* **28** 323
[2] Anagbo P E and Brimacombe J K 1990 *Metallurgical Transactions* **B 21** 637
[3] Alexiadis A 2007 *Applied Mathematical Modelling* **31** 1534
[4] Wang H, Viswanathan N N, Ballal N B and Seetharaman S 2010 *International Journal of Chemical Reactor Engineering* **8** 1
[5] Sano M and Mori K 1980 Trans. *ISIJ.* **20** 668
[6] Mori K, Sano M and Sato T 1979 Trans. *ISIJ.* **19** 553
[7] Szekely J 1979 *Fluid Flow Phenomena in Metals Processing* (New York: Academy Press)
[8] Bradshaw A V and Richardson F D 1965 *The Iron and Steel Institute* 24.
[9] Krishnapisharody K and Irons G A 2007 *Metallurgical and Materials Transactions* **B 38** 367
[10] Krishnapisharody K and Irons G A 2007 *Metallurgical and Materials Transactions* **B 38** 377
[11] Meitz J and Oeters F 1989 *Canadian Metallurgical Quarterly* **28** 19
[12] Krishnakumar K, Ballal N, Sinha P, Sardar M and Jha K1999 *ISIJ international.* **39** 419
[13] Turkdogan E T and Fruehan R J 1999 Fundamentals of iron and steelmaking *The Making, Shaping, and Treating of Steel* vol. 2, ed R J Fruehan (Pittsburgh: The AISE Steel Foundation)
[14] Sigworth G K and Elliott J F 1974 *Metal Science* **8** 298
[15] Bannenbergg N, Bergmann B and Gaye H 1992 *Steel Research* **63** 431

Acknowledgements
The authors thank JSW Steel Limited Salem Works for their support to carry out this project.

IOP Conf. Series: Materials Science and Engineering **314** (2018) 012007 doi:10.1088/1757-899X/314/1/012007

PZT/PLZT - elastomer composites with improved piezoelectric voltage coefficient

K.Harikrishnan[1*,3], D V Bavbande[2], Dhirendra Mohan[2+], B Manoharan[1], M R S Prasad[1] and G Kalyanakrishnan[4]

[1] Zirconium Complex (A unit of NFC, Hyderabad), Dept. of Atomic Energy, Pazhayakayal, Tuticorin.
[2] Glass and Advanced materials division, Materials Group, Bhabha Atomic Research Centre, Trombay, Mumbai, India
[2+] Retired , Bhabha Atomic Research Centre, Trombay, Mumbai, India
[3] On leave from BARC Training School, Bhabha Atomic Research Centre, Trombay, Mumbai, India
[4] Nuclear Fuel Complex, Dept. of Atomic Energy, Hyderabad
[1*]Corresponding author Email: hari_gce@yahoo.co.in

Abstract: Lead Zirconate Titanate (PZT) and Lanthanum-modified Lead Zirconate Titanate (PLZT) ceramic sensor materials are widely used because of their excellent piezoelectric coefficients. These materials are brittle, high density and have low achievable piezoelectric voltage coefficients. The density of the sintered ceramics shall be reduced by burnable polymeric sponge method. The achievable porosity level in this case is nearly 60 – 90%. However, the porous ceramic structure with 3-3 connectivity produced by this method is very fragile in nature. The strength of the porous structure is improved with Sylgard®-184 (silicone elastomer) by vacuum impregnation method maintaining the dynamic vacuum level in the range of -650 mm Hg. The elastomer Sylgard®-184 is having low density, low dielectric constant and high compliance (as a resultant stiffness of the composites is increased). To obtain a net dipole moment, the impregnated ceramic composites were subjected to poling treatment with varying conditions of D.C. field and temperature. The properties of the poled PZT/ PLZT - elastomer composites were characterized with LCR meter for measuring the dielectric constant values (k), d_{33} meter used for measuring piezo-electric charge coefficient values (d_{33}) and piezo-electric voltage coefficient (g_{33}) values which were derived from d_{33} values. The voltage coefficient (g_{33}) values of these composites are increased by 10 fold as compared to the conventional solid ceramics demonstrates that it is possible to fabricate a conformable detector.

Key Words: Dielectric constant, Composites and Sintering

1 Introduction

Piezoceramics are the modern ceramics which, in present-day technology, find many interesting applications due to the unique electro-mechanical coupling associated with these materials. The various applications include pressure sensor, acoustic sensor, accelerometer, gas lighter, ultrasound, sonar, buzzer, actuator etc. The piezoelectric materials are characterized using parameters like d_{33}, d_{31}, g_{33} etc. The d_{33} parameter is the piezoelectric strain coefficient which gives the electric charge per unit area generated by a unit mechanical stress. It also measures the strain induced by a unit electric field in these materials. The 1st and 2nd numbers of suffix 33 or 31 show the direction of electrical and mechanical stimuli respectively, 3 being the direction parallel to electric field applied during poling and 1 being other perpendicular direction. The g_{33} (piezoelectric voltage coefficient) gives the amount of electric field generated on the application of unit stress in these materials. This value is derived by using the relation $g_{33}=d_{33}/(k*\varepsilon_o)$, where ε_o - the permittivity of the material- 8.854×10^{-12} $C.V^{-1}m^{-1}$. This value is used as a figure of merit for the passive piezoelectric devices [1, 2].

The g_{33} value which can be achieved in poled PZT ceramics by varying their composition is of the order of 20-40 x10^{-3} V-m/N [1]. This value can be enhanced by an order of magnitude by using composites made of piezoceramics & polymer. The other advantages of composites are their low density, good compliance and flexibility. Lower density makes them suitable for devices like ultrasound and sonar due to their better coupling with human tissues and water respectively. Better compliance provides damping (lower mechanical Q) in these materials. Due to this, acoustic sensor, pressure sensor etc. using these materials may not require any back up materials. Their flexibility offers the interesting possibility of fabricating a conformable detector [3, 4].

In this study polymeric sponge method was used to make porous ceramic with 3-3 connectivity as it was easier to obtain large amount of porosity using this method [5, 6]. Pores were filled with silicone elastomer by vacuum impregnation. Poling parameters like D.C. field strength, temperature were varied to find their effect on the d_{33} and g_{33} values of the composites.

2 Experimental methods

2.1. Synthesis of ceramics

Two ceramics of the following stoichiometry $Pb_{0.95}La_{0.05}(Zr_{0.58}Ti_{0.42})O_3$ [PLZT] and $Pb(Zr_{0.55}Ti_{0.45})O_3$ [PZT] were prepared by the solid state route. The starting powders, AR grade PbO, ZrO_2, TiO_2, La_2O_3 (Aldrich Corp.) were ball milled for 6 hours, followed by calcination at 800^{0}C for 1 hour. The calcined powders were analysed by X-ray diffraction analysis (Philips make) with the target of Cu-Kα and the radiation wavelength (λ)-1.54A^o. Data collection was performed in the 2 θ range of 20°–60° with a step scan with a step size of 0.02° and counting time of 2s/step. Fig.1 and 2 illustrates the XRD pattern of PZT and PLZT respectively showing the pure perovskite phase formation in these materials. The theoretical density of PLZT and PZT ceramics calculated by XRD analysis were found to be 8.18 g/cc and 7.98 g/cc, respectively. The powders were pressed into 10 mm diameter pellets at 200 MPa using a hydraulic press.

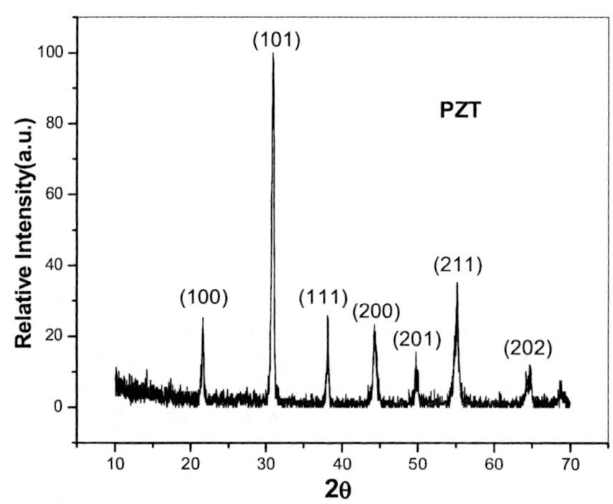

Fig.1 XRD pattern of PZT

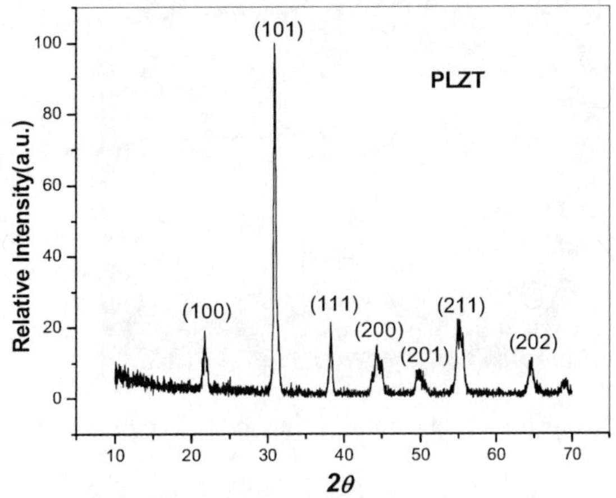

Fig.2 XRD pattern of PLZT.

2.2. Preparation of porous ceramics

To make porous ceramics, slurry of the powders was first prepared. In this preparation 2 g of PVA (Polyvinyl alcohol) was dissolved in 100 cc of hot distilled water. About 25 cc of this solution was added to 45 g of ceramic powder and ball milled for 2 hours to get slurry of good consistency. Round pieces of sponge were dipped in the slurry and ball milled for 2 minutes to ensure the voids of the sponge are filled with slurry. Slurry dipped sponge pieces of each ceramic formulation were further divided into two lots. One lot of each formulation was spun in a centrifuge to remove excess of slurry. In this way four lots, two spun and two non-spun were prepared. Porous ceramics and composites prepared from the spun samples are referred as spun ceramics/composites in this work. Other two lots are referred as non-spun ceramics/composites. The slurry loaded sponge was dried in an oven at 80^0C. The green ceramic pellets and slurry loaded sponge pieces were sintered at 900^0C, 1000^0C and 1200°C for an hour in PbO atmosphere. The rate of heating and cooling in the sintering schedule was kept at 1°C/minute to avoid cracking in the fragile porous structure. The polymeric sponge was burned out during sintering leaving a porous ceramic structure. The density and porosity of the bulk and porous ceramics were calculated using Archimedes Principle. The sintered density of non-spun porous ceramic sintered at 900-1200°C ranged between 40-60% T.D. The density of spun porous ceramic samples varied between 8.5 to 12.5% T.D. The magnified pictures of the porous ceramics are given in Fig. 3 and 4.

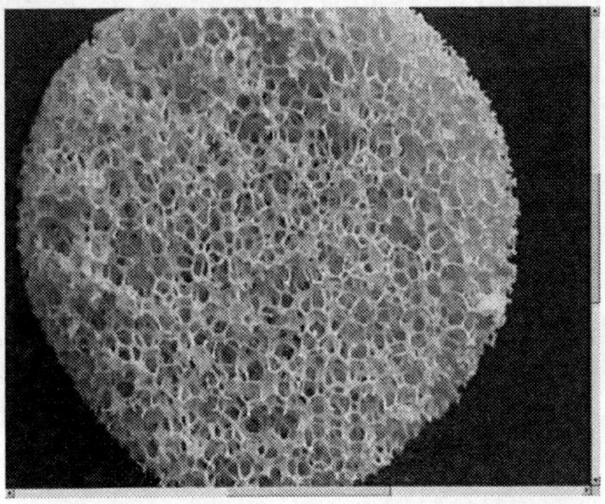

Fig.3 Photograph of porous spun ceramics sintered at 1100^0C (sample dia= 8.8mm)

Fig. 4 Photograph of porous non-spun ceramics sintered at 1100^0C (sample dia = 8.8 mm)

2.3. Impregnation of porous ceramics

The epoxy and hardener of the silicone elastomer (Sylgard® 184, Dow Corning Corp., U.S.A) were mixed in the required ratios. The porous samples were placed in a flask and a vacuum of -670 mm column of Hg was created by a rotary pump. The liquid elastomer was then poured into the flask and vacuum was maintained for ½ - 1 hour. The impregnated samples were then cured at 100°C for 1 hour. The density of the samples was measured by Archimedes Principle. By comparing the density difference between the impregnated composites and density of porous ceramics, the volume fractions of ceramics, elastomer and air in composites were calculated. There were considerable variations in the volume fraction of these phases. For example, the variations in volume fractions of ceramics, elastomer and air phase of composite made from porous PLZT ceramics (non-spun) sintered at 1100°C ranged between 33-35%, 22-39% and 26-44% respectively.

2.4. Electroding

Flat surfaces of sintered bulk ceramics were coated with air-fired silver and heated at 550°C for 30 minutes. For composites, round pieces of aluminum foil of 5 mm and 8 mm diameter were pasted on opposite surface of composite with the help of an epoxy (5 Minute Epoxy, Devcon, USA).

2.5. Poling

Electroded samples of bulk ceramics and composites were given poling treatment to align the dipoles of each domain in a particular direction, so as to get a net dipole moment. Each sample was fixed in a suitable jig and then heated in a stirred silicone oil bath to a temperature ranging from 120^0C to 140^0C. After reaching the temperature a D.C. field of 3-7 kV/mm was applied for 15 minutes. Then the oil bath was allowed to cool to 70^0C with D.C. field on, after which the field was removed.

2.6. Dielectric and piezoelectric constant measurement of composites

The dielectric constant (k) values of the samples were determined using a LCR meter (Stanford Research System 715). The d_{33} values of the bulk ceramics and composites were measured with the help of a d_{33} meter (S5865 APC Inc.). The g_{33} values were derived using the relation $g_{33} = d_{33}/(\varepsilon_0 * k)$.

3 Results and discussions

The properties of the PZT/PLZT- elastomer composites were analyzed with LCR meter used for measuring the dielectric constant values (k), d_{33} meter used for measuring piezo-electric charge coefficient values (d_{33}) and piezo-electric voltage coefficient (g_{33}) values which were derived from d_{33} values prepared in the different processing conditions were given in the Table 1 and 2.

Table1 Dielectric, piezoelectric properties & density of PLZT ceramics & their composites.

Material	Sample	Sintering Temp. ^{0}C/hr.	Poling Temp.^{o}C	Poling Field kV/mm	Dielectric constant (k)	d_{33} x 10^{-12} C/N	g_{33} V-m/N	Density g/cc
PLZT ceramic	PL11	900/1	140	3	870	90	0.012	6.07
	PL12	1100/3	140	3	1200	335	0.032	7.26
	PL13	1200/3	140	3	1090	325	0.034	6.17
Non-Spun Composite	PL21	900/1	120	3	20.5	14.9	0.063	2.31
	PL22	900/1	120	4.5	21	23.8	0.12	2.31
	PL23	1100/3	130	7	20	40.7	0.23	3
	PL24	1100/3	120	3	18.5	27	0.17	3
	PL25	1100/3	135	3	21	51.3	0.27	3
Spun Composite	PL31	1100/3	140	4.5	13.8	14.7	0.12	1.72
	PL32	1200/3	140	4.5	14.7	14.1	0.11	1.73

Table 2 Dielectric, piezoelectric properties & density of PZT ceramics & their composites

Material	Sample ID	Sintering Temp. ^{0}C/hr.	Poling Temp. ^{o}C	Poling Field kV/mm	Dielectric constant (k)	d_{33} x 10^{-12} C/N	g_{33} V-m/N	Densit y g/cc
PZT Ceramic	P11	900/1	140	3	748	117	0.017	6.65
	P12	1100/3	140	3	530	186	0.039	7.55
	P13	1200/3	140	3	463	181	0.044	7.42
Nonspun Composite	P21	1100/3	140	3	12.5	37.6	0.34	2.2
	P22	1200/3	140	3	6.9	32.9	0.54	2.22
Spun Composite	P31	900/1	140	3	6.8	4.7	0.08	1.36
	P32	1100/3	140	3	9.3	9.4	0.114	1.47
	P33	1200/3	140	3	8.7	11.5	0.15	1.68

The k, d_{33} and g_{33} values of bulk PZT and PLZT ceramics fairly matched with the values given in the literature [1, 7]. The dielectric constant (k) values of the bulk PZT samples decrease with increase in the sintering temperature, which is also reported elsewhere [8]. This trend is similar to the dependence of dielectric constant on internal stresses in barium titanate ceramics [9]. In these ceramics, the dielectric constant (k) at the room temperature decreases with increase in their grain size. However, reverse trend is noted for the PLZT ceramics. Such dielectric behavior of PLZT ceramics is similar to the dielectric behavior of PMN (Lead Magnesium Niobate) ceramics, where increase in grain size is associated with rise in the dielectric constant values [10]. The k and d_{33} values of PLZT ceramics are greater as compared to these values in PZT ceramics. However, the g_{33} values of PZT ceramics are on higher side as compared to those of PLZT ceramics. Similar trends are carried forward in composites made from PZT and PLZT ceramics. As the dielectric and piezoelectric constant values of a composite will depend on the amount of the ceramic phase present, k, d_{33} and g_{33} values of the non-spun composites are higher as compared to these values in the spun composites. Higher g_{33} values in the range of 0.34-0.54 V-m/N could be achieved in the non-spun composites made from PZT ceramics. The D.C field and temperature during poling has minor influence on the k-values of the composites, but has pronounced effect on the d_{33} and g_{33} values as is evident from the data history of the samples PL21 to PL25 shown in Table 1. The increase in D.C field strength and temperature is associated with the rise in the d_{33} and g_{33} values of the composites. Out of the two parameters, temperature has a more drastic effect on these values as is clear on comparing the values of samples PL23 and PL25.Variation in density ranging between 1.3 g/cc to 1.7 g/cc for spun composites and 2.3 g/cc to 3.0 g/cc for non-spun composites was obtained. These values are relatively much lower than the density of the bulk ceramics.

4 Conclusions

The lower density composites of PZT/PLZT ceramics impregnated with silicone elastomer could be prepared which is having high g_{33} values about ten times greater than that of bulk ceramics demonstrate that it is possible to fabricate a conformable detector.

PZT is a better candidate material than PLZT to make composites with higher g_{33} values. Non-spun composites have higher g_{33} values as compared to the spun composites. D.C. field and temperature during poling have important role in obtaining composites with higher g_{33} values.

5 References

[1] Bernard Jaffe, William R. Cook, Hans Jaffe 1971 *Piezoelectric Ceramics* (London & New York- Academic Press).
[2] Heartling G H 1999 *Ferroelectric Ceramics History and Technology* J. Am. Ceram. Soc. **82** (4) p 797- 818.
[3] Thomas R. Shrout, Walter A. Schulze and James V. Biggers 1979 *Simplified Fabrication Of PZT/Polymer Composites* Mat. Res. Bull. **14** (12) p 1553-59.
[4] Lewis R W C, Dent A C and Bowen C R 2007 *Network Modelling of 3-3 Piezocomposite Materials, Ferroelectrics* **351** p 216-224.
[5] Newnham R E, Skinner D P and Cross L E 1978 *Connectivity and Piezoelectric-Pyroelectric Composites* Mat.Res.Bull.**13** (5) p 525-536.
[6] Jeannine Saggio, Woyansky, Curtis E. Scott and Minnear W P 1992 *Processing of Porous Ceramics, Am.Ceram.Soc.Bull.* **71** (11) p 1674-82.
[7] Ramji Lal 1994 *Piezoelectric Sensor and Actuator Materials Developed at NMRL for Underwater Applications*: Proceedings of the National Seminar on Materials for Marine Applications, Bombay, pp. 113-131.
[8] Dhirendra Mohan, Sharma A K, Bhattacharya S and Grover A K 2006 *Development of A.E-Sensor Element*: Proceedings of the National Seminar on Advances in Electroceramics, DRDO, Pune, India, pp 236-241.
[9] Buessem W R, Cross L E and Goswami A K 1966 *Phenomenological Theory of high permittivity in fine grain Barium Titanate* J.Am.Ceram.Soc. **49** pp 33-36.
[10] Philippe Papet, Joseph P. Dougherty and Thomas R. Shrout 1990 *Particle and grain size effects on the dielectric behavior of the relaxor ferroelectric Pb(Mg1/3Nb2/3)O3* J.Mater.Res. **5** (12) p 2902-09

Acknowledgement

The corresponding author would like to express his sincere thanks to the management of Nuclear Fuel Complex, Hyderabad for giving the opportunity to publish this manuscript.

IOP Conf. Series: Materials Science and Engineering **314** (2018) 012008 doi:10.1088/1757-899X/314/1/012008

Influence of nitrogen as grain refiner in low carbon and microalloyed steels

B M Hasan[1] and P Sathyamurthy
[1]JSW Steel Limited, Salem Works, Tamilnadu, India

Email: bm.hasan@jsw.in

Abstract. Microalloyed steel is replacing using of low alloy steel in automotive industry. Microalloying elements like vanadium, niobium and titanium are used to enhance the steel property. The current work is focused on using nitrogen as a strengthening element in existing steel grade. Nitrogen in free form acts as solid solution strengthener and in combined form as precipitates acts as grain refiner for enhancing strength. The problem of grain coarsening at high temperature in case carburizing steel was avoided by increasing nitrogen level from 60ppm to 200ppm. Grain size of ASTM no 10 is obtained at carburizing temperature of 950 °C by increasing nitrogen content from grain size no 6 with lower nitrogen. Mostly crankshaft is made from Cr-Mo alloyed steel. At JSW, nitrogen in the level of 130-200ppm is added to medium carbon steel to meet property requirement for crankshaft application

1.Introduction

Microalloyed steels are group of steels which contains small addition of niobium, vanadium and titanium. The development of microalloyed steel started in sixties for flat rolled categories with niobium addition. For forging application microalloying steels was developed in 1980s [1][2]. Vanadium was the primary microalloying element and microstructure consists largely of pearlite. Currently these steel grades constitute around 12-14% of total steel production. The rising strength levels combined with excellent toughness favoured development of microalloyed steels. Microalloyed bar and forging steels are greatly used in automobile applications namely gears, crankshafts, spindle, fasteners, driveline components, connecting rods, spindle etc [3][4][5][6]. There is a high potential for this type of steel in forging application. Property of the steel is dependent on the microstructural aspects. Throughout the year's large work have been carried out to understand the influence of these elements on microstructural aspects like effect of grain refinement, precipitate hardening and recrystallization effects during thermomechanical process. The common feature among these microalloying elements is their affinity towards nitrogen in forming nitrides and carbonitrides and these precipitates is responsible for increased properties. Generally, microalloyed steels have nitrogen content around 60-70ppm. In this work the effect of nitrogen at levels of 130-190 ppm is studied with respect to grain refinement at various temperatures is studied for case carburized steels. For increasing productivity nowadays carburizing is done at higher temperatures of 1000°C [3]. Low carbon steel grades are subjected to carburizing treatment for case hardening are prone to grain coarsening if sufficient amount of grain boundary pinning elements are not present at these high temperatures. Moreover, the development of crankshaft material with increased nitrogen content is discussed in this paper.

2.Strengthening mechanisms

The micro-alloyed steels are developed by means of a suitable combination of chemical composition and thermo-mechanical treatment parameters to achieve good strength without compromising toughness, ductility and formability. Properties of a material depend on the metallurgical features. The yield strength of steel can be increased by one or more of several strengthening mechanisms. [7][8][9] a) Dislocation strengthening: the resistance to dislocation movement due to the obstacles presented to other dislocations. b) Grain-boundary strengthening: grain-boundary hindering of dislocation movement. With fine grains very high strength may be achieved. c) Solid-solution strengthening: the resistance to dislocation movement due to the presence of interstitial or substitutional solute atoms in a crystal lattice. The limitation of this mechanism is imposed by the solubility limit in each alloying system. d) Precipitation strengthening: the resistance to dislocation movement due to the effect of second-phase particles. Second-phase particles like precipitates formed from a supersaturated solid solution which is more present in practice. The limitation of this mechanism is governed by the influence of the size and shape on toughness, not on the strengthening. The dominant mechanism is grain-boundary strengthening, because the influence on the grain size can be accomplished by a modification of chemical composition and process parameters. The stability of austenite grains depends on the number of precipitates rather than their type. Any development in metallurgical aspect is focussed on grain refinement. Grain refinement has played a greater role in achieving higher mechanical properties. The combined effect of grain refinement and precipitation hardening in microalloyed steel replaced conventional C-Mn and Cr-Mo steels. Fine grain size is essential for high strength steels. Figure 1[2] shows with lower grain size better properties can be achieved. Small amount of titanium helps in maintaining the grain size at higher temperature. Figure 2 [2] shows role of microalloying elements in grain refinement, recrystallization and precipitation hardening.

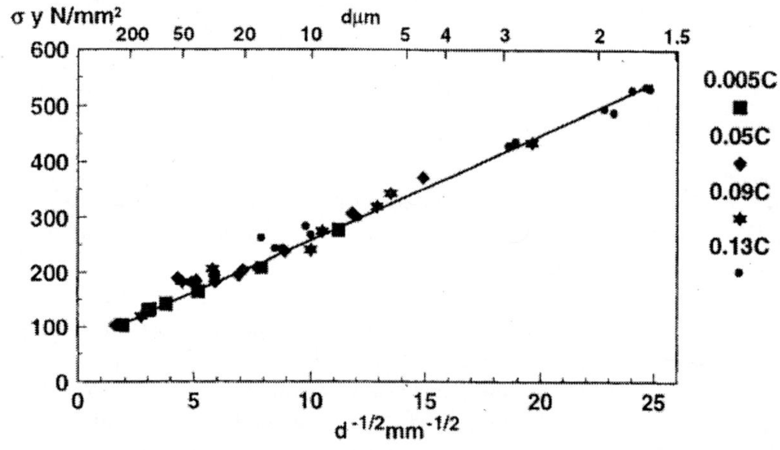

Figure 1. Strength variation with grain size [2]

Table 1. Role of microalloying elements [2]

Microalloy	Precipitation strengthening after hot rolling	Precipitation strengthening after normalizing	Influences recrystallisation during hot rolling	Refines grain size on normalizing	Refines grain size during high temperature austenizing
V	VN,VC	VC	-	VN	-
Nb	NbCN	-	Nb,NbCN	NbCN	-
Ti	TiC	-	-	TiC	TiN

3. Nitrogen in steel

Nitrogen acts as both beneficial and detrimental element in steel based on the product application. Nitrogen causes yield-point phenomena and results in stretcher-strain formation in deep drawing operation. Formation of AlN causes intergranular fracture in continuously cast steels [10]. The blessing effect of nitrogen is it helps in grain refinement forming precipitates with Al, Ti, V and Nb. In general steel nitrogen with Al in steel is used for grain refinement. AlN precipitate along grain boundary and retards grain growth at high temperature. The size of these particles influence effect of grain refinement. AlN precipitates helps in grain refinement up to 900°C. Nitrogen is also beneficial in stabilizing the austenitic phase. Free nitrogen in steel helps in strengthening by solid solution strengthening mechanism as they are interstitial atoms. The behaviour of nitrogen in solid steel has to be understood for its use in steel.

4.Case Carburizing application

4.1Experimental procedure:

JSW steel limited Salem works, India produces long product special steels for automotive applications namely crankshaft, gears, bearings and many through Energy optimizing Furnace-Ladle Furnace-Vacuum Degassing-Caster route. The billets produced from billet caster is hot rolled to get the final product of different sizes. In the current work 30mm dia bars was taken for study. Table 1 shows typical chemistry for conventional and modified case carburizing grade taken for understanding the grain coarsening behaviour at high temperature for different nitrogen levels.

Table 2. Chemical composition of case carburizing grade

	C	Si	Mn	P	S	Ni	Cr	Mo	Al	N
Type1unmodified	0.20	0.27	0.80	0.016	0.004	0.03	1.09	0.20	0.029	90ppm
Type2 Modified	0.20	0.25	0.88	0.010	0.023	00.2	1.05	0.2	0.028	190ppm

5 Samples from both the material are subjected to McQuaid- Ehn test as per ASTM E-112 for understanding grain coarsening at different temperatures of 850°C, 875°C, 900°C, 925°C, 950°C, 975°C, 1000°C, 1025°C and1050°C.

4.2. Results

The samples after the test is polished, etched and grain size is measured under optical microscope. Table 3 shows ASTM grain size no for type 1 steel with carburizing temperature. It can be seen that till 900°C fine grain size of ASTM no 9 to 10 is achieved. With further increase in temperature grain coarsening has started and worsened with increasing temperatures.

Table 3. Grain Size at different temperatures for N=90ppm

Temp°C	850	875	900	925	950	975	1000	1025	1050
Sample 1	10.6	10.3	9.3	Mix 8.2/4	Mix 6.2/4	5.6/4	4.7/3	4.3/3	4.3/3
Sample 2	10.5	9.9	9.3	Mix 8.2/4	Mix 5.8/3	5.7/4	4.8/3	5.3/3	4.5/3
Sample 3	10.7	10.6	9.2	Mix 6.0/3	Mix 6.2/4	5.6/4	4.5	5.2/2	3.9/1
Sample 4	10.7	10.5	10.1	Mix 6.8/4	Mix 5.8/4	5.5/3.5	5.5/3	3.6	4.4
Sample 5	10.7	10.1	10.2	Mix 9.7/1	Mix 4.2/2	5.3/4	3.5	3.7/2	3.8

IOP Conf. Series: Materials Science and Engineering **314** (2018) 012008 doi:10.1088/1757-899X/314/1/012008

Figure 2 shows coarsened grain microstructure of samples subjected to carburizing temperature of 950°C and 1000°C.

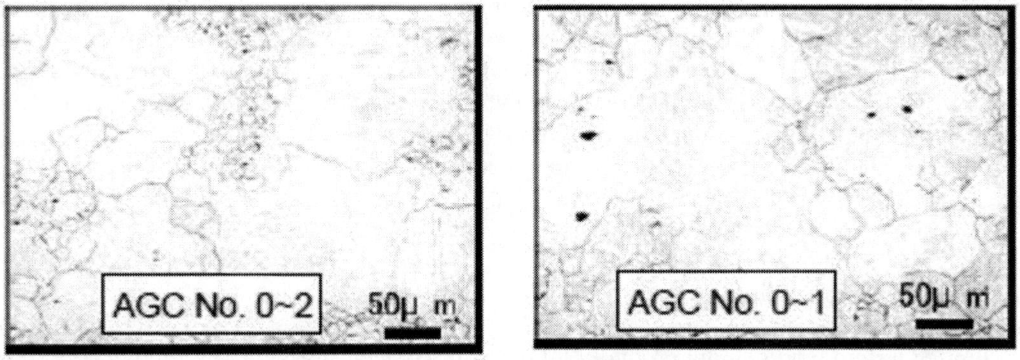

(a) (b)

Figure 2. Microstructure at a)950 °C b)1000°C

Table 4. Grain Size at different temperatures for N=190ppm

Temp°C	850	875	900	925	950	975	1000	1025	1050
Sample 1	10.6	10.6	10.5	10.6	10.6	10.4	Mix10.3/2.0	Mix3.4/1.0	Mix4.9/2.0
Sample 2	10.8	10.8	10.7	10.7	10.8	10.7	Mix10.1/2.0	Mix 2.9/1.0	Mix2.9/1.0
Sample 3	10.8	10.9	10.7	10.6	10.8	10.6	Mix5.3/1.0	Mix2.8/1.0	Mix 3.0/1.0
Sample 4	10.8	10.8	10.8	10.7	10.7	10.5	Mix2.5/0.0	Mix2.2/0.0	Mix3.7/-1.0
Sample 5	10.9	10.7	10.5	10.6	10.5	10.3	Mix1.3/0.5	Mix 2.5/1.0	Mix 1.7/0.5

From table 4 it can be seen that fine grain size of ASTM no 10 are obtained till 975°C. Beyond 975°C grain coarsening have started. The reason is due to formation of more AlN precipitates due to increased nitrogen content.

5.Crankshaft application

In this development the strength of a medium carbon steel is obtained by increasing nitrogen content and adding titanium to existing medium carbon steel. Table 5 gives the chemical composition of existing and modified grade. Medium carbon steels without chromium and molybdenum is not used for crankshaft application. The modified steel grade has lower carbon and manganese compared to existing grade. In this development solid solution strengthening effect of nitrogen and precipitation hardening by formation of TiN is used for strength improvement.

Table 5. Chemical composition of crankshaft grade

	C	Si	Mn	Al	Nppm	Ti
Existing	0.45	0.20	0.70	0.02	75	Trace
Modified	0.40	0.20	0.58	0.002	160	0.01

Table 6 show properties of the modified grade which is comparable to existing grade with reduced carbon and manganese content.

Table 6. Properties for modified grade

Condition	Ultimate tensile strength (Kg/mm2)	Yield strength (Kg/mm2)	Elongation %	Reduction in area %	Charpy Joules	Hardness BHN	Grain Size ASTM No
Rolled	62.5	47.3	32	36	44	179	7
Normalized	63	46.7	32.5	35.5	45	183	7

Figure 3 shows microstructure after heating at different temperature. Even at 1100°C grain size no of 5 is achievable in the modified grade. At forging temperature of 1100°C the modified grade meets industry requirement of grain size no 5. Formation of AlN and TiN precipitates have reduced grain coarsening at high temperatures.

a) 950°C, GS – 7-8 b) 1000°C, GS 5-7 c)1100°C, GS 5-6

Figure 3. Microstructure at different temperature and corresponding grain size

6. Conclusion

The effect of high nitrogen content in two different application of steel grades is studied. Grain coarsening started at 950°C in case carburizing steel grade with nitrogen levels of 90 ppm, whereas fine grains are obtained even at 950°C in these grades of steel having 190 ppm nitrogen. Formation of more AlN precipitates have contributed to grain refinement at higher temperatures. Increased nitrogen content helped in processing the material at higher temperatures. AlN is effective till 900°C and for higher temperatures vanadium and niobium is added which adds to increased cost of production. Here to reduce cost, nitrogen a cheaply available element is used to prevent grain coarsening at higher temperatures.

For crankshaft application nitrogen is used to get the same property of medium carbon steel with reduced manganese and carbon content. So the above work uses nitrogen effectively for grain refinement strengthening and solid solution strengthening.

7.References

[1] Morrison W B 2000 The *Proceedings of the Vanitec Symposium*, (Guilin, China) p25.
[2] Krauss G 2001 Vanadium microalloyed forging steels. *In The Proceedings of the Vanitec Symposium*, Beijing, China.
[3] Van Tyne C J, Matlock D K, Speer J G. 2008 *Microalloyed forging steels.* Advanced Steel Processing and Products Research Center, Colorado School of Mines, Golden, Co p189-97

[4] Lemaitre C, Diericks P and Bittes G 2006 *International conference on new developments in long and forged products: Metallurgy and Applications*, AIST, Warrendale, p.29
[5] Naylor D J. 1998 *Microalloyed forging steels*. In Materials Science Forum Trans Tech Publications Vol. 284, p. 83-94.
[6] Speer J G, Matlock D K. 2007 *Some recent developments in microalloyed bar and forging steels. Microalloyed steel: emerging technologies and applications*, p434-41.
[7] Gladman T 1997 *The physical metallurgy of microalloyed steels*. Maney Publications.
[8] Dieter G E, Bacon D J 1986 *Mechanical metallurgy*. New York: McGraw-Hill.
[9] Courtney T H 2005 *Mechanical behavior of materials*. Waveland Press.
[10] Ward R G. 1962 *An introduction to the physical chemistry of iron & steel making*, Edward Arnold.

Optimization of tribological behaviour on Al- coconut shell ash composite at elevated temperature

R Siva Sankara Raju[1]*, M K Panigrahi[2], R I Ganguly[2], G Srinivasa Rao[3]

[1] Department of Mechanical Engineering, Gandhi Institute of Engineering and Technology, Gunupur, Odisha, India.
[2] Department of Metallurgical Engineering, Gandhi Institute of Engineering and Technology, Gunupur, Odisha, India.
[3] Department of Mechanical Engineering, RVR&JC College of Engineering, Guntur, Andhra Pradesh, India.

Abstract. In this study, determine the tribological behaviour of composite at elevated temperature i.e. 50 – 150 °C. The aluminium matrix composite (AMC) are prepared with compo casting route by volume of reinforcement of coconut shell ash (CSA) such as 5, 10 and 15%. Mechanical properties of composite has enhances with increasing volume of CSA. This study details to optimization of wear behaviour of composite at elevated temperatures. The influencing parameters such as temperature, sliding velocity and sliding distance are considered. The outcome response is wear rate (mm^3/m) and coefficient of friction. The experiments are designed based on Taguchi [L_9] array. All the experiments are considered as constant load of 10N. Analysis of variance (ANOVA) revealed that temperature is highest influencing factor followed by sliding velocity and sliding distance. Similarly, sliding velocity is most influencing factor followed by temperature and distance on coefficient of friction (COF). Finally, corroborates analytical and regression equation values by confirmation test.

1. Introduction

In the current scenario, aluminium metal matrix composites (AMC) is preferred in most application due to lightness and cost effective. Generally, AMCs are fabricated with matrix as aluminium (Al) and reinforcement is ceramics. AMCs has been attributed inherent strength, high strength modules, and better wear resistance with customize properties of thermal due to the addition of ceramics[1–4].
The reinforcements (ceramics) restricted the deformation of Al from load and distributed load uniformly throughout the base matrix with enhances the properties of composite. The preferred reinforcement is stated by several influencing factors would like to anticipated with application, kind of process, cost and availability. Many researchers are interested to use the natural minerals and agro-wastages as reinforcement owing to renewability, inexpensive, moderate recycled, and eco-friendly. However, various agro wastes like fly ash [5,6], red mud [7] and colliery shale[8,9] , rice husk [10], shell char [11–14], and bagasse [15] and jute [16] .
Moreover, still there is a sang in preparation of the composite in term of fabrication and processing of reinforcement [4]. Usually, AMC is fabricated by several ways i.e. liquid metallurgy (*i.e.,* stir or compo casting and infiltration), squeeze casting and powder metallurgy route. Generally, for the mass and low cast production, liquid metallurgy route is most preferred than the other ones.
Aku et al.[14]evaluated microstructure, hardness and density coconut shell ash composites having 3-15% weight volume. They found that by increasing the vol. % of reinforcement, density decreases and hardness values increases. Shen et al. [17] reported on the correlation between hardness and tensile strength of composite, based on the effect of particle size, volume fraction and aging behaviour of prepared with Al-Cu-Mg alloy reinforced with SiC. Mahendra et al.[18] Prepared aluminium (2024) composites reinforced with SiC, fly ash and hybrid of SiC- fly ash at various volume fractions (0, 5, and 10%) by stir casting route. Moreover, the authors reported on mechanical properties i.e. density, tensile strength, hardness, % of elongation and yield strength of all the composites. Certainly, Hybrid composite is exhibit better mechanical properties than the other. Siva et al. [9] investigated the

IOP Conf. Series: Materials Science and Engineering **314** (2018) 012009 doi:10.1088/1757-899X/314/1/012009

mechanical properties, abrasion and frictional behaviour of Al-colliery shale (CS) and Al-Al_2O_3 composites under forged condition by using pin-on –disc set up. They reported that the mechanical properties like ductile, toughness, stiffness, tensile strength and percent of elongation exhibit better in Al-colliery shale than Al-Al_2O_3. The wear behaviour in the forged condition of Al-CS had superior characteristic than Al-Al_2O_3 and base matrix Aibodion and Hassan [19] studied the effect of mechanical and micro-structural properties on Al- Si-Fe reinforced with Silicon carbide of 5-25% of volume. The authors concluded that increase of reinforcement with increasing hardness, tensile strength, and porosity whereas decrease density and impact strength. The composite has improved mechanical properties due to the addition and uniform distribution of reinforcement (SiC) in the ductile matrix. Kok and Ozdin [20] investigated the wear mechanisms of the 2024 aluminium alloy composites reinforced with Al_2O_3. They reported that wear resistance of the composites was greater than the base alloy, and increased with "increasing reinforced particles size", also decreased with "increase of sliding distance, load and the abrasive particle size".

Similarly, few researchers made effort on elevated temperature which is a one of the main influencing factor in Tribology of composite materials. Rashed and Mahmoud [21] has developed A356-SiC composite by the way of rheo casting. The author reported on optimization of wear resistance of composite using artificial neural network (ANN) model. The contributed input parameters were size of SiC, volume fraction of SiC, applied pressure, and temperature. The wear resistance of base metal is inversely proportional characteristic to temperature. The composite exhibited better resistance up to 200 °C then significantly decreased. Gudlur et al.[22], reported on thermo mechanical properties of Al-A12O3 composite. The properties of composite are estimated by numerical method and compared with OOF element code. Sharma et al. [23] investigated wear behaviour of AA7075-E glass fiber composite at elevated temperature up to 200 °C. They exposed composite exhibited better wear resistance than virgin due to delaying transitions wear. From the wear mechanism, severity in composite increased with increasing of applied load and temperature. Delaminated surfaces and debris are corroborating with by SEM-EDX.

Wilson and Alpas[24] prepared two aluminium (A356 and A16061) composites with reinforcing volume of SiC (20%), Al_2O_3 (20%), and Gr (10%). They determined wear performance of developed composites at an elevated temperature up to 450 °C. The tested conditions was minimized to lower load (11.55 N), sliding speed (0.1m/s) in an isothermal atmosphere. The seizure condition of composites (i.e. A356-SiC and A16061-Al_2O_3) is improved from 230 to 450 °C and 190 to 340 °C respectively as compared to base materials. The Gr addition deteriorates strength but enhances friability of composite.

This investigation is mainly intense towards preparation of aluminium based composite reinforced with agro waste i.e. coconut shell ash (CSA). The Al-CSA composite has been prepared by compo casting route with various percentage of CSA such as 5, 10, and 15% of volume. This study deals with investigation of mechanical properties, micro structural with wear resistance of composite at elevated temperature. The wear mechanism of composite has understood by SEM analysis.

2. Materials and Methods

2.1. Materials

In the present work, Al 1100 grade (Table 1) aluminium (HINDALCO, BBSR) is used as the matrix, whereas coconut shell ash is used as reinforcement for the preparation of composites.

Table 1 Chemical composition for Al-1100

Elements	Cu	Zn	Mn	Si	Fe	Al
%	0.05-0.2	0.1	0.05	0.7-0.95	0.35-0.85	Balance

2.2. Preparation of reinforcement and Composites

The list The Al-CSA composites are prepared using compo casting route. Initially, the collected coconut shells (CS) are dried and outer skins is smoothened then, crushed in a jaw crusher followed by a hammer mill in order to produce a coarser form of CS powder. The obtained powder is packed in a graphite crucible and fired in an electric resistance furnace at the temperature of 1320 °C and cooled at furnace temperature to form micro-sized coconut shell ash (CSA) powder. Finally, the CSA powder is sieved in a rotary sieve shaker with the size of ≤ 240 BSS mesh (63 microns). The obtained CSA is treated in tubular furnace under below (argon) atmospheric at a temperature of 1530 °C up to ≈4 hr. The final formed CSA particulate is used as reinforcement.

The compo casting technique has been employed for the preparation of AMCs. The aluminium matrix is pre-heated in a bottom pouring furnace at 450 °C and then raised melting temperature up to 670 °C to keep the matrix alloy in a semi-solid state. On the other hand, the CSA particles are pre-heated in a muffle furnace up to 3 hours for 1100 °C (CSA). Thereafter, the molten metal along with reinforcements is stirred with the help of a motor driven stirrer in the presence of argon gas at a speed of 600 rpm up to 9 minutes. Then, the molten metal is superheated above the liquid temperature at 690 °C and poured into a preheated (300 °C) cast iron mold of the size of 100 × 20 × 40 mm^3.

2.3. Characterization techniques

The as-prepared AMCs were characterized for their structural, morphological and mechanical properties. X-ray diffraction pattern for CSA was recorded using an X-ray diffractometer (Phillips PW-1729) using CuKα radiation of wavelength λ = 0.1541 nm in the scan range 2θ = 10–60°. Morphology of the sample was investigated using scanning electron microscope (SEM with EDXA, JEOL JSM-5600LV). Microstructural examinations (ASTM E-7-95) of samples have been carried out using standard metallography technique (XJL-1000). The composite are performed with Brinell hardness tester with a load of 250 kg and 5 mm ball indenter. The tensile properties of the composites have been tested with Hounsfield tensometer (model: ETM-ER3/ 772/12) at a velocity of 1 $mmsec^{-1}$ with a maximum load of 20 KN following the ASTM standard E8. Pin on the disc wear testing machine (Model: TR-201 LE-PHM 400) is used for conducting the wear test (ASTM-G99) experiments on composites.

2.4. Taguchi approach

Taguchi is a robust design to enhance the productivity during development of components at lower price. It reduces economically variation of productive functions to the customer environment[25,26]. In this study, Taguchi (L9) orthogonal array has been preferred to optimization of number of experiments. The specified input factors along with their levels are shown in Table 2. Moreover, output response is wear-rate (WR) and coefficient of friction (COF). To obtained outcomes, the experiments are designed with specified inputs such as Temperature, sliding distance and velocity. The designs of experiments are shown in Table 3. Furthermore, a statistical analysis of variance (ANOVA) has been performing to known input factor with significant. Ultimately, a corroboration regression equation is verifying the optimal process parameters.

Table 2. Used factors with their levels

S.No	Factors	Units	Levels		
			-1	0	1
1	Temperature (T)	°C	50	100	150
2	Sliding Distance (D)	m	1000	1500	2000
3	Sliding Velocity (V)	m/s	0.5	1	1.5

IOP Conf. Series: Materials Science and Engineering **314** (2018) 012009 doi:10.1088/1757-899X/314/1/012009

Table 3. Experimental design (L9)

Si.No	T	D	V
1	50	1000	0.5
2	50	1500	1
3	50	2000	1.5
4	100	1000	1
5	100	1500	1.5
6	100	2000	0.5
7	150	1000	1.5
8	150	1500	0.5
9	150	2000	1

Table 4. Chemical Composition of CSAp

Elements	Al2O3	CaO	Fe_2O_3	MgO	K_2O	MnO	Na_2O	SiO_2	ZnO
% of Wt.	21.42	2.67	4.42	18.2	0.59	0.25	0.47	48.2	0.32

3. Results and Discussion

3.1. Mechanical properties of Al-CSA composites

The coconut shell ash is subjected to analysis by XRD and details of chemical constitutes as shown in Table 4. Aluminium alloy (ISO 99.0Cu) has been used as the matrix in the present investigation. The density of receiving alloy was 2.71g/cc. Table 5 shows the variation of properties in AMCs. It is clear from the table that with an increase of % reinforcement the tensile strength is increased. The yield strengths are found to be 90, 101, 127, and 142 N/mm2 for pure, 5%, 10%, and 15% of volume in Al-CSA respectively. Similarly, the Hardness value is increased with increase in reinforcement as observed in Table 5, which is due to strain hardening of composite. Similarly, density and elongation decreases with increase of CSA volume in composite. Generally, reinforcement particles (CSAp) are harder, which increases load bearing capability, and restrict dislocation movement of the matrix reduces interspacing, particle movement has been critical.

3.2. Metallographic testing:

The microstructures of the pure and reinforced MMCs are presented as Fig. 1. The metallography images observed bright area specifies matrix and dark area specify reinforcement particles. The distribution of particles in the microstructure is seen to be uniform. It is also observed from Figure that with an increase in the % of reinforcement grain size decreases, which in turn increases the strength and decreases the ductility.

Table 5. Mechanical Properties of Al-CSA-MMCs

% of CSAp	0	5	10	15
VHN	56.3	61.1	78.4	82.1
Density(g/cc)	2.71	2.65	2.59	2.51
UTS(N/mm2)	90	101	127	142
% of Elog.(mm)	28	23	18	14

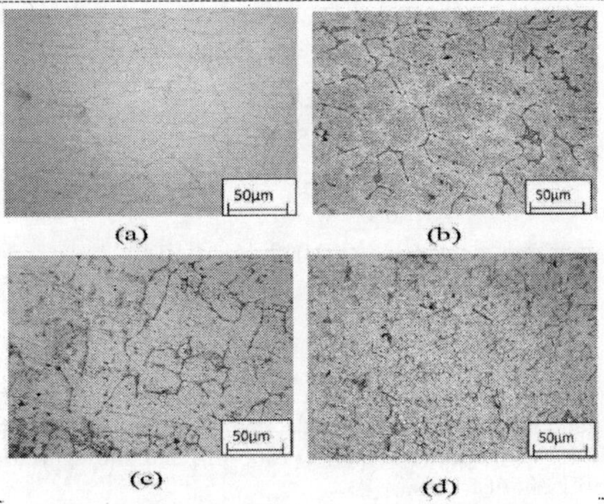

Figure 1. Microstructure for (a) Al-1100 (b) Al-5% CSA (c) Al-10% CSA (d) Al-15% CSA

Table 6. Outcome- response with S/N ratio

Run	WR(mm^3/m)	COF	WR SN ratio (dB)	COF SN ratio (dB)
1	0.002705	0.491	51.35665	6.182794
2	0.00207	0.488	53.68059	6.238726
3	0.001435	0.485	56.86296	6.295021
4	0.00232	0.554	52.69024	5.136078
5	0.001685	0.551	55.468	5.185642
6	0.002955	0.569	50.58885	4.901572
7	0.001935	0.616	54.26638	4.202043
8	0.003205	0.635	49.88344	3.947946
9	0.00257	0.632	51.80134	3.991158

3.3. S/N ratio results

The signal to noise ratio is attained by Taguchi technique. The ratio for each response has been resolute by averaging S/N ratios at the consequent level. The influencing factors such as Temperature, sliding distance, and sliding velocity on WR & COF for Al-CSA composites are detailed in Table 6. The influencing factors with the highest S/N ratio will give in the optimum feature with least variance can determine by the response Tables 7 and Table 8. The sliding velocity is a dominant factor on the wear rate, followed by temperature and sliding distance. Similarly, from Table 8 temperature is a main influencing factor followed by sliding velocity and distance for Coefficient of friction.

Table 7. Response table of wear rate

Level	Temp	SD	SV
1	53.97	52.77	50.61
2	52.92	53.01	52.72
3	51.98	53.08	55.53
Delta	1.98	0.31	4.92
Rank	2	3	1

Table 8. Response table of COF

Level	Temp	SD	SV
1	0.4876	0.5536	0.5647
2	0.5576	0.5576	0.5576
3	0.6276	0.5616	0.5504
Delta	0.14	0.008	0.0143
Rank	1	3	2

The main effect plot (Fig. 2) resulted that wear rate and coefficient of friction are increased with increase of temperature and sliding distance and decreased with sliding velocity. The optimum input factors are temperature (50 ° C), sliding distance (1000 m), and sliding velocity (0.5 m/s) irrespective of the wear rate and coefficient of friction.

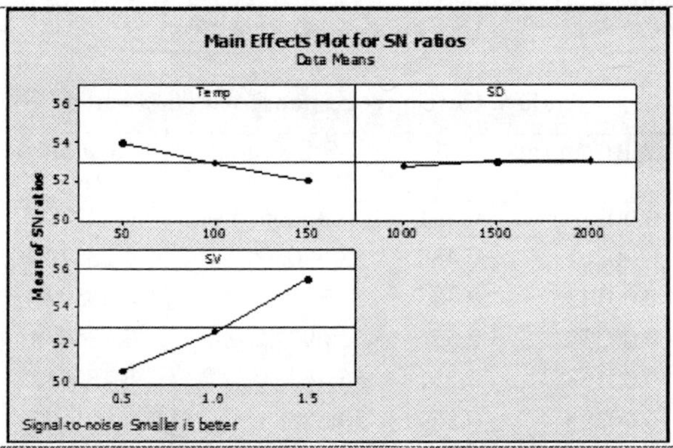

Figure 2. Signal to noise ratio plot for wear rate and coefficient of friction

3.4. Analysis of variance (ANOVA)

ANOVA is a technique of portioning changeability into certain source of disparity and the related degree of freedom in experimentation. The frequency test (F-test) is exploiting in statistics to examine the important effect of input factor, which form the quality characteristics. Table 9&10 shows the result of linear effect on ANOVA analysis of S/N ratio of wear rate (WR) and coefficient of friction (COF). This analysis was carried out for a level of significance of 5%, i.e., for 95% a level of confidence. The last column of the table shows the "percent contribution" (P) of each factor as the total variation, indicating its influence on the result.

Table 9. ANOVA of wear rate

Source	DF	Seq SS	Adj SS	Adj MS	F	P
T	2	5.9056	5.9056	2.9528	35.99	0.027
D	2	0.161	0.161	0.0805	0.98	0.505
V	2	36.5917	36.5917	18.2959	223.01	0.004
Error	2	0.1641	0.1641	0.082		
Total	8	42.8224				
S = 0.2864		R-Sq = 99.6%		R-Sq(adj) = 98.5%		

Table 10. ANOVA of COF

Source	DF	Seq SS	Adj SS	Adj MS	F	P
Temp	2	7.21536	7.21536	3.60768	32096.5	0.001
SD	2	0.01857	0.01857	0.00929	82.61	0.012
SV	2	0.07052	0.07052	0.03526	313.69	0.003
Error	2	0.00022	0.00022	0.00011		
Total	8	7.30467				
S = 0.01060		R-Sq = 97.5%		R-Sq(adj) =96.4%		

4. Multiple linear regression and Confirmation test

Multiple linear regression equations were developed to establish the association amid of input factors and outcome. The value of regression coefficient R^2 (0.996 & 0.975) is in good agreement with the adjusted R^2 (0.985 & 0.964) for the wear rate and coefficient of friction respectively. In view of both values are logically close to unity, the models provide a good justification for the liaison among input factors and outcome response.

The regression equation developed for the wear rate of Al-CSA composite is as follows:

SRN WR = 0.000935 -0.000143 Temp + 0.000131 SD + 0.000228 SV – (1)

The regression equation developed for the coefficient of friction of the Al- CSA composite is as follows:

SRN COF = 0.313 + 0.0715 Temp + 0.00954 SD + 0.0325 SV – (2)

From the observation of Equations (1) & (2) that the temperature (T) plays a major role on wear rate, followed by sliding velocity (V) and sliding distance (D). The coefficient associated with temperature (T) is negative, thus indicating that the wear rate and coefficient of friction decreases with increasing temperature. Conversely, the wear rate increases with increasing sliding distance and sliding velocity because the coefficients allied with these factors are positive. Sliding velocity is a larger effect on wear rate and coefficient of friction compared with the sliding distance according to its coefficient value.

Assessment of confirmation has been performing to corroborate the numerical analysis by choice trial conditions that are dissimilar from those employed in the analysis. The factors utilized in the confirmation test are offered in Table 11.

Table 11. Parameters used in the confirmation test

TEST	T	D	V
1	50	1500	1.5
2	100	1000	0.5

Tentative outcome are evaluate with the estimated values developing from the regression equation. Table 12 shows that the experimental values and calculated values from the regression equation are nearly the same with the least error (±5%). The resulting equations are capable of predicting the surface roughness to an acceptable level of accuracy.

Table 12 Results of confirmation test.

	Test 1			Test 2		
	Model Eq.	Exper.	%Error	Model Eq.	Experi.	%Error
WR	0.00145	0.00132	2.153	0.00274	0.00282	-4.953
COF	0.482	0.481	1.11	0.551	0.554	-3.619

5. Conclusions

- The Al-CSA composites are prepared with compo casting route by various percentage of volume 5, 10, and 15.
- Mechanical properties of composites such as tensile strength and hardness have been increased. Similarly, elongation and density decreased with increasing volume of CSA in Al composite.
- Morphological examination revealed that grain size decreasing with increase of reinforcement, which enhance the hardness of composite.
- Temperature has the highest influence on wear rate and coefficient of friction in the wear behaviour of Al- CSAp composite followed by sliding velocity and sliding distance.
- Optimal wear behavioral conditions, such as temperature (50 °C), sliding velocity (0.5 m/s), and distance (1000 m), can be used to achieve the minimum wear rate and coefficient of friction, Al-CSAp composites.

6. References

[1] Adams D F 1987 *Composites* **18** 261
[2] Miracle D B 2005 *Compos. Sci. Technol.* **65** 2526–40
[3] Rohatgi P K 1993 *Def. Sci. J.* **43** 323–49
[4] Surappa M K 2003 *Sadhana* **28** 319–34
[5] Govindarao R, Ganguly R I, Dash R R, Surya Prakash Rao P, Reddy G S and Singh S K 2015 *Trans. Indian Inst. Met.* **68** 951–8
[6] Rohatgi P K, Daoud A, Schultz B F and Puri T 2009 *Compos. Part A* **40** 883–96
[7] Prasad N, Sutar H, Mishra S C, Sahoo S K and Acharya S K 2013 *Int. Res. J. Pure Appl. Chem.* **3** 59–74
[8] Venkata Siva S B, Ganguly R I, Srinivasa Rao G and Sahoo K L 2015 *J. Eng. Tribol.* **229** 823–34
[9] Venkata Siva S B, Ganguly R I, Srinivasa Rao G and Sahoo K L 2014 *Tribol. - Mater. Surfaces Interfaces* **8** 117–24
[10] Alaneme K K, Akintunde I B, Olubambi P A and Adewale T M 2013 *J. Mater. Res. Technol.* **2** 60–7
[11] Siva Sankara Raju R, Panigrahi M K, Ganguly R I and Srinivasa Rao G 2016 *31st indian engineering congress,kolkata* ed I Nilanjan senpupta (Kolkata: Allied publishers pvt. Ltd) pp 241–5
[12] Siva Sankar Raju R, Srinivasa Rao G and Muralidhara rao M 2015 *Int. J. Conceptions Mech. Civ. Eng.* **3** 17–21
[13] Murali T P, Surappa M K and Rohatgi P K 1982 *Metall. Trans. B* **13** 485–94
[14] Aku S Y, Yawas D S and Apasi A 2013 *Gazi Univ. J. Sci.* **26** 449–57
[15] Aigbodion V S, Hassan S B, Dauda E T and Mohammed R A 2011 *Tribol. Ind.* **33** 28–35
[16] Biswas B, Chabri S, Mitra B C, Bandyopadhyay N R and Sinha A 2016 *J. Inst. Eng. Ser. C* 1–6
[17] Shen Y L, Williams J J, Piotrowski G, Chawla N and Guo Y L 2001 study *Acta Mater.* **49** 3219–29
[18] Boopathi M, Arulshri K P and Iyandurai N 2013 *Am. J. Appl. Sci.* **10** 219–29
[19] Aigbodion V S and Hassan S B 2007 *Mater. Sci. Eng. A* **447** 355–60
[20] Kök M and Özdin K 2007 *J. Mater. Process. Technol.* **183** 301–9
[21] Rashed F S and Mahmoud T S 2009 *Tribol. Int.* **42** 642–8
[22] Gudlur P, Muliana A and Radovic M 2014 *Compos. Part B Eng.* **58** 534–43
[23] Sharma S C, Krishna M, Vizhian P S and Shashishankar A 2002 *Proc. Inst. Mech. Eng. Part D J. Automob. Eng.* **216** 975–82
[24] Wilson S and Alpas A T *Wear* 1996 **196** 270–8
[25] Taguchi G 1986 *Vol. 4* (Tokyo: Asian Productivity Organization.) p 191

[26] Phadke M S 1989 *Vol. 1* (AT & T Bell Laboratories,Holmdel ,New Jersey: P T R Prentice Hall, Englewood Cliffs,New Jersey 07632) pp 1–327

Acknowledgments
Authors were acknowledge to the department of mechanical and metallurgical engineering, Gandhi institute of engineering & Technology, gunupur for encouragement from colleagues, special work by technical staff and providing the permission for publishing work.

IOP Conf. Series: Materials Science and Engineering **314** (2018) 012010 doi:10.1088/1757-899X/314/1/012010

Novel perovskite coating of strontium zirconate in Inconel substrate

G Venkatesh[1], B Blessto[1], C Santhosh Kumar Rao[1], R Subramanian[1], L John Berchmans[2]

[1]Dept. of Metallurgical Engineering, PSG College of Technology, Coimbatore – 641 004, India.

[2]Electropyro Metallurgy Division, CSIR-Central Electrochemical Research Institute, Karaikudi - 630003, India.

Thermal Barrier Coatings (TBC) provides a low thermal conductivity barrier to heat transfer from the hot gas in the engine to the surface of the coated alloy component. $SrZrO_3$ powder are prepared by Sol Gel synthesis method. The synthesized powder sample is characterized by X Ray Diffraction Technique (XRD), Scanning Electron Microscope (SEM) and Transmission Electron Microscope (TEM) and the results are interpreted. The Polycrystalline nature of $SrZrO_3$ is confirmed and lattice spacing are determined in XRD. SEM shows sub-micron sized particles and a fringed pattern is observed in TEM. The IN718 specimen is Wire Cut and Sand Blasted. A $SrZrO_3$ double layer is coated over the Inconel specimen through a Bond Coat made of NiCoCrAlY by Plasma spraying Process and also characterized. SEM analysis of the Coating shows diffusion of Fe, Sr into the substrate.

Keywords: Thermal Barrier Coatings (TBC), Strontium Zirconate, Ceramic coatings, Sol gel synthesis, Atmospheric plasma spraying.

1. Introduction

A protective coating deposited to act as a barrier between the surfaces of the component and the aggressive environment that it is exposed during operation is now globally acknowledged to be an attractive means for reducing damage to the actual component. Thermal Barrier Coatings(TBC) provide a insulation barrier to heat transfer from the hot gas in the engine to the surface of the coated alloy component, whether in the specific applications such as in combustor or the gas turbines [14]. They are provided to protect the base material in various applications such as internal combustion engine bodies, gas turbine blades etc. at very high temperatures thus extending the life of alloy components in the these super temperature regimes by decreasing their surface temperatures [6]. The perovskite Zirconia (ZrO_2) is been used as the best suit for coatings where it is used as the top coat, over a bond coat. As wear resistant coatings are used to prevent wear mainly in textile mill rollers, cylinder liners, pistons, valves, spindles etc. Alumina (Al_2O_3), Titania (TiO_2) and Zirconia (ZrO_2) are the some of the conventional wear resistant coating materials which is presently used for study of novel ceramic coatings for high temperature applications [16].

There are various methods for synthesizing these Perovskite coatings. Molten Salt Synthesis and Sol Gel synthesis are the most commonly used processes. YSZ is the commonly used coating materials due to their thermal properties such as low thermal conductivity, thermal expansion coefficient and an excellent corrosion resistance. CYSZ is an alternate TBCs powder and it has good thermal cycle properties because the addition of ceria makes zirconia more stable at higher temperatures [6]. Calcium titanate synthesized and deposited from conventional powder feedstock by *Jiri Kotlan* et al. is found to be the best choice for plasma spraying when compared with other perovskite structures [12]. Doping with materials such as of Yb_2O_3, Gd_2O_3 are more effective in reducing the specific heat capacity and thermal conductivity for Ceramic powders leading to higher efficiency [2]. The NiAl sprayed coatings are harder than the substrate and hence forms the basis for tribological and wear resistant applications [15].

The method of thermal spraying depends on the type of Thermal type of process employed and consequently flame spraying (FS), high velocity oxy-fuel spraying (HVOF), plasma spraying (PS) etc. are the available choices for material spraying [16]. The study can be extended on degradation of parts and their failure caused due to high relative motion between mating surfaces, degrading environment, peak temperatures

IOP Conf. Series: Materials Science and Engineering **314** (2018) 012010 doi:10.1088/1757-899X/314/1/012010

and repeated stresses [1]. The plasma spraying process can be conducted by varying several process parameters such as oxygen flow rate, spraying speed, fuel gas flow rate, powder carrier gas flow rate, arc current, powder feed rate, nozzle length and stand-off distance [3,4,10]. These parameters have influence on thickness, hardness, microstructure and wear resistance on the coatings [6,9,10]. The powder that is coated fail at different locations due to various reasons. In Optimized functionally graded ceramics spalling occurs at the ceramic and bond coat interface along the thermally grown oxide layer(TGO) whereas in double ceramic layer coatings, it occurs along the bilayer [5]. In double layer coatings failure occurs due to mismatch due to thermal expansion and this can be avoided by formation of a functional gradient coating [7,8].

Based on the literature survey, it is found out that studies on thermal barrier coating using Strontium zirconate has not been explored. Also, several researchers reported only on single layer coating hence in the present study, a novel perovskite double layer has been coated over Inconel substrate and characterization is done.

2. Materials and methods

2.1. Sol gel method

2.1.1. Raw material synthesis

The precursors were chosen based on the processing techniques and the compound formation. The chosen raw materials are $ZrOCl_2.8H_2O$-98% pure, $SrCl_2$-98% pure and EDTA. The precursors were weighted on the basis of stoichiometry for the sol gel reaction and were taken for further process.

2.1.2. Preparation

The weighted materials are taken for the preparatory steps through sol gel method. Initially the required glass wares were cleaned well by distilled water and acetone to ensure the purity. 1.0 M solution of zirconium Oxychloride and Strontium chloride in water is prepared by vigorous stirring. In the clear solution, EDTA is added in the proposed ratio (1:1) to the molten solution under continuous stirring to induce the chelation pH of the solution was increased to 7 and by addition of 5M solution of NaOH, turns the PH till 10. The complete setup was setup at 80°C. This continuous stirring and the effect of the PH converts the solution into gel formation. Then the resultant gel is dried overnight at 350°C for 8 hours in hot air oven for complete removal of water molecules. The obtained mass is grounded to ensure the homogeneity before final calcinations followed by intermediate grinding, the sample is subjected to calcination at 600°C for 4 hours. Resultant powder is grounded well and subjected to sequence of characterization to analyze the desired properties.

2.2. Plasma Coating

$SrZrO_3$ exhibits pseudo tetragonal structure at 750°C and exhibits tetragonal structure at high temperatures of 840°C. The pseudo tetragonal structure creates lattice mismatch between Sr-O atoms, so that coating delaminates easily. So, during the service temperature of the material is subjected to fail at an earlier circumstance. This type of failure requires further modification of coating or a coating interlayer to be sprayed above the bond coat. For this we are going for a multilayer coating with varying feed rates.

The Inconel specimens are wire cut to the desired dimensions and are sand blasted with silica sand to improve its surface roughness for coating adherence. The powder synthesized after conformation of phase purity are then coated onto the Inconel 718 substrate. The coating is done by APS (Air Plasma Spray) technique. The parameters are mentioned in table 1. Initially bond coat of NiCoCrAlY of about 100μm are coated onto the substrate with standard parameters. On top of the bond coat $SrZrO_3$ powder thickness up to 150 μm is deposited with higher powder feed rate and is then cooled. Again, the coating of $SrZrO_3$ powder of thickness 150 μm atop of pervious coated layer with lower powder feed rate. The substrate (Figure 1) along with coating is then cooled and further characterization is done to determine the microstructure of them.

- Substrate - INCONEL 718
- Coating - Strontium Zirconate ($SrZrO_3$) - Double Layer
- Bond Coat – NiCoCrAlY

  Specimen Dimensions (Substrate) - 60mm x 38mm x 6mm

Figure 1. $SrZrO_3$ Coated Inconel Specimen.

Table.1. Various Parameters used in Double Layered Coating.

	Bond Coat	Top Coat 1	Top Coat 2
Current (A)	500-550	500-600	500-600
Voltage (V)	60	60	60
Spraying Distance (mm)	100	100	100
Spraying Speed (mm/sec)	10	10	10
Powder Feed Rate (g/min)	35	35	20
Ar/He Ratio	15/50	15/50	15/50
Oxygen (kg)	4	4	4
Acetylene (kg)	1.8	1.8	1.8
Air	3	3	3

2.3. Characterization

The Crystal structure and Phase of the synthesized $SrZrO_3$ Powder particles is determined using XRD Copper K alpha radiation and Cobalt K alpha radiation and the Phase Purity is confirmed. The surface Morphology of the powder particles are determined using SEM with an EDS attachment in secondary electron mode. The Particle morphology is analyzed using JEOL JEM 2100 High Resolution Transmission Electron Microscope (HRTEM). The Coated specimen is subjected to SEM analysis and the Elemental Distribution along the Top Coat and Bond Coat interface were determined.

3. Results and discussion

3.1. Characterization of $SrZrO_3$ Powders

3.1.1 Phase Analysis-XRD

IOP Conf. Series: Materials Science and Engineering **314** (2018) 012010 doi:10.1088/1757-899X/314/1/012010

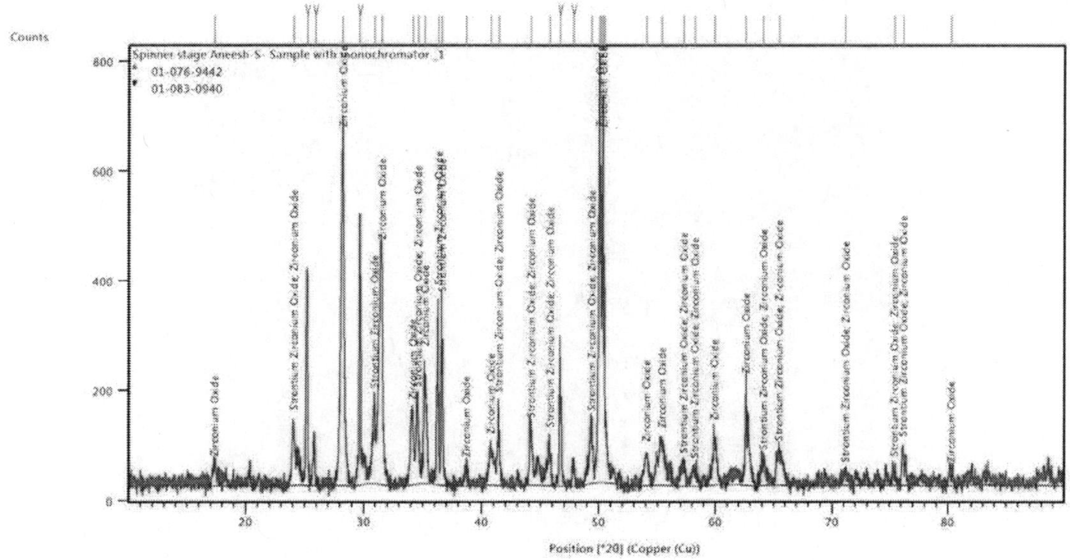

Figure 2. Phase analysis of synthesized $SrZrO_3$

The XRD pattern reveals presence of $SrZrO_3$ phase with Orthorhombic system. The strong diffracted peaks as obtained in Figure 2. represents Strontium Zirconate phases with space group of pbnm thereby the lattice parameters are computed. Existence of the narrow diffracted peaks resembles coarser crystallites which are in the range of few micro meters (µm) which is evident from the SEM analysis. The crystallite size is determined from x-ray line broadening using Debye Scherer's formula, and was found to be 5.20nm.

3.1.2. Scanning Electron Microscope (SEM)

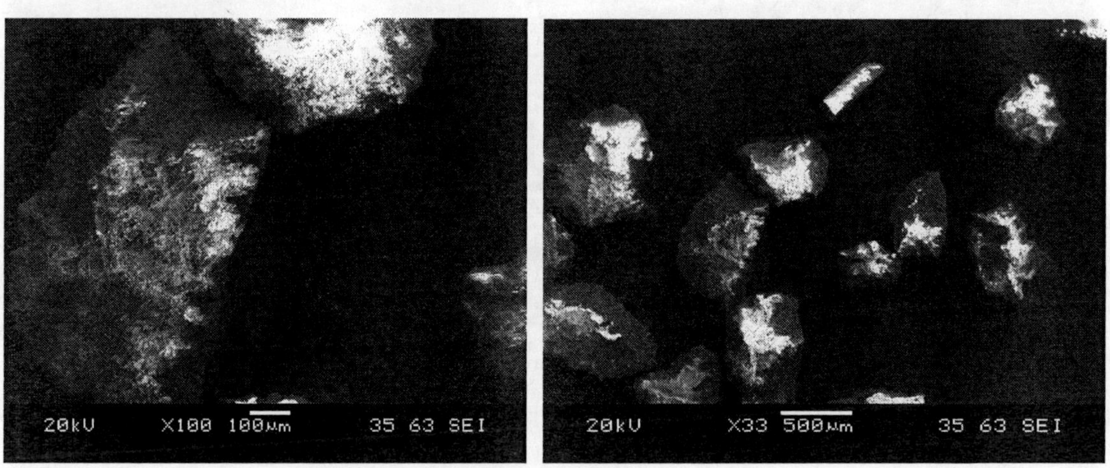

Figure 3. SEM Analysis of $SrZrO_3$ powders.

From figure 3, the particle size is approximately in the sub-micron size with particles being agglomerated. The particle size is in comparison with the particle size calculated from the SEM image.

IOP Conf. Series: Materials Science and Engineering **314** (2018) 012010 doi:10.1088/1757-899X/314/1/012010

3.1.3. Transmission Electron Microscope (TEM)

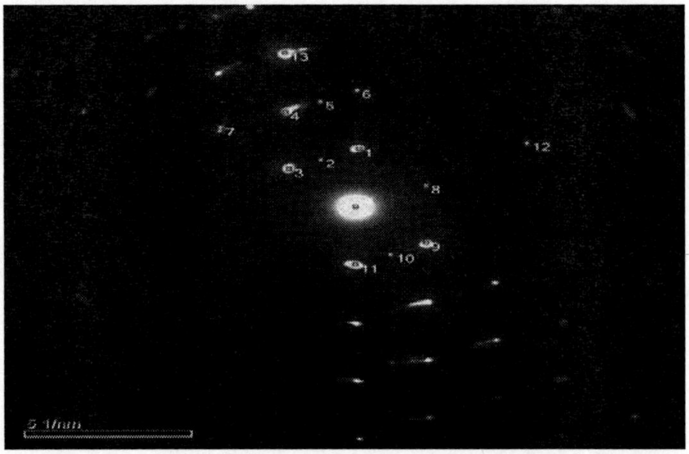

Figure 4. SAED pattern of $SrZrO_3$.

The synthesized $SrZrO_3$ powders are Analyzed in TEM to confirm the presence of its Crystalline nature. The presence of rings confirms the Polycrystalline nature of $SrZrO_3$.

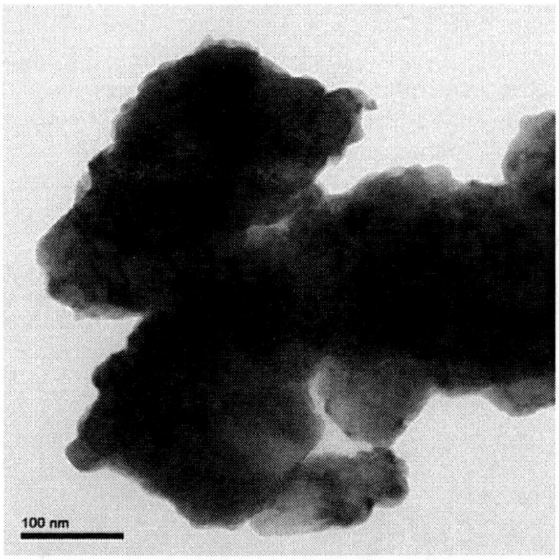

Figure 5. Particle morphology using TEM.

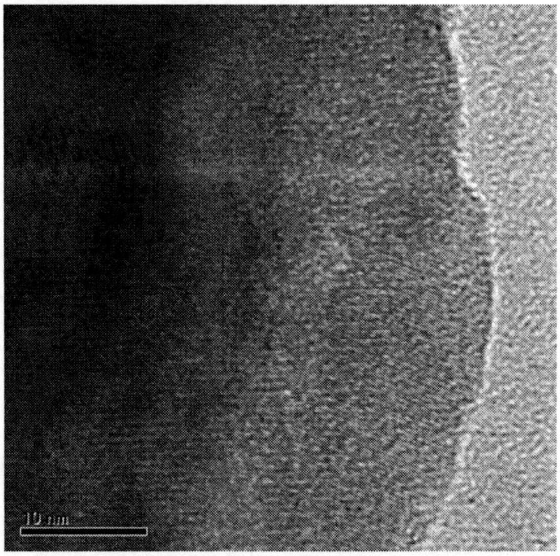

Figure 6. HRTEM image of Strontium Zirconate.

The particle shape and size of $SrZrO_3$ was observed from TEM images (Figure 5,6). The particle size is found to be ultra-fine sized and confirms the morphology as spherical in nature. The particle size computed from the above TEM images ranges between 10nm to 100nm. Particle agglomeration prevents the size reduction factor of $SrZrO_3$.

From SAED pattern (Figure 4) of $SrZrO_3$ reveals the presence of polycrystalline nature of the sample with each diffraction spot corresponding to the reflection from the planes of orthorhombic unit cell respectively [12]. (111), (002), (040) from the orthorhombic unit cell clearly confirms the phase pure formation. The bravais lattice with d-spacing value ranges from 28.23nm to 178.26nm. Hence from the d spacing and the reflection planes, the lattice parameter were calculated and found to be a=5.81nm, b=8.13nm, c=5.69nm.

The HRTEM image (Figure 6) shows the perfect stacking sequence. All atoms are in the [001] direction. The fringe pattern is observed and that justifies the atomic presence over the (111), (200), (040) observed from the SAED pattern. The size ranges from 0.5Å to 1nm.

3.2. Characterization of Coated specimen

3.2.1. Scanning Electron Microscope (SEM)

Figure 7. shows the cross-sectional view of the $SrZrO_3$ powder coated over the Inconel substrate. Presence of NiO or NiO_2 (spinal) and chromia at the TGO layer may lead to Thermal Spalling. The Elemental analysis as shown in Figure 8. shows Cr, Ni which are ingredients of the bond coat.

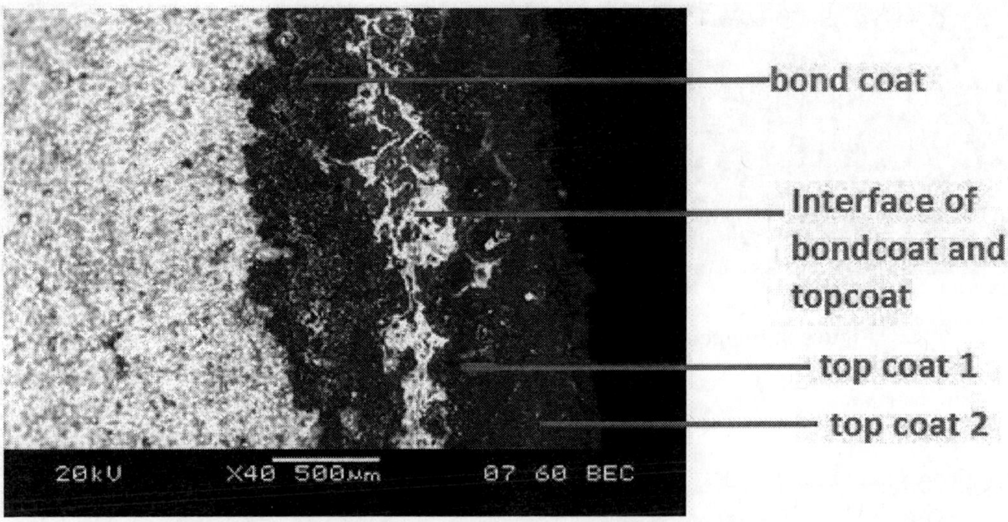

Figure 7. Cross sectional view of sprayed $SrZrO_3$ on Inconel.

From figure 8., the Area analysis shows the variation in concentration of the element across the line. The Area EDS analysis confirms the range of distribution of elements along the interface of the bond and top coat.

IOP Conf. Series: Materials Science and Engineering **314** (2018) 012010 doi:10.1088/1757-899X/314/1/012010

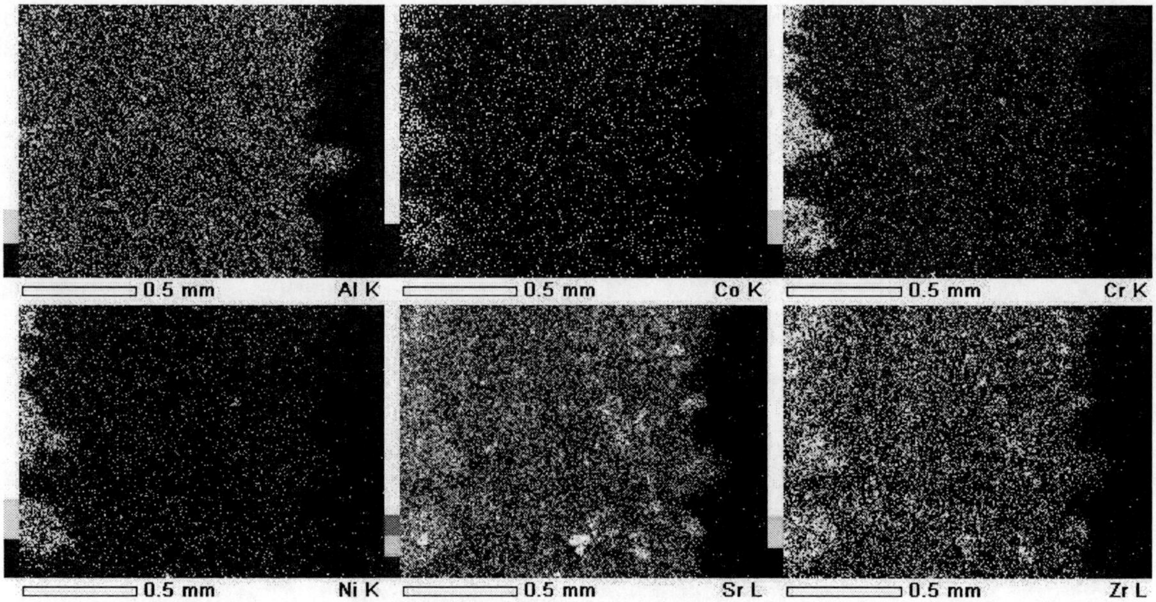

Figure 8. Elemental Distribution along Area at Bond and Top coat interface.

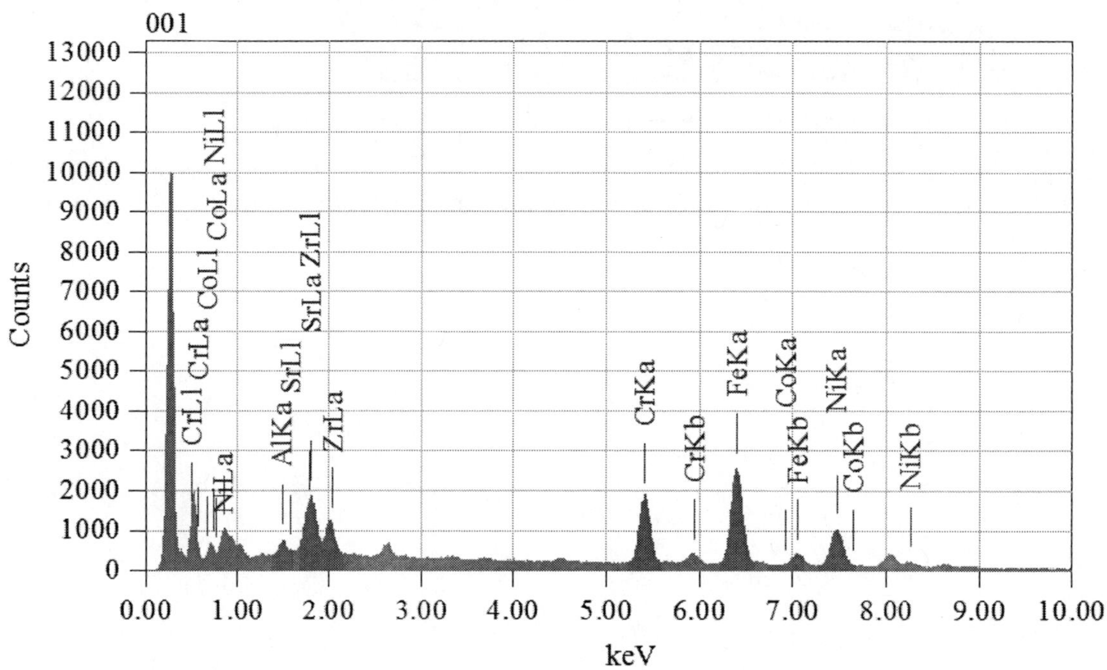

ZAF Method Standardless Quantitative Analysis
Fitting Coefficient : 0.6018

Element	(keV)	mass%	Error%	At%	Compound	mass%	Cation	K
Al K	1.486	2.05	1.43	4.39				1.0023
Cr K	5.411	18.43	2.02	20.44				20.6870
Fe K*	6.398	40.11	2.70	41.41				42.5933
Co K*	6.924	0.43	3.44	0.42				0.4408
Ni K	7.471	24.26	4.40	23.82				24.6054
Sr L	1.806	8.50	3.12	5.59				6.5384
Zr L*	2.042	6.21	3.25	3.92				4.1327
Total		100.00		100.00				

Figure 9. EDS and Chemical Composition of Elements at Inconel and Bond Coat Interface.

EDS shows presence of Ni, Cr, Fe from Substrate and Ni, Co, Cr, Al from the Bond coat as in Figure 9. The presence of Si may be due to the sand blasting process, we can clearly see the difference in concentration of Sr, Cr, Fe, Ni as we move away from the substrate into the coating can be clearly seen in Figure 9. Intermediate to the coating there is a diffused layer wherein the Fe from the substrate and Sr from the top coat are diffused to some extent.

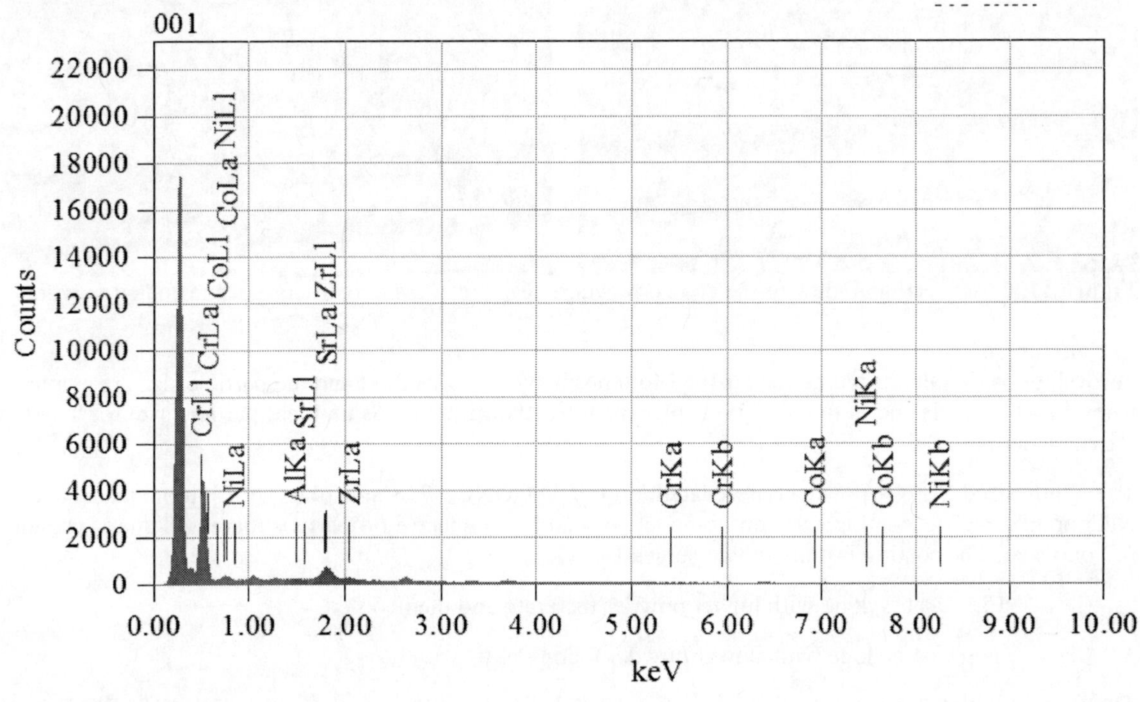

ZAF Method Standardless Quantitative Analysis
Fitting Coefficient : 0.9633

Element	(keV)	mass%	Error%	At%	Compound	mass%	Cation	K
Al K*	1.486	4.89	25.66	14.33				4.6068
Cr K*	5.411	1.46	73.52	2.22				1.5663
Co K*								
Ni K*	7.471	0.42	129.44	0.56				0.4988
Sr L	1.806	59.25	67.74	53.45				72.6495
Zr L*	2.042	33.97	123.21	29.44				20.6786
Total		100.00		100.00				

Figure 10. EDS and Chemical Composition of Elements at Bond coat and Top Coat Interface.

IOP Conf. Series: Materials Science and Engineering **314** (2018) 012010 doi:10.1088/1757-899X/314/1/012010

EDS (Figure 10) shows presence of Ni, Co, Cr, Al from the Bond coat and major content of Sr, Zr from the Top coat. The Cu alloying element in the substrate is also diffused into the top coat because of their high thermal conductivity as the plasma sprayed process is carried at higher temperature. Due to the diffusion of Cu there is a possibility of formation of intermetallic during subsequent thermal process. The whole of diffusion process is in the first coat while the second coat remains unaffected. In single layered coating, the diffusion would have concentrated in top coat that reduces the overall property of the coating, the double layered coating provides nil concentration of diffusion in top coat that can increase the lifetime of the coating and wear property of the substrate in comparison to the single layered coating.

3.2.2 Macrostructure

Figure 11. shows the cross sectional view of the $SrZrO_3$ coating over the Inconel substrate as observed in the image analyzer. The coated surface is porous which may be due to improper melting of large particles during spraying [2]. Figure 12. Gives the macrostructure of the coated powder samples in top view.

Figure 11. Cross sectional view of the coated powders. **Figure 12.**Macrostrucure of the coated powders.

4. Conclusion

Perovskite oxide salt are synthesised using Molten salt synthesis method and the particle size are evaluated using SEM analysis and is found to be approximately 10.6μm which is an ideal particle size for Thermal Barrier Coating(TBC) applications.

The synthesised $SrZrO_3$ powders are coated atop of INCONEL 718 substrate with intermediate layer of NiCr bond coat. The coating was non-homogeneous and the adhesive property is increased due to presence of bond coat. The coating is done in two stages:

(i) First coat is done with higher powder feed rate and then cooled.

(ii) Top coat is done with slower powder feed rate.

The Polycrystalline nature of $SrZrO_3$ is confirmed and lattice spacing are determined in XRD. SEM shows sub-micron sized particles and a fringed pattern is observed in TEM. All atoms are in the [001] direction. The fringe pattern is observed and that gives the atomic presence over the (111), (200), (040).

SEM in Coating shows diffusion of Fe, Sr into the substrate. The Cu alloying element in the substrate is also diffused into the top coat because of their high thermal conductivity as the coating process is carried at elevated temperatures. Due to the diffusion of Cu there is a possibility of formation of intermetallic during subsequent thermal process if ever undergone. In single layered coating, the diffusion of elements is concentrated in immediate top coat that reduces some property of the coating while in the double layered coating the top coat is not subjected to diffusion which increases the lifetime of the coating and wear property of the substrate in comparison to the single layered coating.

Acknowledgments

The authors are grateful to Department of Metallurgical Engineering, PSG College of Technology and PSG Institute of Advanced Studies for the great support.

References

[1] Mishra S C 2012 Analysis of Experimental Results of Plasma Spray Coatings Using Statistical Techniques *Advanced Plasma Spray Applications* 84-96

[2] Ma W, Mack D, Malzbender J, Vaben R and Stover D 2008 Yb_2O_3 and Gd_2O_3 doped strontium zirconate for thermal barrier coatings *Journal of the European Ceramic Society* ***28*** *Issue 16 3071-81*

[3] Wei-Cheng Lih, Yang S H, Su C Y, Huang S C, Hsu I C and Leu M S 2000 Effects of process parameters on molten particle speed and surface temperature and the properties of HVOF CrC/NiCr coatings *Surface and Coatings Technology* 54-60

[4] Rampon R, Marchand O, Filiatre C and Bertrand G 2008 Influence of suspension characteristics on coatings microstructure obtained by suspension plasma spraying *Surface & Coatings Technology* 4337–42

[5] Chaohui Wang, You Wang, Shan Fan, Yuan You, Liang Wang, Changlong Yang, Xiaoguang Sun and Xuewei Li 2015 Optimized functionally graded $La_2Zr_2O_7$/8YSZ thermal barrier coatings fabricated by suspension plasma spraying *Journal of Alloys and Compounds*

[6] Muhammet Karabas, Emre BAL, Ayşe KILI and Yılmaz Taptık 2016 Effect of Air Plasma Spray Parameters on the Properties of YSZ and CYSZ Thermal Barrier Coatings T*he Australian Ceramic Society*

[7] Caoa *X Q*, Vassenb R, Tietzb F and Stoever D 2006 New double-ceramic-layer thermal barrier coatings based on zirconia–rare earth composite oxides *Journal of the European Ceramic Society* 247–251

[8] Zhenhua Xu, Limin He, Rende Mu, Shimei He, Guanghong Huang and Xueqiang Cao 2010 Hot corrosion behavior of rare earth zirconates and yttria partially stabilized zirconia thermal barrier coatings *Surface & Coatings Technology* 3652–61

[9] Ozkan Sarikaya 2005 Effect of some parameters on microstructure and hardness of alumina coatings prepared by the air plasma spraying process *Surface & Coatings Technology* 388–393

[10] Khalida M, Mujahida M, Nusair Khanb A, Rawatc R S and Mehmood K 2013 Effect of arc current on microstructure texturing and wear behavior of plasma sprayed CaZrO3 coatings *Ceramics International* 2293–2302

[11] Panadda Niranatlumpong, Chalermchai Sukonkhet and Kittichai Ninon 2015 Loss of Y from NiCrAlY powder during air plasma spraying *Surface & Coatings Technology* 277–281

[12] Jiri Kotlana, Zdenek Palaa, Radek Musaleka and Pavel Ctibor 2016 On reactive suspension plasma spraying of calcium titanate *Ceramics International* 4607–15

[13] Mishra S C and Praharaj S 2011 Deposition of Plasma Spray Coating using industrial waste with low grade ore mineral **18** No2 199-206

[14] Gosipathala Sreedhar, Masroor Alam M D and Raja V S 2009 Hot corrosion behaviour of plasma sprayed YSZ/Al2O3 dispersed NiCrAlY coatings on Inconel-718 super alloy *Surface & Coatings Technology* 291–299

[15] Chaithanya M 2001 processing & characterization of nickel – aluminide coating on metal substrates *Thesis*

[16] Lalitendu tripathy 2010 Analysis and prediction of plasma sprayed alumina-Titania coating deposition using neural computation *Thesis*

IOP Conf. Series: Materials Science and Engineering **314** (2018) 012011 doi:10.1088/1757-899X/314/1/012011

Synthesis of second phase hybrid ceramics using two different bio-source and a comparative study on their morphological characterization

Adhitya Karivaratharajan[1], **Sidharth Baskaran**[1], **K. Thillairajan**[2]

[1] Metallurgical Engineering, PSG College of Technology, Coimbatore – 641004
[2] Assistant Professor, Metallurgical Engineering, PSG College of Technology, Coimbatore – 641004

Abstract. Ceramics are generally synthesized with various sources and methods. The most common method for synthesis of ceramics with reduced cost and energy is SOL-GEL method. Combustion synthesis is also a most widely used method for ceramic synthesis. In general, ceramics have enhanced hardness and dimensional stability even at elevated temperatures. For this reason, they are used in the production of refractories, thermal barrier coatings, chemical resistant coatings, wear resistant coatings, and also as reinforcement material to produce metal matrix composites and polymer matrix composites. This work concentrates on the comparison of morphological characterization of such reinforcement particles synthesized from different sources. The particles size range varying from 7 µm to 250 µm with flaky and spongy structures are observed in the ash of Vicia faba. However, the ash of Cocos nucifera resulted in fibrous structure with a diameter of 50 µm to length above 600 µm, particles size ranging from 10 µm to 70 µm, micro tubes of diameter 3.6 µm to length of 150 µm. The EDX and XRD analysis of Vicia faba showed the presence of carbon as the major element with a few other elements.

1. Introduction

Ceramics are synthesized as glasses, poly crystals, single crystals, and in many forms dictated by their end use, including fine powders, monoliths, thin films, and composites. It paved way for several inventions like ultrahigh-strain single crystal piezo electrics, new high frequency dielectrics, phosphors intercalation compounds, ductile layered structural ceramics, and thermo electrics.

Ceramics are generally classified as traditional ceramics and synthetic ceramics. Thermal conductivity is low for ceramics in most of the cases, although there can be some ceramics with a temperature dependent conductivity and may go into superconductivity range at elevated temperatures. In this work, seeds of Vicia faba and leaves of Cocos nucifera is taken as the sources for synthesis of ceramics. This paper concentrates more on morphological study of synthesized particles and compare the obtained results.

2. Experimental Work

2.1. Materials and Pre-Treatment

In this work, seeds of a plant named Vicia faba are used. The seeds collected from the farmer after through checking for quality. The leaves of Cocos nucifera were also collected from the same farm which is actually a waste. Pre-treatment is given to the seeds and leaves by washing them in running water to remove the unwanted contaminants from them. Followed by washing, seeds alone are boiled in water and milk and it is not done for leaves. This is because the seeds in general, contains organic compounds with very low boiling point and could vaporize even at a low temperature. Seeds of Vicia faba are bought from farms directly for combustion synthesis. Pretreatment of seeds are made to avoid unwanted evolution of gases during combustion. Seeds are collected from the farms and are dried under sun for two days. After drying, seeds are washed in running water and dried under sun shade. After washing, seeds are boiled in water for one hour. Then again it is boiled in milk (if milk is not available dissolve 25G of milk powder in 100ml of water) for one hour. Followed by drying under the sun. After drying these seeds, they are checked manually for moisture. If moisture is still present the seeds are roasted to remove the moisture content. After roasting the seed is ready for burning.

Initially 500 g of seed is taken. After pre-treatment and drying, the weight of seed is found to be 419.392 grams. This indicates that the removal of excess moisture and some organic impurities from the seed has taken place. In one part of the pre-treatment, the seed is taken as it is without powdering and pre-treated. During burning excessive evolution of gases are found giving unpleasant smells. In another section, powdered seed is taken for pre-treatment.

The organic compounds are generally classified into water soluble and fat soluble. So that, while boiling in water and milk, the organic compounds are also removed accordingly. Whereas the leaves do not contain such compounds and the impurities present in leaves generally vaporize on burning. The leaves were roasted at a temperature of 300 °C for 2 hr to remove the surface moisture and unwanted volatile compounds. Once roasting is done, leaves are ready for burning.

2.2. Burning of Leaves and Seeds:

The seeds after drying under the sun are weighed before burning. After weighing the required amount of seed, they are filled in the crucible and burnt in muffle furnace at 750° C for 3 hr to produce the ash. Then the crucible with ash is removed from the furnace and cooled in air. After sometime, ash samples are collected and stored. Ash particles may get attached to each other due to a weak mechanical bonding. This bonding is broken by simply subjecting them to manual crushing. Leaves are burnet at (i) 800° C for 2 hr and (ii) 950° C for 2 hr. Once burning is done, ash from each condition are separated and allowed for cooling.

3. Results and Discussion

3.1. Visual observations

The time taken for drying and burning is reduced and also evolution of gases is highly minimized or suppressed when the seed is used in powdered condition. Also when we use powdered seed we can burn comparatively high quantity of seeds. Hence efficiency is increased to some extent by using powdered seeds. When the seed was burned at 500° C, blackish product is obtained which contains some unburned seeds. This is due to the presence of volatile impurities that were unreacted during burning process. This unburned seed is found at the bottom of the crucible. It implies that sufficient amount of air is not available at the bottom layer for burning. Hence again the seed is burned at 750° C for 3 hours in the

IOP Conf. Series: Materials Science and Engineering **314** (2018) 012011 doi:10.1088/1757-899X/314/1/012011

muffle furnace with intermediate manual agitation by doing this the bottom layer comes to the top and the top comes to the bottom, Which results in uniform burning of seeds. While burning at 750° C a mixture of greyish and black powder is obtained which is taken for characterization. While burning leaves of Cocos nucifera, no such evolution of gas is observed and also the leaves burned completely and formed blackish grey ash. The obtained ash is observed like micro tubes and small particles. They are further subjected to sieve analysis to study the size distribution of the particles. Since the greyish colour ash was completely burned, it is taken for characterization.

3.2. Sieve analysis

The sieve analysis of burned seed infers that the more amount of fine powders are obtained when the seeds are burned after crushing instead of burning as it is. Also it the reduction in burning time is observed in powdered seed. On sieving the burned ash of Cocos leaf, it is observed that when the burning temperature increases the size of the particle is reducing. This is observed because of removal of volatile impurities at elevated temperature.

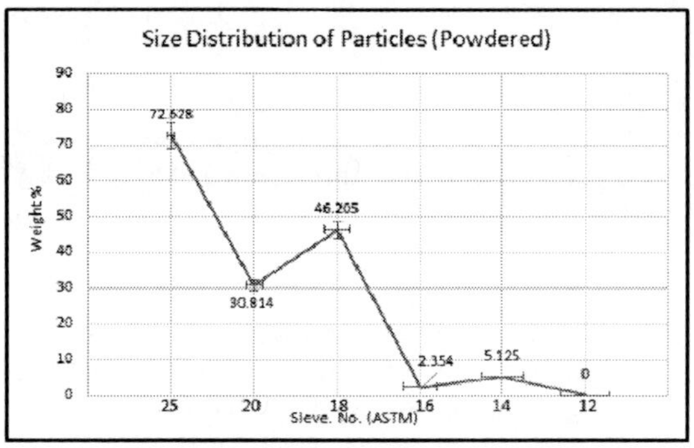

Graph 1: Size distribution of powdered seed after burning

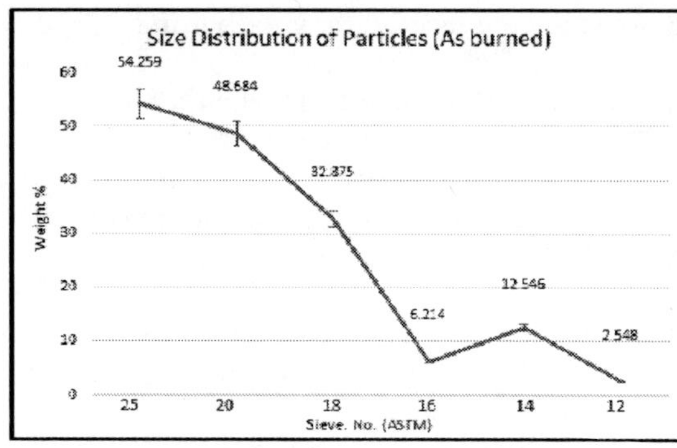

Graph 2: Size distribution of un-powdered seed after burning

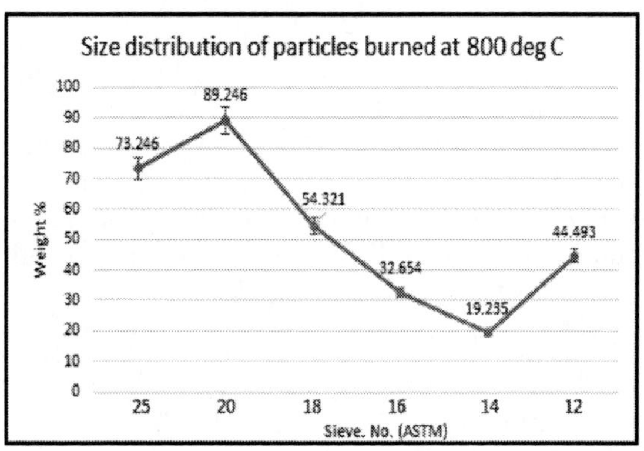

Graph 3: Size distribution of leaves burned at 800°C

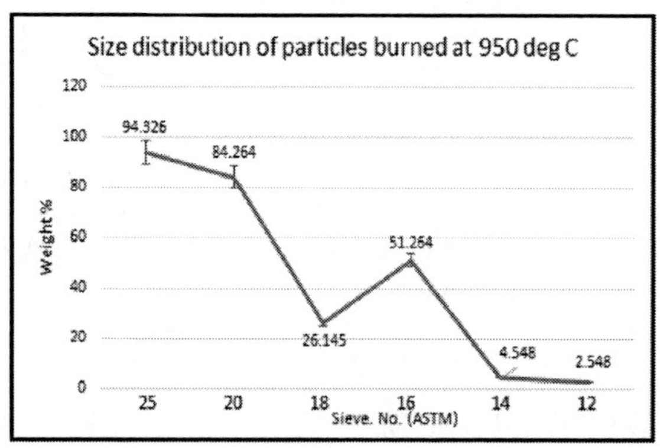

Graph 4: Size distribution of leaves burned at 950°C

3.3. SEM Analysis

The SEM images of the burned seed with varying size is represented in the figures 1, 2, 3, and 4, particles ranging from coarser to very fine. The maximum particle size of 210 micron meter and minimum of 7 micron meter is observed. Coarse particles formed spongy structure (cauliflower structure), size

IOP Conf. Series: Materials Science and Engineering **314** (2018) 012011 doi:10.1088/1757-899X/314/1/012011

ranging from 102 μm to 210 μm with high amount of porosity which is generally not preferred because ceramics are naturally hard and hence it has apparent density and tap density more or less equal. So compact ability of ceramics is difficult, so it is difficult to compact ceramics. When porosity is present in high amount the compact ability of the ceramics will further reduce and high energy is required for compaction. If it is used for reinforcement, this porosity may be an initiation size for crack growth and also it accelerates the crack propagation thereby affecting the properties of the composite.

Figure 1: Coarse particles

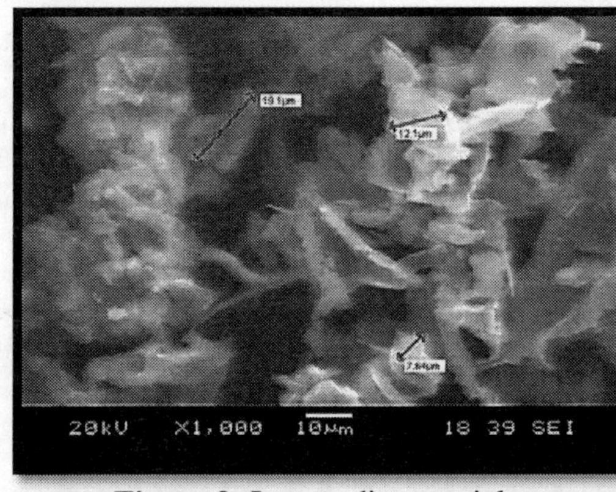

Figure 2: Intermediate particles

Figure 3: Micro particles

Figure 4: Fine particles

Coarse particles formed spongy structure (cauliflower structure), size ranging from 102 μm to 210 μm with high amount of porosity which is generally not preferred because ceramics are naturally hard and hence it has apparent density and tap density more or less equal. So compact ability of ceramics is difficult, so it is difficult to compact ceramics. When porosity is present in high amount the compact ability of the ceramics will further reduce and high energy is required for compaction. If it is used for reinforcement, this porosity may be an initiation size for crack growth and also it accelerates the crack propagation thereby affecting the properties of the composite.

Intermediate particles formed flaky structure (fern like structure) with size ranging from 40-90 μm. This particles has less porosity than coarser particles. The compact ability of the intermediate particles are preferably good because they have a considerably difference between apparent density and tap density. But they have sharp corners, this sharp corners may act as initiation of tri axial stress in a reinforced composite. The development of tri axial stress is not preferred as it failures the material much earlier of its working lifetime. So to blend this corners, ball milling is preferred before reinforcement which will also results in uniformly distributed particles with respect to size and shape of the particle.

Micro particles formed flaky structure (rose petal like structure) with a size ranging from 15-40 μm with moderate amount of porosity. In this microstructure needle like structure is observed size ranging from 20-30 micron meter. This elongated structure has good compact ability and moderate advantage in reinforcement.

In very fine particles, flaky structure (cluster of leaves like structure) with particle size ranging from 7-20 μm is observed. In this structure blunt flakes are observed instead of sharp corner edges. The porosity is also comparatively low. The difference between apparent and tap density is relatively high and hence it has good compact ability. These particle are also preferred for reinforcement because of the presence of blunt corners instead of sharp. But the uniformity in distribution of particles with such morphology is low. Hence it is preferred to undergo milling to attain uniform particle size and shape distribution. Few particles are cylindrical in structure, few are acicular, and few are rounded structure.

The SEM images of the burned seed with varying size is represented in the figures 5, 6, 7, 8, 9, 10, and 11. Where figures 5, 6, and 7 represents the SEM image of leaves burned at 800 deg Celsius and figure 8, 9, 10 and 11 are the SEM images of leaves burned at 950 deg Celsius.

Figure 5: Fine particles

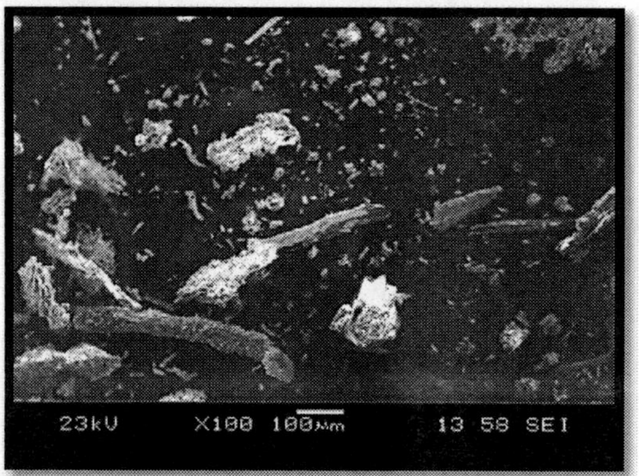

Figure 6: Intermediate particles

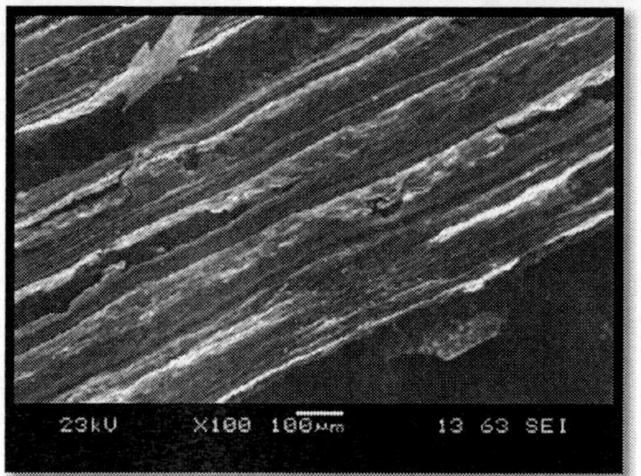

Figure 7: Coarse particles

In this, particles of size ranging from 10 μm to 100 μm is observed. The particles are rod like and flaky structured. Few dendrites are also observed. The intermediate particles are observed under SEM. The morphology of the particles shown rod like structure with diameter of 50 μm to length of 600 μm. Few particles of 50 to 200 μm are also observed. The final particles remained in the sieve is taken and analysed under SEM. The formation of tubes of diameter 100 μm and length above 1000 μm is observed.

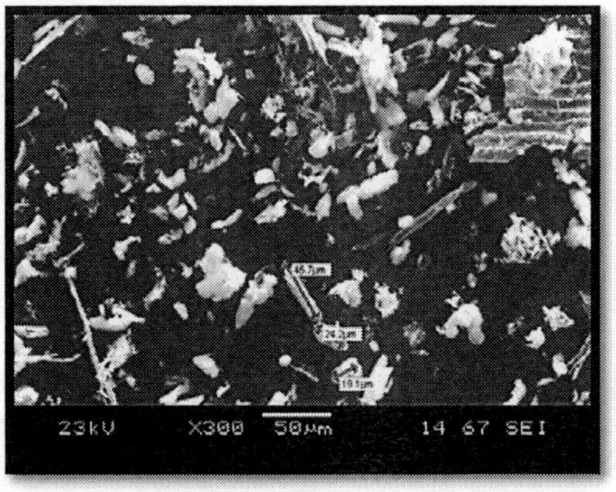

Figure 8: Fine particles

Figure 9: Intermediate particles

Figure 10: Fibres

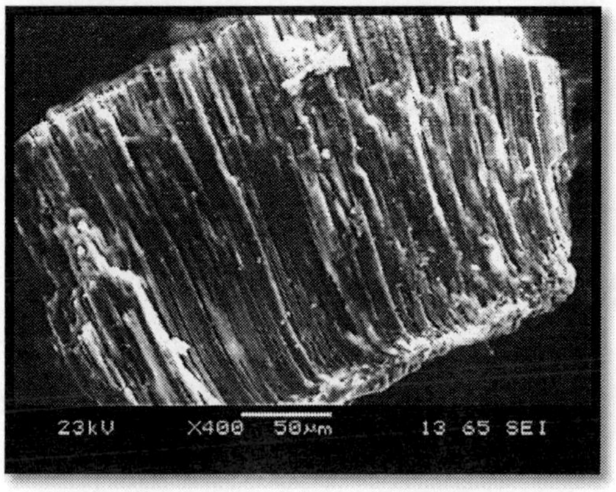

Figure 11: Clustered tubes

Particles of size ranging from 10 µm to 50 µm are observed. The particles formed are cylindrical in nature with blunt ends. Few particles are formed like thin needles with diameter of 5 µm. In figure 9, particles of size ranging from 50 µm to 100 µm are observed. The morphology of the particles are rod like structure. The whiskers of diameter ranging from 20 µm to 50 µm is observed in figure 10. The whiskers are fractured at some part which may be occurred during handling of the particles. Particles of size ranging 300 µm are seen in figure 11. Upon magnifying further, fine tubes formation is observed with diameter of 1 to 5 µm.

3.4. EDS Analysis

Table 1. EDX analysis of burned powders in % mass are shown below.

S1 – Burned seed, S2– Leaves burned at 800 deg Celsius.

	% C	% O	% P	% Mg	% Fe	% K
S1	65.69	8.55	3.63	0.65	5.38	16.47
S2	81.23	4.62	1.99	0.18	3.21	8.77

EDX analysis confirmed the presence of carbon as the major element in the obtained ash. Other than carbon, few other elements are present in a considerable amount.

3.5. Methanol Test

To verify the presence of carbon in organic form, the ash is dipped in methanol solution. The ash precipitated down indicating the presence of organic carbon in both ash of seed and leaves as shown above. Generally organic carbon are more reactive than inorganic carbon. Also the bond nature of organic compound make it highly active at elevated temperature.

4. Conclusions

The ash of both source is characterized and compared. The ash of the seed resulted in particles of flaky structure. New structures like rosette structure, cluster leaves structure and cauliflower like structures are observed. The composition analysis of the burned ash infers the presence of carbon as the major element. The particles of size ranging from 7 μm to 250 μm are observed.

The ash of burned leaf shows various particles morphology like whiskers, micro tubes, rods, fine tubes and bars. The particle size ranging from 10 μm to 250 μm is observed. Tubes of 100 μm diameter and 1000 μm length is observed. Few particles are observed with dimensions of 500 μm, but they are composed of fine tubes of diameter 1 to 5 μm.

Methanol test verifies the presence of carbon is organic in nature. In addition to carbon, other elements like phosphorous, potassium, magnesium, copper, iron are also observed.

5. Acknowledgment

Authors are grateful to thank Dr. M. Kavitha, Mr. G. Venkatesh for their support during the work. We also thank all who helped directly or indirectly during this work.

6. References

[1] Effects of breadfruit seed hull ash on the microstructures and properties of Al–Si–Fe alloy/breadfruit seed hull ash particulate composites. C.U. Atuanya etal.

[2] Naresh P. Development and characterization of metal matrix composite using red mud an industrial waste for wear resistant applications, Ph.D. thesis, Department of Mechanical Engineering, National Institute of Technology, Rourkela 769 008, India; 2006, p. 23–34.

[3] Besterci M, Slesar M, Jangg G. Structure and properties of dispersion hardened AlAl4C3 materials. Powder Metallurgy International 1992; 24:2732.

[4] Introduction to Materials Science, Chapter 13, Structure and Properties of Ceramics

[5] Intermetallic phase particles in 6082 aluminium alloy G. Mrówka-Nowotnik*, J. Sieniawski, M. Wierzbiñska Department of Materials Science, Rzeszów University of Technology, ul. W. Pola 2, 35-959 Rzeszow, Poland * Corresponding author: E-mail address: mrowka@prz.edu.pl

[6] J.E. Hatch Ed., Aluminium. Properties and Physical Metallurgy. ASM Metals Park, Ohio 1984.

[7] L.F. Mondolfo, Aluminium Alloys: Structure and Properties. Butterworths, London-Boston, 1976.

[8] Synthesis and characterization of Aluminum – Silicon – Silicon carbide composite, J P Pathak J K Singh and S Mohan, Centre of advanced study, Department of Metallurgical Engineering, Institute of technology,

Banaras Hindu University, Varanasi 221 005, Indian journal of engineering and material sciences, Vol 13, pp 238 – 246.

[9] C.D. Marioara, S.J. Andersen, J. Jansen, H.W. Zandbergen,The influence of temperature and storage time at RT on nucleation of the '' phase in a 6082 Al-Mg-Si alloy, Acta Materialia 51 (2003) 789-796.

[10] N.C.W. Kuijpers, W.H. Kool, P.T.G. Koenis, K.E. Nilsen, I. Todd, S. van der Zwaag, Assessment of diffrent techniques for quantification of -Al(FeMn)Si and -AlFeSi intermetallics in AA 6xxx alloys, Materials Characterization 49 (2003) 409-420.

[11] J.E. Hatch Ed., Aluminium. Properties and Physical Metallurgy. ASM Metals Park, Ohio 1984.

[12] S. Zajac, B. Bengtsson, Ch. Jönsson, Influence of cooling after homogenization and reheating to extrusion on extrudability and final properties of AA 6063 and AA 6082 alloys, Materials Science Forum 396-402 (2002) 399-404.

[13] G. Mrówka-Nowotnik, J. Sieniawski, Influence of heat treatment on the micrustructure and mechanical properties of 6005 and 6082 aluminium alloys, Journal of Materials Processing Technology 162-163 (2005) 367-372.

[14] Development of a new aluminium matrix composite reinforced with iron oxide (Fe3O4) E. Bayraktar*, D. Katundi Supmeca/LISMMA-Paris, School of Mechanical and Manufacturing Engineering, France* Corresponding author: E-mail address: bayraktar@supmeca.fr Received 29.11.2009; published in revised form 01.01.2010

[15] Shadakshari R, Dr.Mahesha K, Dr.Niranjan H Carbon Nanotube Reinforced Aluminium Matrix Composites – A Review International Journal of Innovative Research in Science, Engineering and Technology *Vol. 1, Issue 2, December 2012* Copyright to IJIRSET www.ijirset.com 206.

IOP Conf. Series: Materials Science and Engineering **314** (2018) 012012 doi:10.1088/1757-899X/314/1/012012

Recent developments in turbomachinery component materials and manufacturing challenges for aero engine applications

Srinivas G[1], Raghunandana K*[2] and Satish Shenoy B[3]

[1]Assistant Professor, Dept of Aero & Auto Engineering, Manipal Institute of Technology, Manipal University, Manipal, Udupi, Karnataka, India-576104.
[2]Professor, Dept of Mechatronics Engineering, Manipal Instituteof Technology, Manipal University, Manipal, Udupi, Karnataka, India-576104.
[3]Professor, Dept of Aero & Auto Engineering, Manipal Instituteof Technology, Manipal University, Manipal, Udupi, Karnataka, India-576104.

Abstract. In the recent years the development of turbomachinery materials performance enhancement plays a vital role especially in aircraft air breathing engines like turbojet engine, turboprop engine, turboshaft engine and turbofan engines. Especially the transonic flow engines required highly sophisticated materials where it can sustain the entire thrust which can create by the engine. The main objective of this paper is to give an overview of the present cost-effective and technological capabilities process for turbomachinery component materials. Especially the main focus is given to study the Electro physical, Photonic additive removal process and Electro chemical process for turbomachinery parts manufacture. The aeronautical propulsion based technologies are reviewed thoroughly where in surface reliability, geometrical precession, and material removal and highly strengthened composite material deposition rates usually difficult to cut dedicated steels, Titanium and Nickel based alloys. In this paper the past aeronautical and propulsion mechanical based manufacturing technologies, current sophisticated technologies and also future challenging material processing techniques are covered. The paper also focuses on the brief description of turbomachinery components of shaping process and coating in aeromechanical applications.

1. Introduction

In the present trend the aeromechanical propulsion industry response for turbomachinery sector like steam turbines, gas turbines engines and –stationary turbines and all air breathing engines, aircraft engine starting device turbochargers are also consistently developing due to increasing global necessity for vitality and flexibility. These turbomachinery components need to develop keeping environmental safely constrain in usage of oil, gas and coal together in producing the levels of carbon dioxide, nitrogen oxide levels to be maintained according to the international air standard environmental defined safety marginal levels. Thus, it is required to produce the environmental free highly sophisticated effective turbomachinery structures.

Especially in aeronautical engineering sector day to day according to IATA (International Air Transport Association) 2016 year calculation the growing rate drastically changed to 7.25 % in both passenger aircrafts and cargo aircrafts. This value virtually almost twice within a decade according to International global market forecast 2016. Adding to this the maintenance, repair and Overhaul are also playing a vital role increasing demand. Further the market growth, future challenges faced by aircraft turbomachinery components manufacturing organizations are also increasing, as upcoming aircraft engine turbomachinery components need also be cleaner due to official rules and regulations, quitter and most importantly more fuel efficient. According to the survey conducted by the Airbus engines council for aviation goals for 2020, plan to reduce carbon monoxide levels 55%, and 75% Nitrogen oxide and also noise levels reduction by 30%. Finally in precipitously understood that there is huge research gap need for selecting good design in the process selecting materials, extensive manufacturing methods, repair technologies able to phase present challenging impending difficulties for turbomachinery structures.

2. Turbomachinery components challenges and manufacturing capabilities

Especially in aeronautical propulsion industry, the turbomachinery components like fan, axial flow compressor, centrifugal compressor and turbines required to handle large amount of mass flow of air and hot gas at higher velocities. Therefore the turbomachinery components required to have characteristics like to hold maximum temperature, maximum clear-cut strength and uniform resilient resources. Machining of these component edges really too difficult to cut by method of conventional methods. Conventional methods has drawback that it takes more metal removal rate, low precession due to great cutting forces leads to more tool cost due to more wear subsequently low process efficiency[1]. Therefore, use of economical and highly sophisticated technologies is of more interest.

The aircraft engines preferred more Titanium and Nickel based alloys (see figure 1). The temperature capability of these composite materials improved continuously through different manufacturing techniques and new alloy composition [2]. Single crystal materials gives much better properties compare to polycrystalline materials as it gives superior creep properties and also best suitable for high temperature applications like aero engines [3]. The use of such advanced titanium and Nickel based composites (used in aero engine turbines and compressors), matrix composites (especially polymer matrix composites [4] used in aircraft turbofan engines) required the more sophisticated advanced manufacturing strategies.

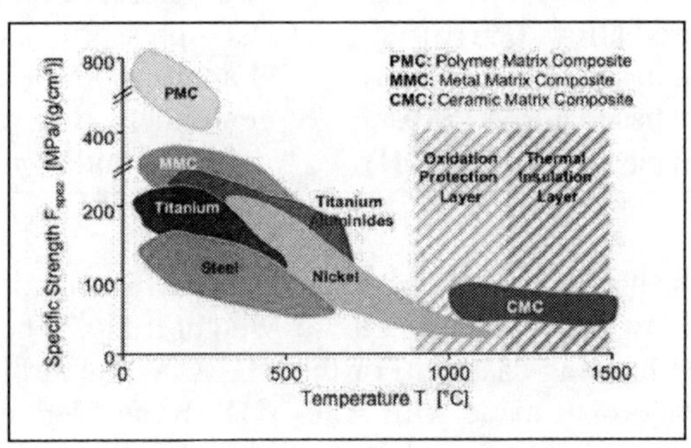

Figure 1: Representation of material requirement between strength and temperature for aero engines. [86].

All the aircraft turbofan engines have additional features like after burner with bypass provision to increase propulsive efficiency [5-7]. These additional features enhance the thermal effectiveness, the enhancement further can improved by incorporating blade cooling materials, maximum temperature resilient and low weight carrying materials like single crystal, smart graded materials, ceramic materials and ceramic matrix metal materials [8-12]. Further it is from the literature that Nickel based alloys gives more predominant outcomes in improvising thermal efficiency and propulsive efficiency in turbofan engines [13]. Additionally light weight material engine structures, aerodynamic shaped smart trailing edge structures require more sophisticated techniques which also improves the thrust development and high efficiency enhancement [97].

In order to overcome the difficulties currently phasing in aircraft engine component manufacturing design for manufacturing (DM) and single process tools as well as innovative primary manufacturing and process chains for repairs necessary [14]. For example considering the milling process in compressor blade electro chemical based manufacturing the metal removal rate reduced the risk of blisk gap so that saved the manufacturing time as well as maintained high precision [15]. Economic capabilities have to be assessed especially in aero engines components to produce in large volumes through electro chemical manufacturing techniques [16].

Table1: Materials used in turbomachinery components [90-91].

Material	Turbomachinery component
Titanium Based alloy (HK40)	Housing
Titanium Based alloy HK50	Housing
CAFA7	Housing
Alloy 282 (cast)	Housing
Waspaloy	Disk
Udimet 720	Disk
Alloy 718	Disk
A-218	Disk
Rene 41	Disk
CMSX-8	Turbine Blade
PWA 1483	Turbine Blade
Rene N4	Turbine Blade
CMSX-4	Turbine Blade
Nickel Based alloys (316-SS)	Compressor blades
Nickel Based alloys (Heyness 230)	Compressor blades
Nickel Based alloys (Alloy 625)	Compressor blades
Nickel Based alloys (PE16)	Compressor blades
Chromium based alloys (316ss)	Fan, rings
Chromium based alloys (310ss)	Fan inlet guide vanes
Chromium based alloys (800H)	Turbine outlet guide vanes.

Various materials used in turbomachinery components listed in table 1. The compressor hub housing made with HK 50/40 due to its high strength to weight ratio. The CAFA7 also used for high temperature resistant engine housing casting. For the industrial gas turbine housing purpose Alloy 282 (Cast) used. The stator blade disk made with Alloy 718, Rene 41 due to its high stiffness and high temperature resistance capacity in aero engines. Similarly the nickel based alloys like 316-SS, Heyness 230, Alloy 625 and PE 16 are used for axial flow compressor stages at higher Mach number ranges. Chromium based alloy grades like 316ss/310ss uses axial flow fan in turbofan engines. In the recent years especially turbine outlet guide vanes chromium based alloy grade 800H using.

The main objective of this paper is to give holistic view of more sophisticated manufacturing technologies and economically more compatible in producing large volumes. Especially concentrating earlier, current and future modern techniques like electro discharge machining, electro chemical machining and laser beam process for turbomachinery structures. More focuses on the explanation of shaping process not including the coating and welding process applications. All the aero engine turbomachinery components manufactured and designed by titanium and nickel based alloys, therefore this paper more described about the analysis of titanium and Nickel based alloys.

3. Advanced processes in turbomachinery components

3.1Turbomachinery components manufacturing process through electrochemical machining

The major advantage with electrochemical machining features are maximum material cut rate with virtually less instrument wear. The tool used for electrochemical machining generally high cost intensive one, therefore during the production process the wearing rate very much minimum also this process is best suitable for massive production of turbomachinery components like centrifugal compressor, small scale axial flow compressors and turbines etc [92]. This electrochemical machining technology is an alternative manufacturing technology for turbomachinery components. Through this type of manufacturing process can achieve excellent workpiece surface quality at high metal removal rate without any heat exaggerated zones / strain hardening regions and white layers [17-19].

The final product specifications after completion of electro chemical manufacturing depends on the particular configuration of the bulk composite material. The limitation with this type of manufacturing process is to ensure that material must not be mechanically damages or no thermally damaged otherwise during the machining process rim zones may form over the surface of the material which leads to internal flaws inside the material. The study of internal flaws in alloys through electrochemical process can achieve by crystallographic orientations method [20]. The study of the waviness on the titanium and nickel based alloys studied in detail after specific electrochemical treatments [19, 21].

Figure 2: Image of the experimental setup for blade EDM slicing [21].

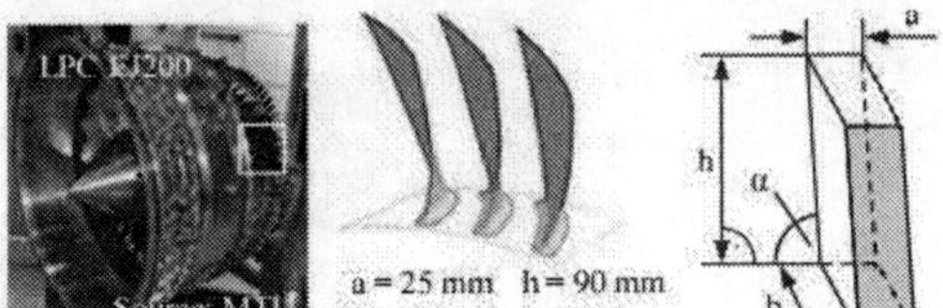

Figure 3: Simplified gap geometry from 3rd LPC stage of EJ 200 [18, 24]

The figure 2 [21, 32] shows for Ti-6Al-4V higher wavelength observed at 5μm due to quicker dissolution in the material. These electrochemical techniques have more significance because it depends purely on iterative cathode designing process and knowledge based approach. Before starting the electrochemical process the model will verify with the theoretical and numerical error based on the consistency then the workpiece further contrived to the actual size, the same flow model is well described in the figure 3 [18, 24]. The process of involving different engineering stream applications called Multiphysics approach, where it involves different physical aspects is well known [22-23]. All aeronautical problems related to aircraft components manufacturing much complicated especially related to engine components. Materials which is used for these components highly classy therefore require too many design factors. To solve all the design factors require Multiphysics with advanced computing facilities [93]. The electrochemical machining process also connects with high computational facilities so that simulations can possible to complete in few hours even any critical geometrical features of turbomachinery components [25].

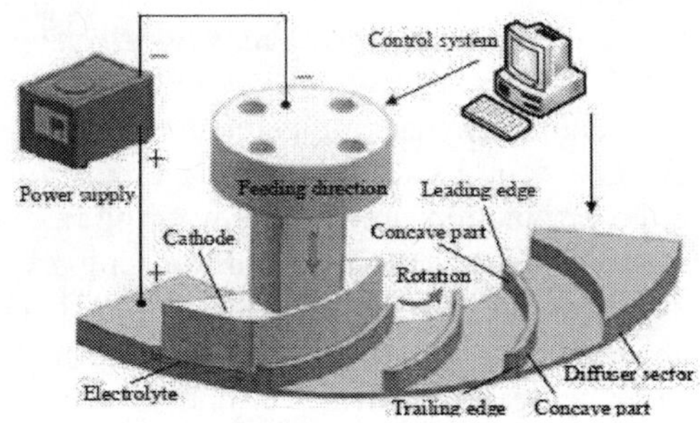

Figure 4: representation of electrochemical machining process various physics coupling modelling [33, 34].

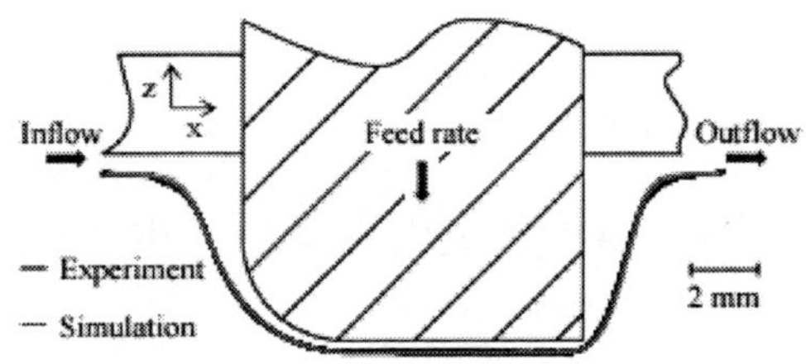

Figure 5: Representation workpiece Figure modelling dependency [25].

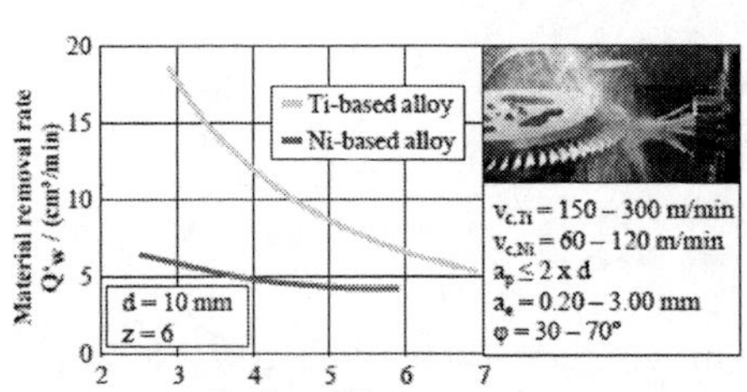

Figure 6. Material removal rates for aero engine alloys [30]

The figure 4 [23] shows complicated blade geometry being solved by Multiphysics approach using high computation facility with better accuracy with minimum time. Based on the geometrical features the profile further modified using inverse simulation method to reduce the computational time for finishing the machining process [27]. The figure 5 [25] shows the three dimensional compressor blade cutting edge by implementing electrochemical process through multiphysics approach. Further the multiphysics simulation results are also well matching at rounded cuboid region of the blade with the experimental data, the error was very minimum 10^-08 and gap was 300 μm [27]. The multiphysics flow simulation also solve many problems like thermal coated metal barrio problems and fluid flow heat transfer using the computer aided cathode design [94].

In the electro chemical machining one more advanced technology is called "pulsed electro chemical turbomachinery component machining". The word 'pulsed' means 'precision' which is most important all the components of aircraft engine [30]. Between the electrodes to workpiece minimum gap maintained to vibrate with designed frequency so as the tool touches the workpiece at desired level so that it goes on remove the material from the work surface [28-29] as shown in figure 6 [30]. The recent pulsed electrochemical machining device tools permits polishing, roughing, and finishing on one single stage [31].

Figure 7: Representation of 2 vibrating cathode tool electrodes of blisk geometries [39,40].

Figure 7 shows the modern electrochemical pulsed machine having total 8 axis degrees of freedom, in which four axis given to tool axis and remaining four axis given to workpiece operating by two fluctuating cathode[32]. This type of machine tool is locked system generally operates by through self-directed electrolyte management. The more advantage with this type of machine for absolutely titanium based alloys. The cathode guns operating at different temperatures ranges according the geometrical profile of the blade so as to reach good smooth workpiece [33-35].

Figure 8: Representation of electro chemical machining process applications[35,87].

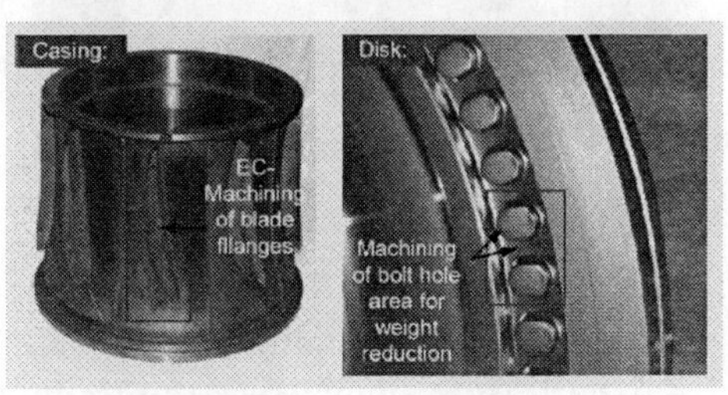

Figure 9: Representation of jet engine casing and disk machining process using electrochemical machining [40].

Figure 8 [35,36, 87] is the best example of electro chemical machining process for aero engine compressor blade and gas turbine blade. The machining completed according to the geometrical and aerodynamics features defined in the simulations. In this paper the materials type used for blade is X12CrNiWTiB16-13 at the rate of 2.5 cm^3/min. Few papers also described about nickel based alloys (material type: Inconel 718, at the rate of 2.1 cm^3/ min) and titanium based alloys (material type: Ti6242, at the rate 3.9 cm^3/min) [35-36]. The metal removal rate is very much minimized also maintained high accuracy of surface roughness (Ra=0.8 µm), good accuracy (o.1mm) and estimated the total savings through the manufacturing 30% in comparisons to traditional methods of cutting procedures [37, 95]. Tooling costs mainly reduced because of continuous production operation. The blade shroud and hub surfaces smoothening also much free through this electrochemical process [38].

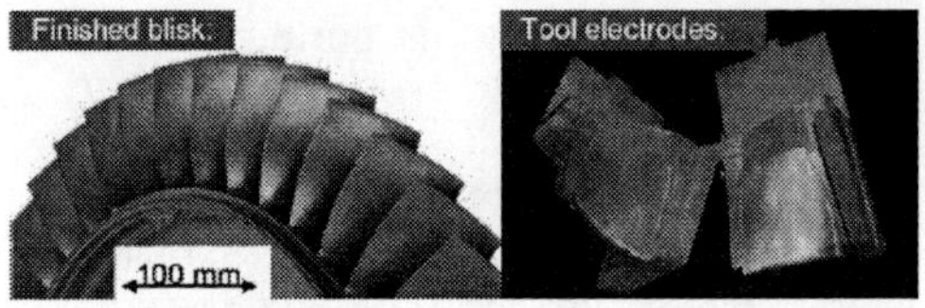

Figure 10: Representation of subsonic low pressure turbachinery blisk using electro chemical machining. [40, 43].

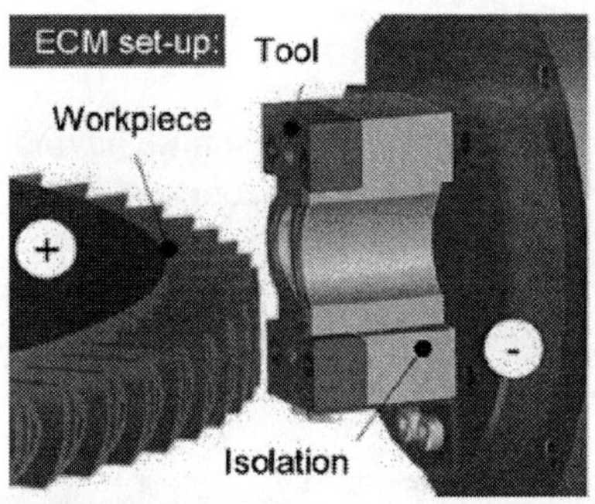

Figure 11: presentation of Electro chemical machining on aero engine gas turbine blades [39].

Figure 9 describes the steam gas turbines machining process representation using direct current electrochemical process. The material used titanium and nickel based alloys (X22CrMoV211), the feed rate maintained in this process from 3mm /min [39].

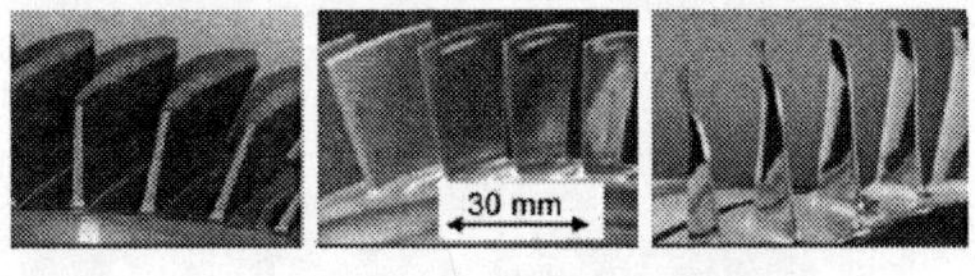

Figure 12: Representation of polished technique on turbomachinery components using electro chemical machining [44, 32].

Figure 10 represents the aircraft engine casing machining process (size 350x350 mm) operated at the voltage of 20 V and current 12 KA [40]. New attempts also made providing currents of 12-15 KA, feed rate takes from 1 mm per minute and the voltage range from 13 V to 18 V the component finishes within 5 minutes of time per blade [40-42]. Especially like low speed axial flow compressor two electrodes used to perform the machining process where the gap maintained minimum 100 mm per cycle so that, blade grooves well maintained with high precession shown in figure 11 [40, 43].

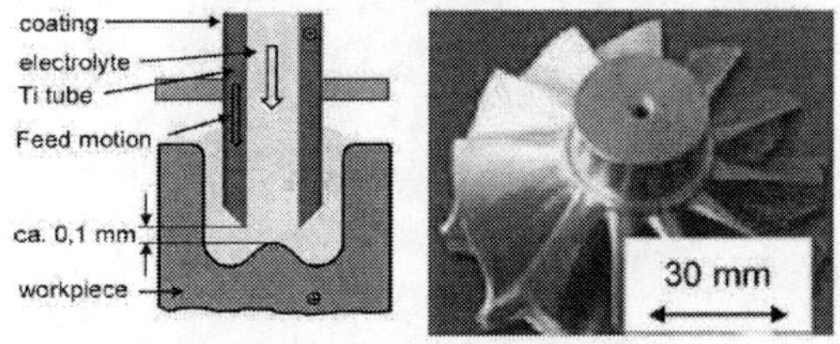

Figure 13: Representation of centrifugal compressor drilling the cooling holes through electro chemical machining [40, 45].

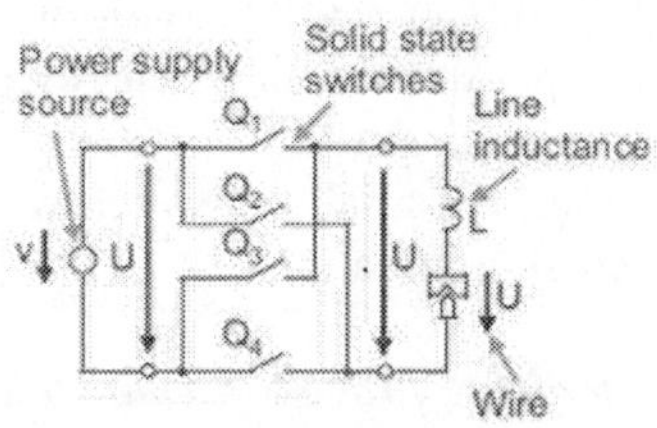

Figure 14: Representation of advanced transistor based generator [60].

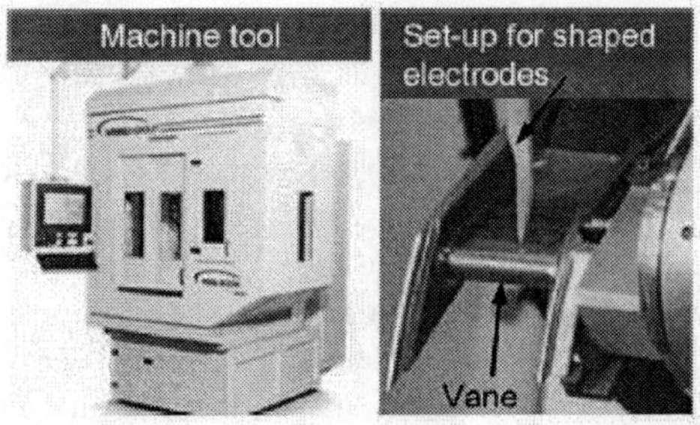

Figure 15: Representation of High speed electro discharge machining and set up [88]

In order to have better look with shining the blades undergo removal of oxide particles through oscillating polishing shown in figure 12 [32,44]. Further the complex geometries like aircraft gas turbine blades, applications of turbo charger disk machined successfully through electrochemical process [36]. In order to sustain the high temperature releasing from the aircraft combustion chamber, the turbine blade cooling system plays vital role to increase the performance of the turbine. The electrochemical process also make small holes in the blade to pass the coolant in it. Figure 13 represents the blade cooling system arranged through electrochemical drilling (0.3 to 5 mm) operation principle [45]. In the blade spherical holes can be created by incorporating the feed rate reduction and tool higher degree of freedom swirl within the air flow. Much more clear information regarding blade swirl creation process, modelling and performance parameters can be found several journal publication papers [46-49, 96].

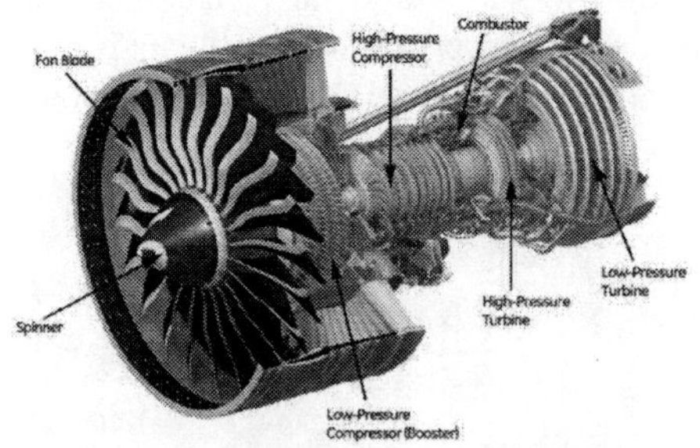

Figure 16: Representation of various components of machined capabilities on aero engine through elector discharge machining. [Courtesy: Pinterest-The world's catalogue of ideas]

There are many more innovations need to invent in electrochemical machining process especially drilling maintaining the high aspect ratio number that too with mixed approach adapting with cathodes is inevitable. To know more about these parameters one can review article by Sen and Shan detailed study on micro and macro electrochemical drilling process [50-52]. This process has a high significance with lower metal removal rate, especially new innovative metal composite difficult to cut titanium alloys in industrial aeromechanical applications.

3.2Turbomachinery components manufacturing process through electro discharge machining

This type of machining process quite new especially in aeronautical applications in 1955's. The electro discharge machining either wire based or die-sink based device, however recently several new innovations brought into the aeromechanical industrial applications [53-55]. For aerospace applications there are other numerous innovations brought like drilling the holes in turbine blades and radial compressors through electro discharge machining [56]. Through this electro discharge machining process it is possible to polishing (swirl grinding) the engine casings (material: Ti–6Al–4V) surface [57-59]. Recent year's transistor based ultra-fast recovery electro discharge machines introduced in which very low level inductances incorporated up to 0.5 μH, current slopes up to 600 A/μsec (see in figure 14) and metal removal rate of 600 mm^2 / minute. This type of facility more useful in machining the grove surface of the blade in axial flow compressor and turbines, see [60-62]. The figure 15 represents the high ultra-speed electro discharge machine set up for blade electrodes to finish the grove vane. The figure 16 represents the industrial based gas turbine engine different components machined by electro discharge machining process. Some articles are also discussed in detail mathematical model for electro discharge machining process for calculating the performance parameters like improving workpiece surface roughness, metal removal rate etc. [63-65]. Later the electro discharge machines with the multi degrees of freedom facility introduced for die-sink based titanium alloys applications [66-70]. Later the wire based electro discharge machines introduced, in this machining process instead of solid tool the continuously travelling wire used to cut the material. Despite from few decades the research and development conducting turbomachinery arc manufacturing extremely limited. The main reason for this type of machining require electro erosion need to maintain at lower metal removal rate, at lower cost for high usage of titanium and nickel based alloys. An example. In this example we can see that there are footnotes after each author name and only 5 addresses; the 6th footnote might say, for example, 'Author to whom any correspondence should be addressed.' In addition, acknowledgment of grants or funding, temporary addresses etc might also be indicated by footnotes.

3.3 Turbomachinery components manufacturing process through Additive manufacturing

This method is absolutely suitable for mass production at higher precession, lower metal removal and shorter time processing technique. Additive manufacturing technique employed in small scale prototype like rapid prototype, rapid manufacturing and rapid tooling so that scaling can maintain at any small value at higher accuracy. This additive manufacturing technology also much useful efficiently in repair applications, powder metal based applications. The power metal based small scale components machined based the principle of electron beam melting or selective laser melting. This technology basically requires advanced process knowledge and also supplementary post treatment phases [71]. All the aeronautical aero engines small components and turbomachinery devices started working from 1990's effectively. Apart from the prototyping the small scale production increased by incorporating the increased deposition rates on the workpiece at the rate 10 cm^3/min [72-73]. After the machining all the components required to undergo according the standards defined by the international civil aviation organisation air certification, therefore the components should undergo the quality certification through different levels.

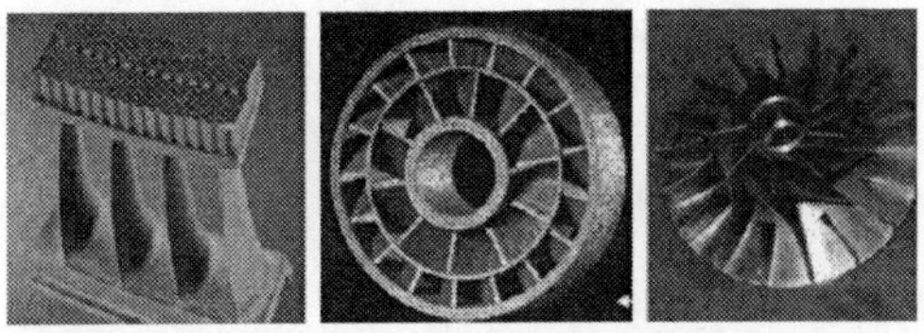

Figure 17: Representation of combustor, impeller and vane segment prototype with honey comb steel structure using additive manufacturing technology [79, 89, 80].

The certifications levels are prescribed indetail especially for this additive manufacturing for all titanium based alloys [74-75]. In aero engine component machining, wherever the complicated geometrical features encounters like vane segment, impeller swirl angle, combustor swirler (see in figure 17) there new possibilities of additive manufacturing plays a vital role to keep the significant cost and weight reduction [76-80]. According to the 2016 year National Science Report of NASA, the organisation also using additive manufacture technique to build various small scale products in large amount for space launch systems. Especially more focus giving to selective laser melting technique and electron beam melting techniques because of its high precession and low cost process [81]. Very Recently the General Electric (GE) Company announced that by the time 2020 year the additive manufacturing process going to give more than one lakh small and medium type of aircraft engine components. Especially focusing more on engine twisted blades, fuel nozzles edges and fuel injector members as their geometrical features so critical in size and difficult to machining at high accuracy [82]. Through the additive manufacturing process, the GE company going to save approximately 24 percent of the amount saving according their initial estimation [83]. To improve the high accuracy machining in different applications like casings, tubes, brackets, vanes, tubes, nozzles, shrouds, liners the research going on through adaptive manufacturing technique [84-85].

4. Conclusions

Through this paper the applications and technical capabilities of different turbomachinery components manufacturing techniques like electro chemical, electro discharge and additive manufacturing process have been studied. Through this study the technical capabilities of aero engine suitable applications and benefits have been investigated of its procedure. Especially in the case of problems in cutting, high advanced metal elimination rate, coolant admission rate, acceptable surface integrity and greater geometrical precision is addressed. All together clear research gap is also identified for every technique with suitable example of aero engine turbomachinery components. This paper also gives a broad description how to use all the techniques effectively for titanium and nickel based alloys with lower cost and further guides to researchers to move into a better way in all aeromechanical applications.

IOP Conf. Series: Materials Science and Engineering **314** (2018) 012012 doi:10.1088/1757-899X/314/1/012012

References :

[1] Soo SL, Hood R, Aspinwall DK, Voice WE, Sage C (2011) *Machinability and Surface Integrity of RR1000 Nickel Based Superalloy*. CIRP Annals – Manufacturing Technology, Volume **60,** PP:89–92.
[2] Reed RC (2006) *The Superalloys*, Cambridge University Press, Cambridge.
[3] Burger M, Koll L, Werner EA, Platz A (2012) *Electrochemical Machining Characteristics and Resulting Surface Quality of the Nickel-Base Single- Crystalline Material LEK94*. Journal of Manufacturing Processes **14**:62–70.
[4] Aspinwall DK, Dewes RC, Mantle AL (2005) *The Machining of γ-TiAl Inter-metallic Alloys*. CIRP Annals – Manufacturing Technology **54(1)**:99–104.
[5] Bradley A (2010) *Engine Design for the Environment*, RAeS, Hamburg. June.
[6] Jefferies R (2013) *Continuous Lower Energy, Emissions and Noise (CLEEN) Program*. USACA Spring Association Meeting, **21** May.
[7] Walther R (2012) *Recent Challenges in Air Breathing Propulsion*. Proceedings of the 14th International Symposium on Transport Phenomena and Dynamics of Rotating Machinery.
[8] EleaonoraFerraris (2016) Shaoing of engineering ceramics by electro, chemical and physical process, CIRP Annals – Manufacturing Technology, Volume **65,** PP:01–24.
[9] FAA (2012) Continuous Lower Energy, Emissions & Noise (CLEEN) Technologies *Rolls-Royce Programm Overview*. November 8, www.Rolls-royce.com.
[10] Sieber J (2014) Langfristige Sicherung des Luftverkehrs durch neue Antrieb-stechnologien und alternative Brennstoffe. Retrieved from www.mtu.de.
[11] Steffens K, Walther R (2003) Driving the Technological Edge in Airbreathing Propulsion. Retrieved from www.mtu.de.
[12] Steffens K, Wilhelm H (2013) Werkstoffe, Oberfla¨chentechnik und Fertigungs-verfahren fu¨r die na¨chste Generation von Flugtriebwerken. Retrieved from www.mtu.de.
[13] Henne J (2013) Gearing Up for High Volume GTFTM Production. 21st ISABE Conference, Busan, Korea.
[14] Bußmann M, Kraus J, Bayer E (2005) An Integrated Cost-Effective Approach to Blisk Manufacturing. Retreived from www.mtu.de.
[15] Veselovac D (2013) Process and Product Monitoring in the Drilling of Critical Aero Engine Components, (Dissertation) WZL RWTH Aachen University.
[16] Bayer E (2014) Fertigungsprozesse als Schlu¨sseltechnologie zur Realisierung des Geared Turbo Fan. Retreived from www.mtu.de.
[17] Klocke F, Ko¨ nig W (2007) Fertigungsverfahren 3: Abtragen, Generieren und Lasermaterialbearbeitung, Springer, Berlin. ISBN 3-540-23492-6.
[18] Klocke F, Zeis M, Klink A (2012) Technological and Economical Capabilities of Manufacturing Titanium- and Nickel-Based Alloys Via Electrochemical Ma-chining (ECM). Key Engineering Materials 504–506:1237–1242.
[19] Klocke F, Zeis M, Klink A, Veselovac D (2013) Experimental Research on the Electrochemical Machining of Modern Titanium- and Nickel-Based Alloys for Aero Engine Components. Procedia CIRP 6:369–373.
[20] Steffens K, Platz A, Buckl F, (2004) Feinbearbeitungsverfahren – Schlu¨ ssel-technologien fu¨ r moderne Luftfahrtverdichter. MTU Aero Engines. www.mtu.de.
[21] Baumga¨rtner M (2013) Entwicklung des elektrochemischen Senkens (ECM) und der mechanischen Bearbeitung von Titanaluminiden, Abschlussbericht BMBF-Forschungsvorhaben.
[22] Bogoveev NA, Firsov AG, Filatov EI, Tikhonov AS (2001) Computer Support for ''All-Round'' ECM Processing of Blades. Journal of Materials Processing Tech-nology 109:324–326.
[23] Klocke F, Zeis M, Harst S, Klink A, Veselovac D, Baumga¨rtner M (2013) Modeling and Simulation of the Electrochemical Machining (ECM) Material Removal Process for the Manufacture of Aero Engine Components. Procedia 8:265–270.
[24] van Tijum R, Pajak T (2008) The Multiphysics Approach: The Electrochemical Machining Process, Presentation During COMSOL Conference Hannover.
[25] Klocke F, Zeis M, Harst S, Herrig T, Klink A (2013) *Analysis of the Simulation Accuracy of Electrochemical Machining Processes Based on the Integration Level of Different Physical Effects*. M. Scripts Precision and Microproduction Engineering 7, Fraunhofer IWU, pp- 165–170.
[26] Klocke F, Zeis M, Harst S, Klink A, Veselovac D, Baumga¨rtner M (2013) Modeling and Simulation of the Electrochemical Machining (ECM) Material Removal Process for the Manufacture of Aero Engine Components. Procedia 8:265–270.

[27] Hinduja S, Kunieda M (2013) Modelling of ECM and EDM Processes. CIRP Annals – Manufacturing Technology 62(2):775–797.
[28] Lauwers B, Klocke F, Klink A, Tekkaya AE, Neugebauer R, Mcintosh D (2014) Hybrid Processes in Manufacturing. CIRP Annals – Manufacturing Technology 63(2):561–583.
[29] Schubert A, Hackert-Oscha¨tzchen M, Meichsner G, Zinecker M, Edelmann J (2011) Precision and Micro ECM with Localized Anodic Dissolution. in Slabe JM, (Ed.) TECOS Slovenian Tool and Die Development Centre, Celje: Proceedings of the 8th International Conference on Industrial Tools and Material Processing Technologies, 193–196. ISBN: 978-961-6692-02-1.
[30] Burger M, Platz A, Werner E (2007) Herstellung/Nachbearbeitung von Turbi-nenblisks durch Pra¨zises Elektrochemisches Bearbeiten, TUM. Retrieved from http://www.wkm.mw.tum.de/forschung/postergalerie/.
[31] EMAG's ECM/PECM Machines (2013) Retrieved from http://www.onlineamd.com/Article.aspx?article_id=132602.
[32] Platz A, Feiling N (2013) Precise Electrochemical Machining of Nickel Base Integrated Blade Compressor Rotors, Precision and Microproduction Engineer-ing 7, Fraunhofer IWU, Chemnitz23–32. ISBN 978-3-942267-95-3.
[33] ECM/PECM Technologie Polieren (2011) Entgraten, 3D-Konturen, EMAG ECM.
[34] Gmelin T (2013) PECM – Anwendungsmo¨glichkeiten und Grenzen im Ver-gleich zur Funkenerosion. 9. Fachtagung Funkenerosion, WZL, RWTH, Aachen.
[35] Baumga¨rtner M (2010) Herstellung von Turbinenkomponenten, series of lec-tures ''Ausgewa¨hlte Kapitel der Turbomaschinen'', AKTM, RWTH Aachen Uni-versity.
[36] Leera raju A (2016) *state of the art Review on Micro Electro Discharge machining,* Procedia Technology **25** pp 1281-1288.
[37] M Kliuev (2016) *EDM drilling and shapind of cooling holes in Inconel 718 turbine blades*. 18th CIRP Conference on Electro Physical and Chemical Machining (ISEM XVIII) **42**, pp 322-327.
[38] Baumga¨rtner M (2010) Herstellung von Turbinenkomponenten, series of lec-tures ''Ausgewa¨hlte Kapitel der Turbomaschinen'', AKTM, RWTH Aachen Uni-versity.
[39] Electrochemical Machining (2013) Retrieved from http://www.koepperninternational. Com
[40] Elektrochemisches Abtragen (2009) Verein Deutscher Ingenieure VDI-Richtlinie 3401-1(Entwurf) .
[41] Klocke F, Ko¨ nig W (2007) Fertigungsverfahren 3: Abtragen, Generieren und Lasermaterialbearbeitung, Springer, Berlin. ISBN 3-540-23492-6.
[42] Steffens K, Wilhelm H (2013) Werkstoffe, Oberfla¨chentechnik und Fertigungs-verfahren fu¨r die na¨chste Generation von Flugtriebwerken. Retrieved from www.mtu.de.
[43] Steffens K, Platz A, Buckl F, (2004) Feinbearbeitungsverfahren – Schlu¨ ssel-technologien fu¨ r moderne Luftfahrtverdichter. MTU Aero Engines. www.mtu.de.
[44] Bußmann M, Bayer E (2013) Market-Oriented Blisk Manufacturing – A Chal-lenge for Production Engineering, MTU Aero Engines.
[45] High-Tech Development and Manufacturing for Aero and Industrial Gas Turbine Components, (2010), Sulzer ELDIM.
[46] Pattavanitch J, Hinduja S (2012) Machining of Turbulated Cooling Channel Holes in Turbine Blades. CIRP Annals – Manufacturing Technology 61:199–202.
[47] Ali S, Hinduja S, Atkinson J, Pandya M (2009) Shaped Tube Electrochemical Drilling of Good Quality Holes. CIRP Annals – Manufacturing Technology 58:185–188.
[48] Jain VK, Chavan A, Kulkarni A (2007) Experimental and Analytical Study of Contoured Holes by Shaped Tube Electrochemical Drilling Process. Proceed-ings of the 15th Int Symp on Electromachining, Pittsburgh, USA, 315–318.
[49] Wang MH, Zhub D (2009) Simulation of Fabrication for Gas Turbine Blade Turbulated Cooling Hole in ECM Based on FEM. Journal of Materials Processing Technology 209:1747–1751.
[50] Bilgi DJ, Jain VK, Shekhar R, Mehrotra S (2004) Electrochemical Deep Hole Drilling in Super Alloy for Turbine Application. Journal of Materials Processing Technology 149:445–452.
[51] Thanigaivelan R, Arunachalam RM, Karthikeyan B, Loganathan P (2013) Electrochemical Micromachining of Stainless Steel with Acidified Sodium Nitrate Electrolyte. Procedia CIRP 6:352–356.
[52] Sen M, Shan HS (2005) A Review of Electrochemical Macro- to Micro-Hole Drilling Processes. International Journal of Machine Tools & Manufacture 45:137–152.
[53] Ho KH, Newman ST (2003) State of the Art Electrical Discharge Machining (EDM). International Journal of Machine Tools and Manufacture 43(13):1287– 1300.
[54] Kunieda M, Lauwers B, Rajurkar KP, Schumacher BM (2005) Advancing EDM Through Fundamental Insight into the Process. Annals of CIRP 54(2):64–87.

[55] Pandey A, Singh S (2010) Current Research Trends in Variants of Electrical Discharge Machining: A Review. International Journal of Engineering Science and Technology 2(6):2172–2191.

[56] Toller DF (1983) Multi-Small Hole Drilling by EDM. Proceedings of the 7th International Symposium on Electromachining (ISEM VII), 12–14 April, Bir-mingham, UK, 147–156.

[57] Bache MR, Tuppen SJ, Voice WE, Lee HG, Aspinwall DK (2009) Novel Low Cost Procedure for Fabrication of Diffusion Bonds in Ti6/4. Materials Science and Technology 25(1):39–49.

[58] Lee HG, Simao J, Aspinwall DK, Dewes RC, Voice W (2004) Electrical Discharge Surface Alloying. Journal of Materials Processing Technology 149:334–340.

[59] Okane M, Goto A (2008) Development of MS Coating for Aircraft Engine Parts. Mitsubishi Electric Advance Mechatronics 123. ISSN 1345-3041.

[60] D'Amario R (2008) Method and Apparatus for Generating Machining Pulses for Electrical Discharge Machining, European Patent EP 1719570.

[61] Han F, Wachi S, Kunieda M (2004) Improvement of Machining Characteristics of Micro-EDM Using Transistor Type Isopulse Generator and Servo Feed Control. Precision Engineering 28:378–385.

[62] Winbro Group Technologies (2014) Series 800 Laser & EDM Datasheet.

[63] Hewidy MS, El-Taweel TA, El-Safty MF (2005) Modelling the Machining Parameters of Wire Electrical Discharge Machining of Inconel 601 Using RSM. Journal of Materials Processing Technology 169:328–336.

[64] Ramakrishnan R, Karunamoorthy L (2008) Modeling and Multi-Response Optimization of Inconel 718 on Machining of CNC WEDM Process. Journal of Materials Processing Technology 207:343–349.

[65] Sarkar S, Mitra S, Bhattarcharyya B (2005) Parametric Analysis and Optimi-zation of Wire Electrical Discharge Machining of g-Titanium Aluminide Alloy. Journal of Materials Processing Technology 159:286–294.

[66] Hascalik A, Caydas U (2007) Electrical Discharge Machining of Titanium Alloy (Ti–6Al–4V). Applied Surface Science 253:9007–9016.

[67] Kao JY, Tsao CC, Wang SS, Hsu CY (2010) Optimisation of the EDM Param-eters on Machining Ti–6Al–4V with Multiple Quality Characteristics. Inter-national Journal of Advanced Manufacturing Technology 47:395–402.

[68] Sivakumar KM, Gandhinathan R (2013) Establishing Optimum Process Pa-rameters for Machining Titanium Alloys (Ti–6Al–4V) in Spark Electrical Discharge Machining. International Journal of Engineering and Advanced Tech-nology 2(4):201–204.

[69] Wardono B, Ismail MFB, Liew PJ (2011) An Analysis of EDM Die Sinking Parameters on Ti–6Al–4V. Proceedings of the International Conference and Exhibition on Sustainable Energy and Advanced Materials (ICESEAM2011), Solo, Indonesia, 3–4 October, 367–373.

[70] Yuan R, Wei B, Luo Y, Zhan Y, Xu W, Lamphere M (2010) Advancement in High Speed Electro-Erosion Processes for Machining Tough Metals. Proceedings of the 16th International Sympsium on Electromachining (ISEM), 19–23 April, Shanghai, China, 207–210.

[71] Kruth JP, Levy G, Klocke F, Childs THC (2007) Consolidation Phenomena in Laser and Powder-Bed Based Layered Manufacturing. CIRP Annals 56(2):730–759.

[72] Wohlers T, Gornet T (2011) History of Additive Manufacturing. in Wohlers T, (Ed.) Wohlers Report: Additive Manufacturing and 3D Printing State of the Industry Annual Worldwide Progress Report, Wohlers Associates, Fort Collins, CO.

[73] GKN Aerospace Capabilities (2010) Retrieved from www.gknaerospace.com.

[74] Gasser A, Kelbassa I, Wissenbach K, Witzel J, Go¨ bel M (2012) Additive Manu- Facturing in Turbo-Engine Applications, AKL/EU Innovation Forum, Aachen.

[75] Witzel J, Schrage J, Gasser A, Kelbassa I (2011) Additive Manufacturing of a Blade-Integrated Disk by Laser Metal Deposition. Proceedings of 30th Inter-national Congress on Applications of Lasers & Electro-Optics, Orlando, USA, 250–256.

[76] Richter KH (2010) Using the Laser for Build-Up. Established and Emerging Applications at MTU Aero Engines. International Laser Technology Congress AKL'10, Aachen.

[77] Hess T (2012) Aero Engine Parts Made of SLM – Status Quo and Future Potential. International Laser Technology Congress AKL'12 Aachen.

[78] Lakomiec M (2013) Serienfertigung von Triebwerksteilen mittels Laserstrahl-schmelzen, VDI Fachkonferenz Additive Manufacturing – Vom Prototypen bis zur Großserie mit generativen Fertigungsverfahren, Duisburg.

[79] Kelbassa I, Albus P, Dietrich J, Wilkes J (2008) Manufacture and Repair of Aero Engine Components Using Laser Technology. Proceedings of the 3rd Pacific Conference on Application of Lasers and Optics, 208–212.

[80] Xue L, Li Y, Wang S (2011) Direct Manufacturing of Net-Shape Functional Components/Test-Pieces for Aerospace, Automotive and Other Applications. Proceedings of the 30th ICALEO, 23–27 October, Orlando, USA, 479–488.

[81] Hubscher B (2012) NASA's Space Launch System Using Futuristic Technology to Build the Next Generation of Rockets.

[82] Wohlers TT (2012) Wohlers Report 2012: Additive Manufacturing and 3D Printing State of the Industry: Annual Worldwide Progress Report, Wohlers Associates, Fort Collins, CO.

[83] Joyce D (2012) GE Aviation. Barclays Capital Industrial Select Conference.

[84] Additive Manufacturing at GE (2011) Loughborough. July 12.

[85] Allen J (2011) The Potential for Aero Engine Component Manufacture Using Additive Layer Manufacturing. Aerodays Conference, Madrid.

[86] Adam Ruszaj, Jozef Gawlik, Sebastian skoczpiec "Electrochemical machining – Special equipement and applications in aircraft industry" Journal of management and production engineering review, volume .7, Number 2, June 2016, pp 34-41. DOI: 10.1515/mper-2016-0015.

[87] Ho KH, Newman ST (2003) State of the Art Electrical Discharge Machining (EDM). International Journal of Machine Tools and Manufacture 43(13):1287– 1300.

[88] NCMT (2013) Deep-Hole EDM Drilling of Turbine Components is Seven Times Faster. Retrieved from http://www.ncmt.co.uk.

[89] Dilba D (2011) Schnelle Schichtarbeit. MTU Report 2011(1):20–25. http:// www.mtu.de/de/take-off/report/archive/2011_1.pdf.

[90] Dr Mark Anderson (2012) Materials, Turbomachinery and Heat exchangers for supercritical CO2 systems. Reactor concept s RD & D . Project No. 09-778.

[91] James Keiser et al Characterization of Turbomachinery Materials Exposed in 750°C, 20 MPa sCO2 slides. ORNL is managed by UT-Battelle for the US Department of Energy.

[92] Shailendra Kumar Bohidar (2013), Advanced Materials used for different components of Gas Turbine, International Journal of scientific research and management (IJSRM), Volume 1, Issue 7, Pages 366-370. Website: ISSN 2321-3418.

[93] Joel Bretheim (2012), A Review of Power-Generating Turbomachines, Proceedings of the 2012 ASEE North Central Section Conference, American Society for Engineering Education.

[94] S. Luque (2017) Scaling of Turbine Metal Temperatures in Cooled Compressible Flows – Experimental Demonstration of a New Theory, Journal of Turbomachinery. Received June 15, 2016; Accepted manuscript posted January 25, 2017. doi:10.1115/1.4035831 Rolls-Royce plc.

[95] Alistair john (2017), Alleviation of shock-wave effects on a highly loaded axial compressor through novel blade shaping, Proceedings of ASME Turbo Expo 2016: Turbomachinery Technical Conference and Exposition. doi:10.1115/1.4035833.

[96] J P. bons (2017), a simple physics-based model for particle rebound and deposition in turbomachinery, journal of turbomachinery. Received January 10, 2017; accepted manuscript posted february 8, 2017. doi:10.1115/1.4035921 copyright (c) 2017 by ASME.

[97] Martin C Arenz (2017) Development and Application of a Fast-Response Total Temperature Probe for Turbomachinery, Journal of Turbomachinery Copyright VC 2017 by ASME MAY 2017, Vol. 139 / 051010-1.

IOP Conf. Series: Materials Science and Engineering **314** (2018) 012013 doi:10.1088/1757-899X/314/1/012013

DRY SLIDING WEAR CHARACTRISTICS OF ALUMINUM 6061-T6, MAGNESIUM AZ31 AND ROCK DUST COMPOSITE

R Balachandar[1], R Balasundaram[1], G Rajkumar[1]

[1] Department of Mechanical Engineering, K.Ramakrishnan Collece of Engineering, Trichy-621112, Tamilnadu.

Abstract. In recent years, the use of aluminum composite is gaining popularity in a wide range of applications like automobiles, aerospace and constructions (both interior & exterior) panels etc., due to its high strength, low density characteristics. Various reinforcing materials are used with aluminum 6061-T6 in order to have better mechanical properties. The addition of 0.3% of magnesium AZ31 will increase the ultimate tensile strength by 25 %. The reinforcement of rock dust will decrease the density. Hence, in order to have an advantages of magnesium AZ31 and rock dust, in this work, these two constitutes are varied from 1% to 2% on the base material of Al6061-T6 in stir casting. To evaluate the wear characteristics, Pin on disc is used in these composites. The input parameters are speed, time & load. The output response is wear. To minimize the number of experiments, L9 orthogonal array is used. The test results showed that a composite of 97% of Al (6061-T6), 1% Mg (AZ31) & 2 % of rock dust produced less wear. To find the best value of operating parameter for each sample, ANN-GA is used.

1. Introduction:

A composite is when at least two different materials are joined together to make a common and one of a kind material. Composites are made up of individual materials referred to as constituent materials. The composites are classified i) Polymer Matrix Composite (PMC) ii) ceramic Matrix Composite (CMC) and iii) Metal Matrix Composite (MMC). Aluminum composites have high strength to weight ratio and high corrosion resistance properties. These days these composites are broadly utilized as a part of many designing fields and specifically in settings in which weight decrease is a basic figure, for example, aviation applications. Even though unmodified aluminum is a standout amongst the most boundless components on earth, it is too soft and flexible. Therefore, it is frequently consolidated with a wide range of components, such as Cu, Si, Mg, Zn, Mn to form compounds. Al 6061-T6 has a place with the aluminum arrangement 6,000 with silicon and magnesium as important alloying segments. These compounds are formed by warmth treatment (in present case T6) and they demonstrate great formability, workability, weldability and shaving. The Rock dust powder as a strengthening phase can improve the wear resistance of aluminum alloy materials. In order to have the advantages of Aluminum, Magnesium and Rock dust, an Al MMC's is prepared & dry sliding wear characteristics are analyzed in this work.

2. Literature survey

The table-1 shows the importance and the noteworthy research work carried on Al 6061-T6.

Table 1. Important & note worthy contributions on Al 6061.

S1.No	Author	Year	Material	Remarks
1	Wei Guo *et al.*	2016	Al 7A04 alloy & Mg AZ31	The thickness of intermetallic compounds interlays consisted of Mg & Al. The experiments authors showed that by intermetallic layers, the tensile strength of the joint increased and its maximum value is 96 MPa.
2	Zilong Zhao, *et al.*	2016	Al 5052 & MgAZ31	In this work, the authors made magnesium aluminum composite panels by asymmetric metal packaging and studied rolling temperature, holding time, and high temperature heat treatment. It is concluded that the new composite panels' elongation is 24% as compared to traditional magnesium aluminum composite panels' elongation.
3	Ratna Sunil, *et al.*	2016	1.Pure Mg, nano-hydroxypatite (32 nm) 2. AZ31, Al_2O3 3. AZ61, SiO2 4. AZ91, Al2O3 5. ZM21, SiC, B_4C	The Process parameters, for example, apparatus rotational and travel speeds assumes a vital part in creating deformity free mix zone in contact mix welding of magnesium composites.
4	Gali *et al.*	2016	AISI 52100, Mg	The surface/near surface parts confined on an Al-Mg blend in the midst of a 10 pass hot moving timetable using work moves with a surface harshness (Ra) estimations of 0.1 μm and 1.1 μm was investigated.
5	Jufu Jiang, *et al.*	2016	Al 7075 & β –Sic(80 nm)	The stress–strain curve at 600 °C is same as that of 500 °C. The increase in SiC particles leads to increasing Peak stress & steady stress.
6	Taherzadeh Mousavian, *et al.*	2016	A356, SiC	Hot-rolling aluminum structure composites strengthened with multimodal estimated SiC. The fracture enhanced SiC consolidation.
7	Kimura, *et al.*	2015	Pure Mg, Pure Al	The welded joint had the intermetallic compound layer (IMC) at the weld interface. The end result showed that the fracture passed off by means of thermal strain in radial and/or circumferential directions.

8	Mojtaba, *et al.*	2015	Al 606, AZ31	Dispersion welding of Al to Mg compounds becomes accomplished without utilizing an interlayer. The ideal holding parameters included T = 440 °C, P = 29 MPa and t = 60 min. The bond quality at a temperature of 440 °C was accomplished as 42 MPa.
9	Ansary Yar, *et al.*	2009	A356+Mgo (50nm)	A Composite of 1.5 Vol% MgO was fabricated at 850 °C and test results demonstrated that properties, for example, hardness, quality and strength are enhanced when contrasted with different examples. Moreover, sturdiness of composites, for the most part diminished by expanding the substance of MgO.

From the literature, it is shown that the use of aluminum composite has widely attracted by many researchers. The application of Magnesium & rock dust as reinforcement with Al 6061-T6 has gotten less attention to the researchers. Hence, in this work, the Al 6061-T6 is taken as base material and Mg AZ31 & rock dust are taken as reinforcement.

3. Methods & manufacturing:

The table 2- 4 shows the compositions of Al 6061-T6, Mg AZ31 & Rock dust respectively.

Table 2. AL 6061-T6 composition.

Element	Al	Si	Zn	Mg	Mn	Cr	Fe	Cu	Ti
Percent	98	0.4-0.8	0.25	0.8-1.2	0.15	0.04-0.35	0.7	0.15-0.4	0.15

Table 3. Mg AZ31 composition.

Element	Al	Zn	Mn	Fe	Si	Ni	Cu	Be	Mg
Percent	3.08	0.76	0.15	0.005	0.01	0.002	0.001	0.0001	95.9929

Table 4. Rock dust composition.

Element	SiO_2	Al_2O_3	CaO	Fe_2O_3	MgO	Na_2O	TiO_2	K_2O	Others
Percent	51	18.4	10.2	9.29	5.0	2.1	0.78	0.59	Bal

3.1. Manufacturing process

In this work, stir casting is used to fabricate the proposed composite consisting of Al 6061-T6, Mg AZ31 and Rock dust. The Stir Casting process is the best method to fabricate composites for a few volumes of production. Aluminum 6061-T6 compound was mix thrown with magnesium AZ31, rock dust as support in order to figure Al 6061-T6/Mg AZ31/Rock dust composites.

The hotness of the heater was absolutely measured and precisely controlled (±1°C) by thermocouples with the controller. The experimental setup up have 1 HP motor to run the stirrer at various speeds. The Al 6061-T6 is heated upto 720^{0}C and then solid material of Mg AZ31 is added. The strrier is kept rotating at constant speed. Then the preheated rockdust power is added at a uniform rate to molten metal of Al & Mg. Throughout the experiment the batch is kept at 720 °C. Then molten metal is cast into required shape. There are three test specimens were prepared and are shown in Table 5.

Table 5. Composition of proposed composite.

Test Specimen	Aluminum (%Wt)	Magnesium (%Wt)	Rock Dust (%Wt)
T1	97	2	1
T2	97	1	2
T3	96	2	2

Figure 1. Schematic diagram of stir casting and molded composite metal.

4. Experimental work

A pin on disc tribometer comprises of a stationary pin under a connected load in contact with a rotating disc. The pin on disc test has demonstrated valuable in giving a basic wear and friction test for low friction coatings. The disc material is made of EN134 Carbon.

IOP Conf. Series: Materials Science and Engineering **314** (2018) 012013 doi:10.1088/1757-899X/314/1/012013

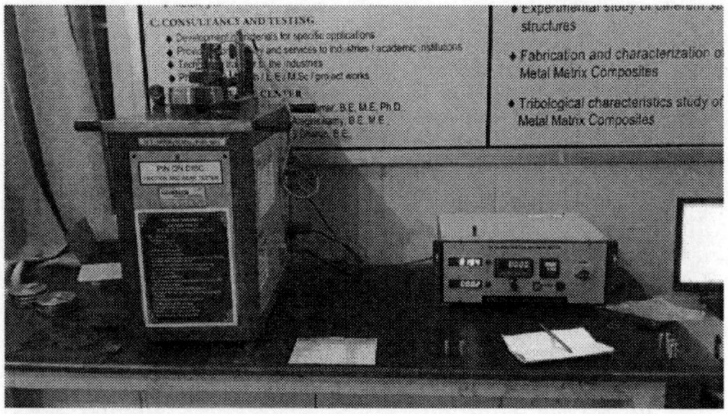

Figure 2. Pin on disc wear testing machine

4.1. Selection of machining parameters

The wear tests are carried out all 3 specimens. To minimize the number of trial runs L9 Ortothogonal array is used. The input parameters are Speed, Load and Time. The output response is wear. The table 6 shows the coded value along with experimentation.

Table 6. Test result of dry sliding wear.

Sl.No.	**INPUT PARAMETRS**						**Wear for specimen (microns)**		
	Coded Value			**Actual Value**					
	Speed	**Time**	**Load**	**Speed (rpm)**	**Time (Minute)**	**Load (N)**	**T1**	**T2**	**T3**
1	**A**	**A**	**A**	150	5	20	138	205	204
2	**A**	**B**	**B**	150	10	40	107	230	222
3	**A**	**C**	**C**	150	15	60	196	320	299
4	**B**	**B**	**C**	300	10	60	189	300	294
5	**B**	**C**	**A**	300	15	20	101	139	117
6	**B**	**A**	**B**	300	5	40	158	94	94
7	**C**	**C**	**B**	450	15	40	195	72	68
8	**C**	**B**	**A**	450	10	20	212	84	72
9	**C**	**A**	**C**	450	5	60	289	145	183

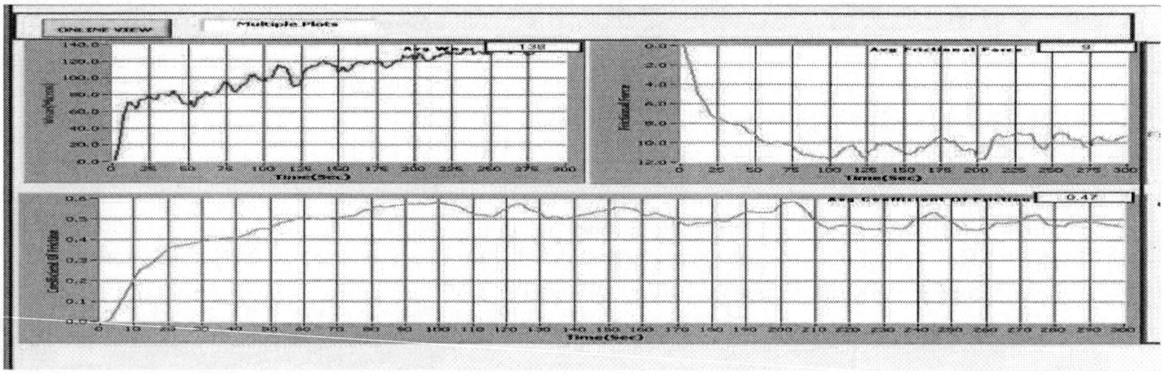

Figure 3. Graph of wear result for the specimen T1 with Speed, Load&Time.

From the figure 3, shows the average value of wear, friction force and co-efficient of friction for the test specimen T1 with speed (150), Load (5) & Time (20).

4.2 .Analysis of test specimen: T1

Table 7 shows the composite Al (6061-T6) 97%, Mg (AZ31) 2% & Rock dust 1%.

Table 7. ANOVA for specimen T1.

Source	Sum of Squares	DEGREES OF FREEDOM	Mean Square	F value	P-value Prob> F	
Mod155el	26792.67	6	4465.44	25.07	0.0388	Significant
A-SPEED	15505.93	1	15505.93	87.06	0.0113	
B-Load	137.52	1	137.52	0.77	0.4722	
C-Time	58.34	1	58.34	0.33	0.6249	
Residual	356.22	2	178.11			
Cor Total	16058.01	5				

$$\text{Wear} = 195.63492 + 0.77143\times\text{speed} - 10.55238\times\text{load} - 2.60000\times\text{time}$$ ..... Equation 1

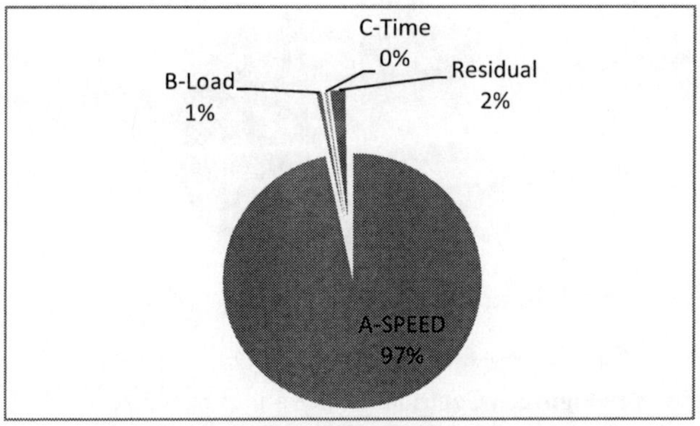

Figure 4. Contribution of various parameters.

From the contributions, the model A-speed (97%), B-load (1%) & C-time (0%).

4.3. Analysis of test specimen: T2

Table 8 shows composite Al (6061-T6) 97%, Mg (AZ31) 1%, Rock dust 2%.

Table 8. ANOVA for specimen T2.

Source	Sum of Squares	DEGREES OF FREEDOM	Mean Square	F value	P-value Prob> F	
Model	54542.33	3	18180.78	6.66	0.0338	Significant
A-SPEED	34352.67	1	34352.67	12.58	0.0165	
B-Load	18928.17	1	18928.17	6.93	0.0464	
C-Time	1261.50	1	1261.50	0.46	0.5270	
Residual	13657.89	5	2731.58			
Cor Total	68200.22	8				

$$W\ a = 186.55556 - 0.50444 \times \text{speed} + 2.80833 \times\ a\ + 2.90000 \times\ i$$ ..... Equation 2

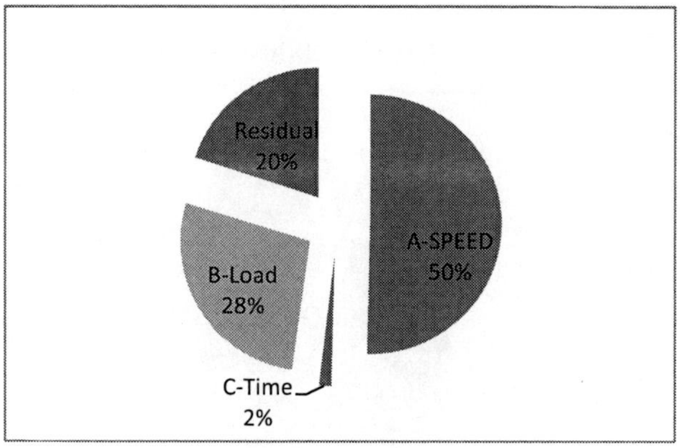

Figure 5. Contribution of various parameters.

From the contributions, the model A-speed (50%), B-load (28%) & C-time (2%).

IOP Conf. Series: Materials Science and Engineering **314** (2018) 012013 doi:10.1088/1757-899X/314/1/012013

4.4 Analysis of test specimen: T3

Table 9 shows composite 3 Al (6061-T6) 96%, Mg (AZ31) 2%, Rock dust 2%.

Table 9. ANOVA for specimen T3.

Source	Sum of Squares	DEGREES OF FREEDOM	Mean Square	F value	P-value Prob> F	
Model	51383.67	3	17127.89	6.4895	0.0355	Significant
A-SPEED	26934	1	26934	10.2049	0.0241	
B-Load	24448.17	1	24448.17	9.2631	0.0286	
C-Time	1.5	1	1.5	0.0006	0.9819	
Residual	13196.56	5	2639.311			
Cor Total	64580.22	8				

$W\ a\ = 177.88 - 0.4466\times \quad + 3.191\times\ a\ + 0.1\times\ i$ ..... Equation 3

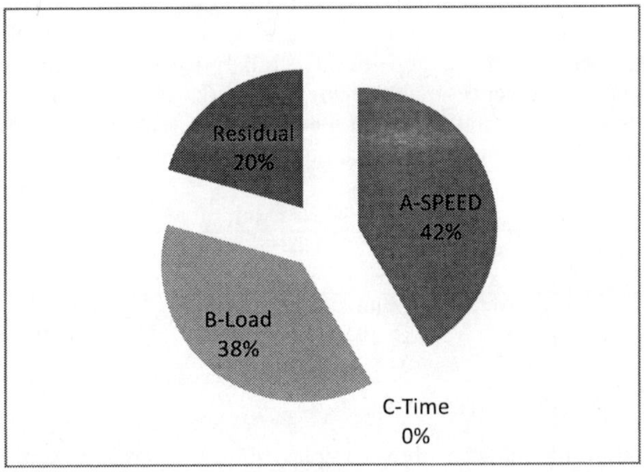

Figure 6. Contribution of various parameters.

From the contributions, the model A-speed (42%), B-load (1%) & C-time (0%).

IOP Conf. Series: Materials Science and Engineering **314** (2018) 012013 doi:10.1088/1757-899X/314/1/012013

5. ANN-GA Approach: Using Software –Neural Power

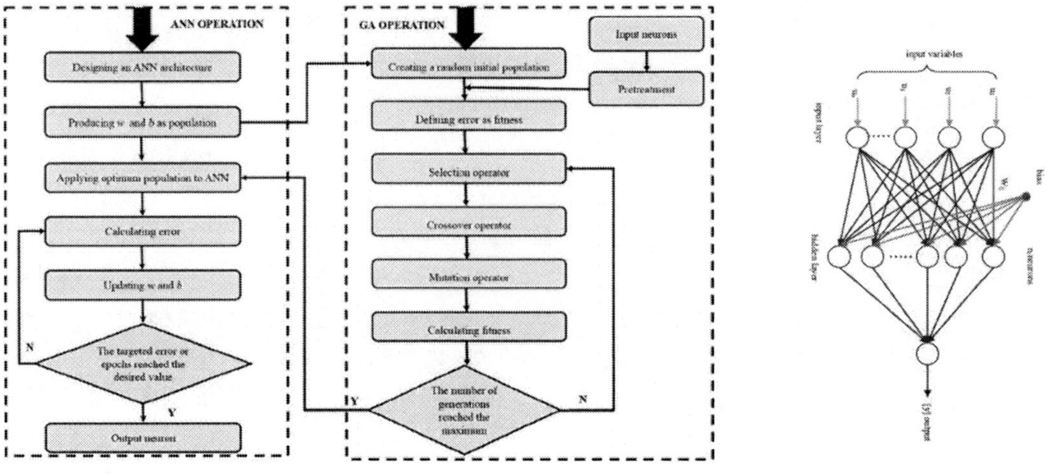

Figure 7. ANN-GA Approach.

Artificial Neural Networks (ANNs) widely applied to deterministic models to find the best value of operating range. In this, Back Propagation Neural Network (BPNN) is used to solve non-linear problems. However, while training BPNN it may converge into local extreme and convergence may be slow. To overcome these problems and improve reliability of network with genetic algorithm is used to achieve global convergence quickly. In this work, Neural power is used to obtain best parameter. The input to the neural power is wear values of each sample. The output is best parameter.

The table-12 shows the best operating parameter for each sample obtained from ANN-GA.

Table 10. Wear result table.

Sl. No	Wear samples	Speed (rpm)	Time (minutes)	Load (Newton)	Wear (micron)
1	Sample 1	234.84	14.99	22.43	99.16
2	Sample 2	449.49	5.00	24.20	59.13
3	Sample 3	434	14.99	20.002	63.14

On investigating the result values of three samples, sample 2 has low wear rate compared to others and corresponding parameters are Speed - 449.949 rpm, Time- 5.00 minutes, Load - 24.20 N, the wear value obtained is 59.13 microns.

6. Conclusion

In this work, the reinforcing materials, magnesium AZ31 and rock dust are used with aluminum 6061-T6 to find the wear characteristics. The mg and rock dust are varied from1 % to 3 % and with Al 6061-T6 during stir casting process. The aluminum and magnesium are taken as solid metal and rock dust as taken as fines (30 μm) powdered form. The temperature of a crucible is maintained around 500 – 800 degrees and stirred continuously. The molten metal is converted into a solid work piece 120 mm lengths with 20 mm diameter.

The wear tests are carried on pin on disc under dry condition as per ASTM G 99-95 measures. To limit the quantity of experimentation ANOVA is utilized. The model developed form ANOVA is significant for the entire three test specimen. The test result shows that the sample-2 (Al (6061-T6) 97%, Mg (AZ31) 1%, Rock dust 2%) posses minimum wear (59.13) as compared to other two specimens. From the iteration plots, it is observed that average contributions of parameters are the speed (63%), load (23.33%), and time (0.67). To find the best value of operating parameters for each sample ANN-GA is used. For the further extension, the above test specimen can be machined using any one of the non-traditional machines.

7. Reference

[1] Wei Guo, Guoqiang You, Guangyu Yuan and Xiuli Zhang November 2016 *Microstructure and mechanical properties of dissimilar inertia friction welding of 7A04 aluminum alloy to AZ31 magnesium alloy* Journal of alloy compound Volume 695 pp 3267-3277.

[2] Zilong Zhao, Qiang Gao, Junfeng Hou, Ziwei Sun and Fei Chen September 2016 *Determining the microstructure and properties of magnesium aluminum composite panels by hot rolling and annealing* Journal of Magnesium and Alloys Volume 4 Issue 3 Pages 242–248.

[3] Ratna Sunil B, Pradeep Kumar Reddy G, Hemendra Patle and Ravikumar Dumpala March 2016 *Magnesium based surface metal matrix composites by friction stir processing* Journal of Magnesium and Alloys Volume 4 Issue 1 pp 52-61.

[4] Gali O A, Shafiei M, Hunter J A and Riahi A R November 2016 *The influence of work roll roughness on the surface/near-surface microstructure evolution of hot rolled aluminum–magnesium alloys* Journal of Materials Processing Technology Volume 237 pp 331-341.

[5] Valavan D, Balasundaram R and Rajkumar G November 2016 *Experimental Design and Analysis of Carbon fiber Reinforced Composite Drive* Shaft Asian Journal of Research in Social Sciences and Humanities Volume 6 Number 11 pp 167-178 DoI:10.5958/2249-7315.2016.01183.7.

[6] Jufu Jiang, Gang Chen and Ying Wang November 2016 *Compression Mechanical Behaviour of 7075 Aluminium Matrix Composite Reinforced with Nano-sized SiC Particles in Semisolid State* Journal of Materials Science & Technology Volume 32 Issue 11 pp 1197-1203.

[7] Kimura M, Fuji A and Shibata November 2015 S *Joint properties of friction welded joint between pure magnesium and pure aluminium with post-weld heat treatment* Materials & Design Volume 85 pp 169-179.

[8] Taherzadeh Mousavian R, Azari Khosroshahi R, Yazdani S, Brabazon D and Boostani A F January 2016 *Fabrication of aluminum matrix composites reinforced with nano- to micrometer-sized SiC particles* Materials & Design volume 89 pp 58-70.

[9] Mojtaba Jafarian, Alireza Khodabandeh and Sahebali Manafi January 2015 *Evaluation of diffusion welding of 6061 aluminum and AZ31 magnesium alloys without using an interlayer* Materials & Design Volume 65 pp 160-164 .

[10] Ansary Yar A, Montazerian M, Abdizadeh H and Baharvandi H R September 2009 *Microstructure and mechanical properties of aluminum alloy matrix composite reinforced with nano-particle MgO* Journal of alloy and compounds volume 484 Issues 1-2 pp 400-404.

Study of Dimple Effect on the Friction Characteristics of a Journal Bearing using Taguchi Method

A.Amar Murthy[1], Dr. Raghunandana *[2]

[1] Department of Mechanical and Manufacturing Engineering, Manipal Institute of Technology, Manipal.

[2]Department of Mechatronics Engineering, Manipal Institute of Technology, Manipal.

Abstract. The effect of producing dimples using chemically etched techniques or by machining process on the surface of a journal bearing bushing to reduce the friction using Taguchi method is investigated. The data used in the present analysis is based on the results obtained by the series of experiments conducted to study the dimples effect on the Stribeck curve. It is statistically proved that producing dimples on the bushing surface of a journal bearing has significant effect on the friction coefficient when used with light oils. Also it is seen that there is an interaction effect between speeds-load and load-dimples. Hence the interaction effect, which are usually neglected should be considered during actual experiments that significantly contributes in reducing the friction in mixed lubrication regime. The experiments, if were conducted after Taguchi method, then the number of experiments would have been reduced to half of the actual set of experiments that were essentially conducted.

1. Introduction

The relation between the friction coefficient and the velocity along with the viscosity and pressure can be graphically expressed by Stribeck curve. Fig.1 shows the curve at three different regimes. To improve and enhance the life of the components used in mechanical systems, reduction in wear and friction plays a crucial role. The profile of the journal bearing surface contributes in reducing the friction apart from developing the required hydrodynamic pressure. Modifying and optimizing the surface topography of the mating surfaces therefore became a subject of interest for many researches leading them to actively investigate in this area. Fogg et al [1] showed that proper control on friction surface, speed and load considerably improved the efficiency, consistency and helped in energy conservation. Out of the different methods available, texturing of journal surface has great effect in reduction of friction and enhancing the various performance properties in tribological related systems. Sinanoglu et al [2] disclosed that the load carrying capacity of a bearing with surface texturing effects can be improved. A number of researches examined on surfaces, proved the potential advantages of altering the surface topography, for example, lower coefficient of friction and spillage issues in mechanical seals, enhanced load capacity in sliding and axial bearings, and additionally diminishes the metal to metal contact and reduced surface destruction. In the present study, an attempt is made to statically analyse the effect of different shaped dimples on improving the frictional characteristics of a journal bearing using Taguchi method.

* Corresponding Author

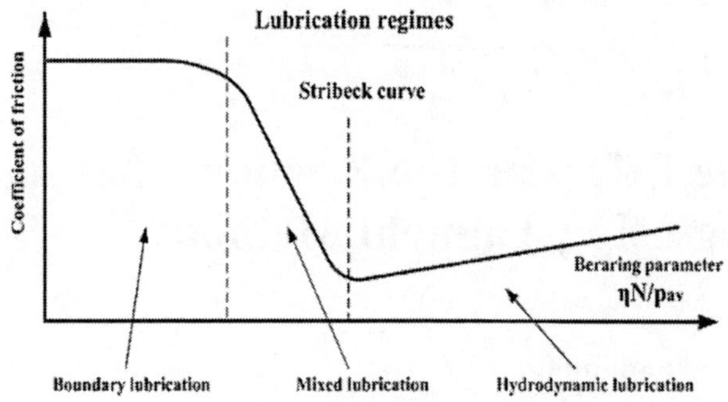

Figure 1. Stribeck curve

Etsion [3] experimentally investigated that there is significant improvement in wear resistance, load capacity, friction coefficient obtained by forming regular micro-dimples on the surface of the bearing. Laser Surface Texturing (LST) was used to obtain the desired texture on the surface. Cupillard et al. [4] numerically studied the pressure buildup due to textures in an inclined slider bearing. Numerical simulations were performed for laminar, steady, and isothermal flows. Their studies concluded that the performance of the bearing with textures was increased. Tala-Ighil et al. [5] established a numerical FDM model to study the textured surfaces with different shapes of micro cavities and at different locations of the texture zone. Their studies revealed that change in the surface texture improves the performance of the bearing.A comparative study was carried out in terms of load carrying capacity for the commonly used dimple shapes for parallel slider gas-lubricated bearings by Qiu, et al. [6]. Their results resolved that the ellipsoidal texture shape was able to give higher load carrying capacity compared to that of all the other types. Simulation were done using the compressible Reynolds equation. Meng, et al. [7] numerically investigated using fluid-structure-interaction method the load capacity of a compound dimpled surface bearing and showed that higher load carrying ability and reduced friction coefficient of the compound dimple. It was seen that the size, geometry and position/interval of compound-dimples play an important role. Sinanoglu et al. [8] studied experimentally and using ANN the topology effect on surface of the shaft on the pressure developed and load carrying capacity. The smooth and the textured surface performances were tested. The textured trapezoidal surface showed a high pressure build up and thus the load capacity. Yushan et al. [9] experimentally discovered a new method of arranging oiling dimples on the surface of axial bearings. 15 to 16 percent reduction in friction was achieved for a patterned surface.

2. Methodology

The flow chart of the methodology adapted for the present research work is as shown in fig.2 is in accordance with the one that is described in Research Methodology by Buckley and Chang [10]. Primarily the experiment is conducted in the laboratory. Later the results obtained by the simulation are used as secondary data for the statistically analysis using Taguchi approach of design of experiments. The data used in the present analysis is based on the results obtained by a series of experiments conducted to study the dimples effect on the Stribeck curve.

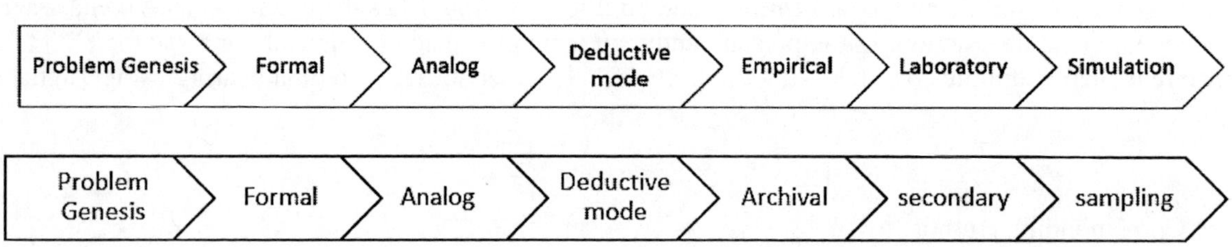

Figure 2. Flow chart of the methodology adopted

The specifications of the various elements used in the experimentation [1] are tabulated from table 1 to table 4.

Table 1. Machined bushing

Bushing	Dimple diameter (mm)	Dimple depth (mm)	Bur diameter (mm)
Conventional	0	0	0
Machined 1	2	0.165	6.223
Machined 2	4	0.448	9.373

Table 2. Etched bushing

Bushing	**Dimple Shape**	**Dimple Depth (mm)**	**Etching time (h)**
Etched 1	4 mm diameter circle, 360^0	0.130	1.0
Etched 2	4 mm diameter circle, 180^0	0.130	1.0
Etched 3	4 mm: 1 mm ellipse, 360^0	1.040	8.0

Table 3. Oil Properties

Oil Type	**Viscosity (cSt)**		**Specific Gravity at 15 ^{0}C**
	40 ^{0}C	100 ^{0}C	
SAE 30	93	10.8	0.890
GT32	31.3	5.25	0.877

(a) (b)

Figure 3. Dimpled bushings (a) machined dimpled bushings (b) etched dimpled bushings

The variation of friction coefficient in different bearing types for various loading condition with SAE 30 and GT32 oil were plotted [5] and is as shown in figure 4 to figure 10.

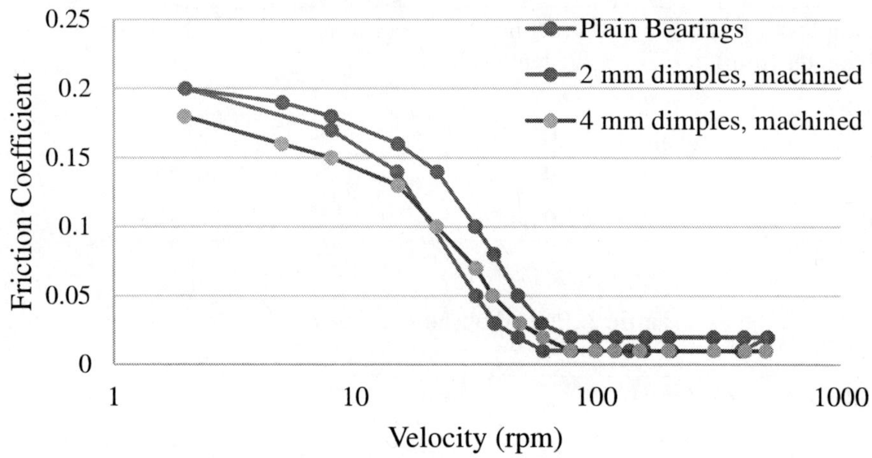

Figure 4. Coefficient of friction v/s velocity (oil – SAE 30 and Load - 667 N)

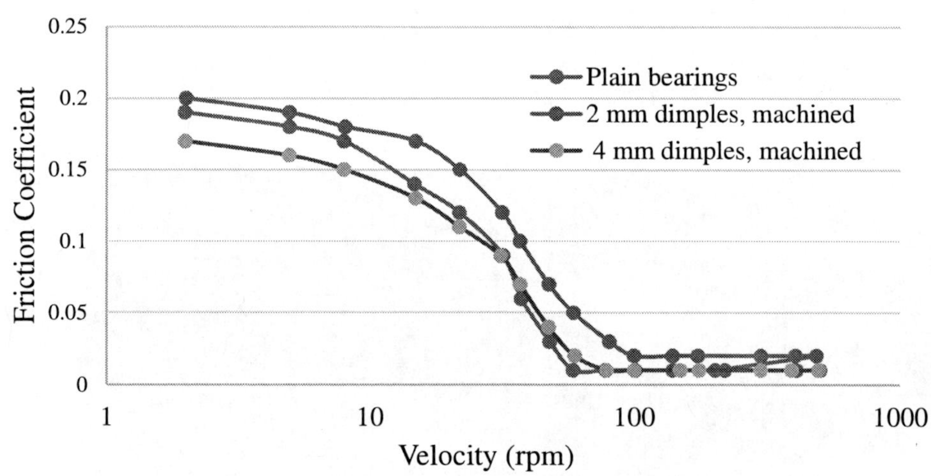

Figure 5. Coefficient of friction v/s velocity (oil – SAE 30 and Load - 890 N)

IOP Conf. Series: Materials Science and Engineering **314** (2018) 012014 doi:10.1088/1757-899X/314/1/012014

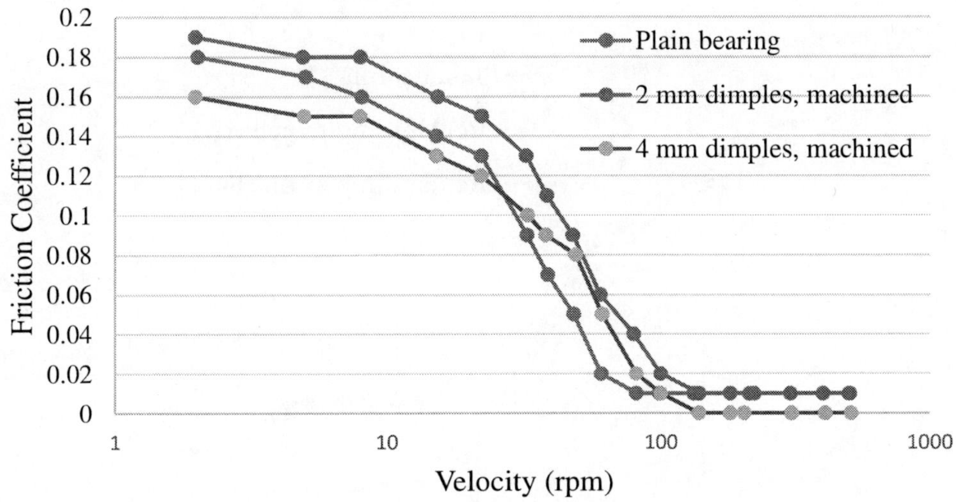

Figure 6. Coefficient of friction v/s velocity (oil – SAE 30 and Load - 1112 N)

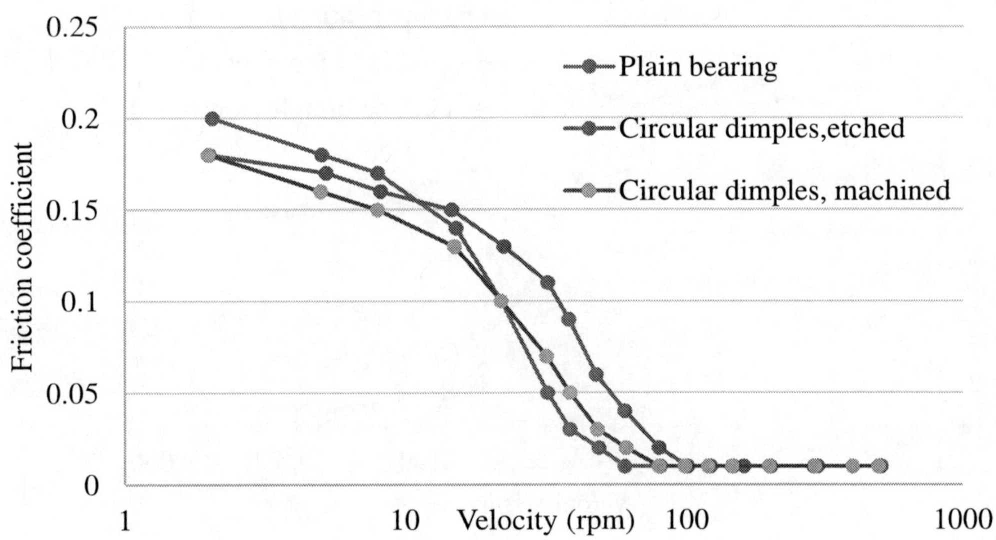

Figure 7. Coefficient of friction v/s velocity (oil – SAE 30 and Load - 667 N)

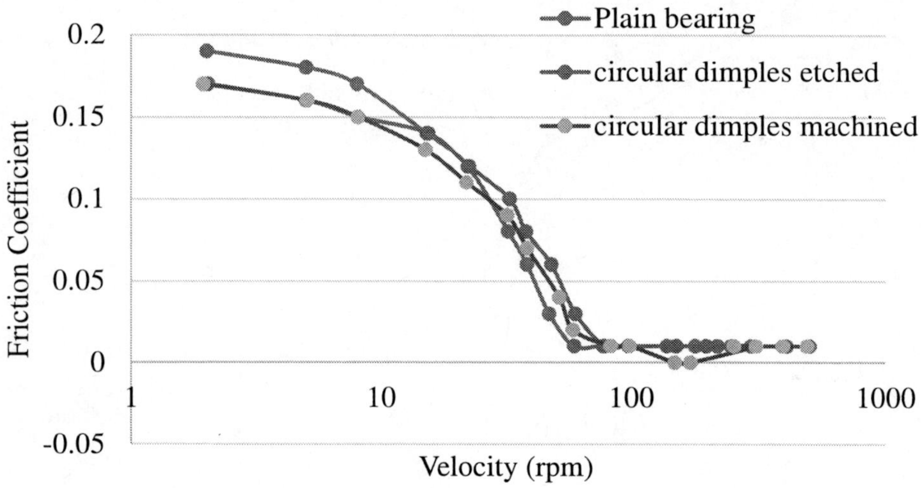

Figure 8. Coefficient of friction v/s velocity (oil – SAE 30 and Load - 890 N)

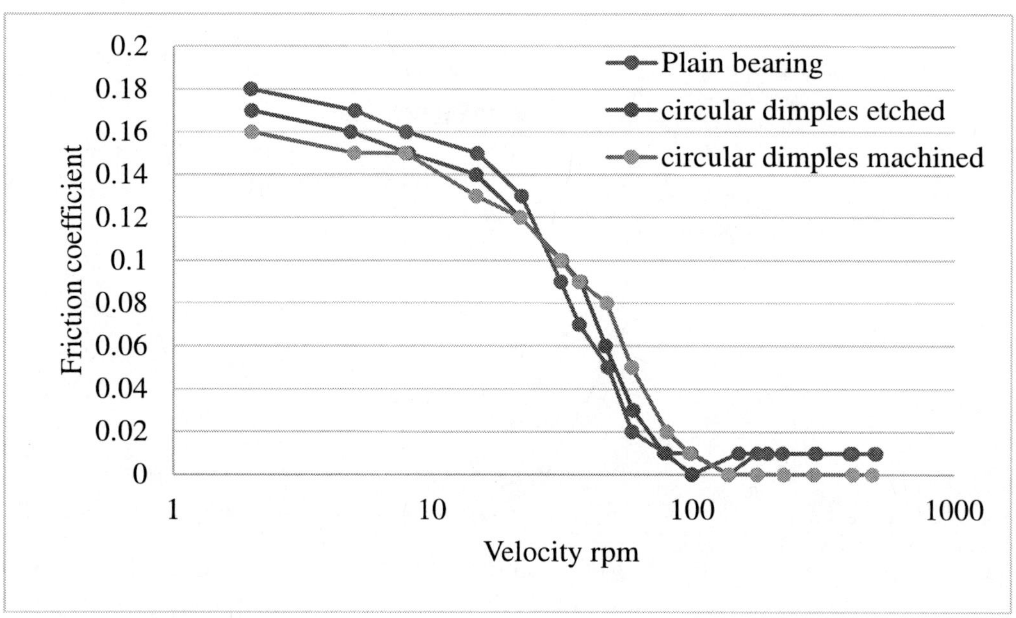

Figure 9. Coefficient of friction v/s velocity (oil – SAE 30 and Load - 1112 N)

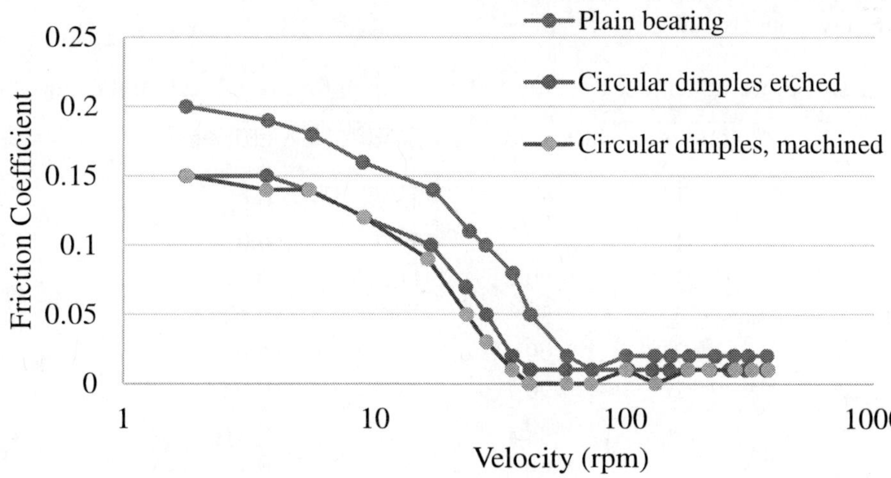

Figure 10. Coefficient of friction v/s velocity (light oil – GT32 Oil and Load - 667 N)

2.1 Taguchi method

The factors that affect the bearing performance are speed, load, surface texture of the bearing, shaft surface texture, geometry of the bearing, viscosity of the oil, temperature of the oil, and type of fluid used as lubricant. During operating conditions of the bearings, it has to be taken care that all these parameters are well balanced so as to produce high load carrying capacity, reduced friction and wear which results in highly stable system.
In the present work, three main independent factors namely speed, surface texture on the bearing bush and load are considered. These are set at three different levels to study their significance on dependent variable i.e., coefficient of friction using Taguchi method.

Case 1: *Comparison of friction co-efficient of circular dimples with 2 mm and 4 mm diameter with that of the plain bearing for (SAE 30 Oil)*

Three levels and three factors are considered. The details are tabulated in table 4.

Table 4: Setting of factors and levels

Factors	Level 1	Level 2	Level 3
Dimples (type)	1	2	3
Speed (rpm)	30	60	90
Load (N)	445	890	1112

Here dimples are designated as 1 – Plain bearing; 2 – 2 mm circular dimples (machined); 3 – 4 mm circular dimples (machined). For each run, data for coefficient of friction is recorded. In a typical case say for 3 factors and 3 levels each one may have to perform , $N = L^F = 3^3 = 27$ experiments. But using

The L9 orthogonal array, as given by Taguchi, the number of experiments or runs to be performed would be 9 which covers all the possible random combination. Table 5 shows the combinations as generated by Minitab for given levels and factors

Table 5. Combinations as generated by Minitab for given levels and factors

Dimples	Speed	Load	Co-efficient of friction
1	30	667	0.03
1	60	890	0.01
1	90	1112	0.01
2	30	890	0.12
2	60	1112	0.06
2	90	667	0.02
3	30	1112	0.11
3	60	667	0.02
3	90	890	0.01

Case 2: *Comparison of friction co-efficient of etched and machined circular dimples with that of the plain bearing for (SAE 30 Oil)*

Table 6 gives the details of three levels and three factors are considered for finding the friction coefficient effect on the etched and machined type of dimple bearings. Table 7 shows the combinations as generated by Minitab for given levels and factors

Table 6: Setting of factors and levels

Factors	Level 1	Level 2	Level 3
Dimples (type)	1	2	3
Speed (rpm)	30	60	90
Load (N)	445	890	1112

1 – Plain bearing; 2 – circular dimples (etched) ; 3 – circular dimples (machined)

Table 7. Combinations as generated by Minitab for given levels and factors

Dimples	Speed	Load	Co-efficient of friction
1	30	667	0.05
1	60	890	0.01
1	90	1112	0.01
2	30	890	0.10
2	60	1112	0.03
2	90	667	0.01
3	30	1112	0.10
3	60	667	0.02
3	90	890	0.01

IOP Conf. Series: Materials Science and Engineering **314** (2018) 012014 doi:10.1088/1757-899X/314/1/012014

Case 3: *Comparison of friction co-efficient of elliptical etched dimples with that of the plain bearing for light oil (GT32 Oil)*

L16 array is used in this case. Two types of dimples, three loading conditions and three different speeds are considered the details are tabulated as shown in table 8. Since with lighter oils the friction performances can be distinguished better for dimple bushing, the oil is now changed to light oil (GT32 oil). Dimple types are 1 – Plain bearing; 2 – Elliptical etched dimple. Table 9 shows the combinations as generated by Minitab for given levels and factors

Table 8: Setting of factors and levels

Factors	Level 1	Level 2	Level 3
Dimples (type)	1	2	
Speed (rpm)	30	60	90
Load (N)	445	890	1112

Table 9. Combinations as generated by Minitab for given levels and factors

Dimples	Speed	Load	Co-efficient of friction
1	30	445	0.225
1	30	560	0.210
1	30	667	0.200
1	60	445	0.160
1	60	560	0.140
1	60	667	0.125
1	90	445	0.110
1	90	560	0.090
1	90	667	0.054
2	30	445	0.150
2	30	560	0.150
2	30	667	0.160
2	60	445	0.100
2	60	560	0.125
2	60	667	0.120
2	90	445	0.060
2	90	560	0.090
2	90	667	0.120

3. Results and Analysis

3.1 Case 1: Comparison of friction co-efficient of circular dimples with 2 mm and 4 mm diameter with that of the plain bearing for (SAE 30 Oil)

Data for coefficient of friction are tabulated in table 5 for case 1. The Taguchi analysis is carried out and the statistical results obtained are tabulated in table 10. The response table for case 1 is shown in table 11

Table 10: Analysis of Variance for Means

Source	DF	Seq SS	Adj SS	Adj MS	F
DIMPLES	2	0.003800	0.003800	0.001900	8.14
SPEED	2	0.008867	0.008867	0.004433	19.00
LOAD	2	0.002067	0.002067	0.001033	4.43
Residual Error	2	0.000467	0.000467	0.000233	
Total	8	0.015200			

Table 11: Response Table for Means

Level	DIMPLES	SPEED	LOAD
1	0.01667	0.08667	0.02333
2	0.06667	0.03000	0.04667
3	0.04667	0.01333	0.06000
Delta	0.05000	0.07333	0.03667
Rank	*2*	*1*	*3*

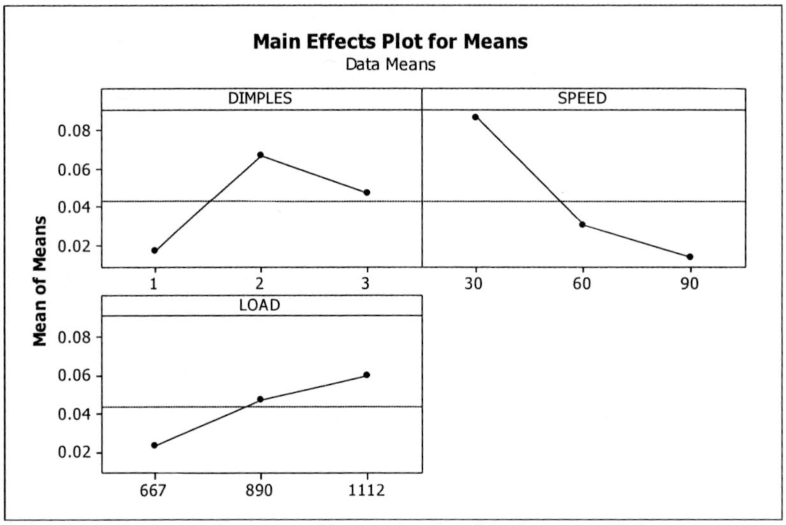

Figure 11. Mean effects plot for friction coefficient

From the above figure 11 it is seen that dimples of type 3 i.e. 4mm circular shaped dimples machined type results in lower value of friction coefficient than compared to the type 2 i.e. 2 mm circular dimples machined type. Also from the response table 11 it is seen that, creating dimples have effect on friction coefficient.

3.2 Case 2: Comparison of friction co-efficient of etched and machined circular dimples with that of the plain bearing for (SAE 30 Oil)

Table 12: Analysis of Variance for Means

Source	DF	Seq SS	Adj SS	Adj MS	F	P
DIMPLES	2	0.0009556	0.0009556	0.0004778	3.31	0.232
SPEED	2	0.0094889	0.0094889	0.0047444	32.85	0.030
LOAD	2	0.0006222	0.0006222	0.0003111	2.15	0.317
Error	2	0.0002889	0.0002889	0.0001444		
Total	8	0.0113556				

Table 13: Response Table for Means

Level	DIMPLES	SPEED	LOAD
1	0.02333	0.08333	0.02667
2	0.04667	0.02000	0.04000
3	0.04333	0.01000	0.04667
Delta	0.02333	0.07333	0.02000
Rank	2	1	3

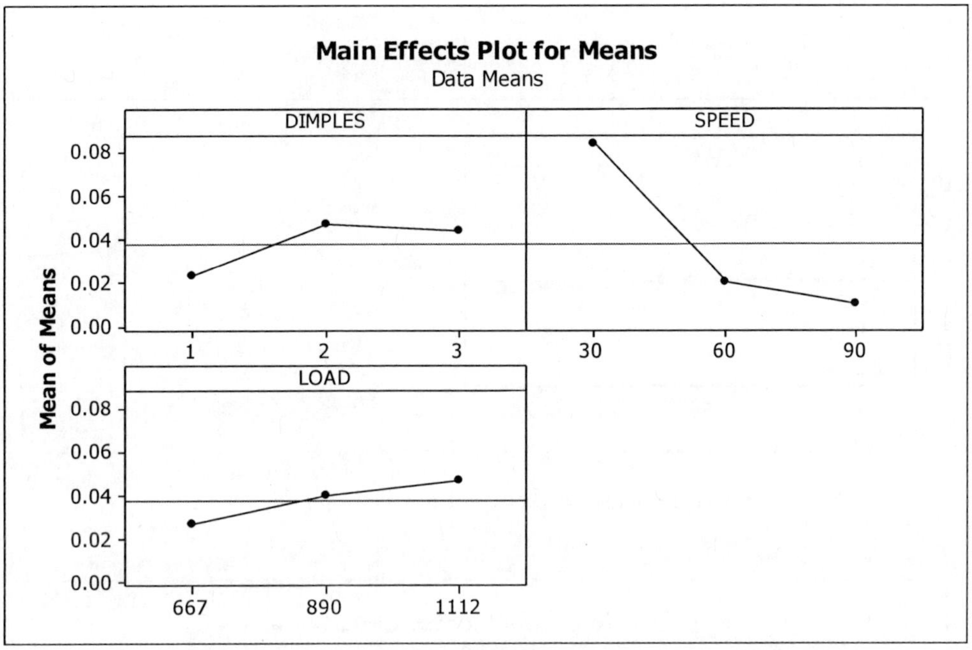

Figure 12. Mean effects plot for friction coefficient

From the above figure 12 it is seen that dimples of type 3 i.e. circular dimples machined type results in lower value of friction coefficient than compared to the type 2 i.e., circular dimples etched type. Also from the response table 13 it is seen that, creating dimples have effect on friction coefficient.

*3.3 **Case 3:** Comparison of friction co-efficient of elliptical etched dimples with that of the plain bearing for light oil (GT32 Oil)*

Table 14: Analysis of Variance for Means

Source	DF	Seq SS	Adj SS	Adj MS	F	P
DIMPLES	1	0.0031734	0.0031734	0.0031734	4.93	0.046
SPEED	2	0.0273434	0.0273434	0.0136717	21.24	0.000
LOAD	2	0.0000751	0.0000751	0.0000376	0.06	0.944
Error	12	0.0077257	0.0077257	0.0006438		
Total	17	0.0383176				

Table 15: Response Table for Means

Level	DIMPLES	SPEED	LOAD
1	0.14600	0.18250	0.13417
2	0.11944	0.12833	0.13417
3		0.08733	0.12983
Delta	0.02656	0.09517	0.00433
Rank	**2**	**1**	**3**

IOP Conf. Series: Materials Science and Engineering **314** (2018) 012014 doi:10.1088/1757-899X/314/1/012014

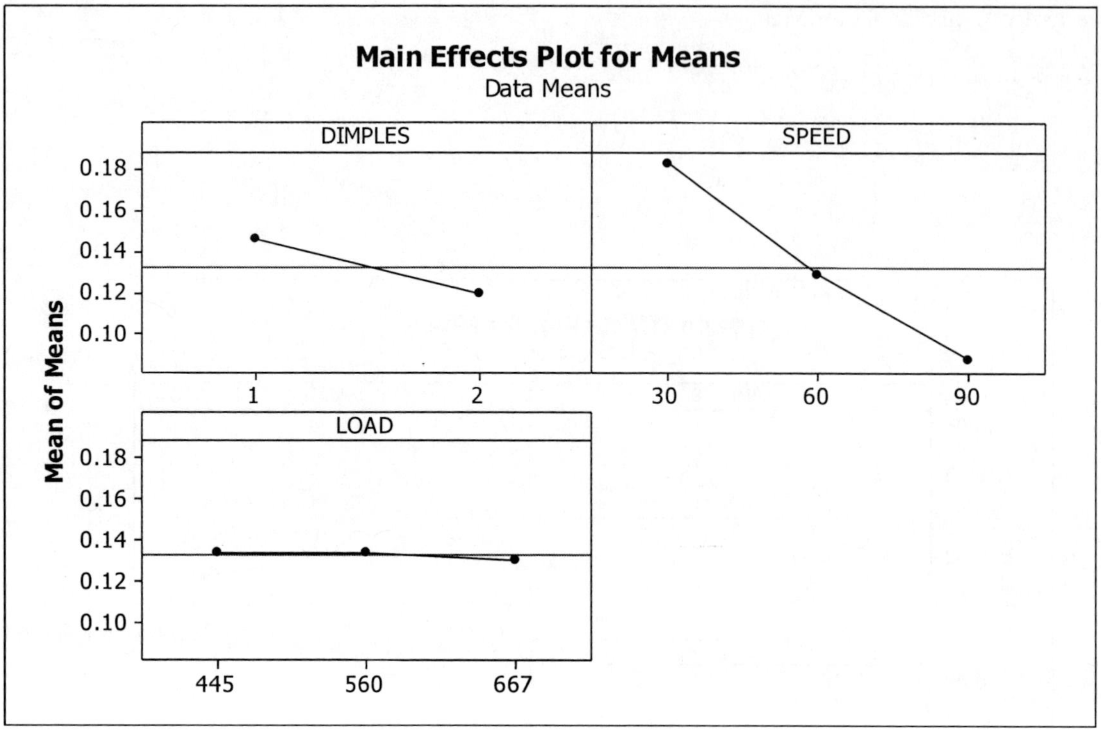

Figure 13. Mean effects plot for friction coefficient

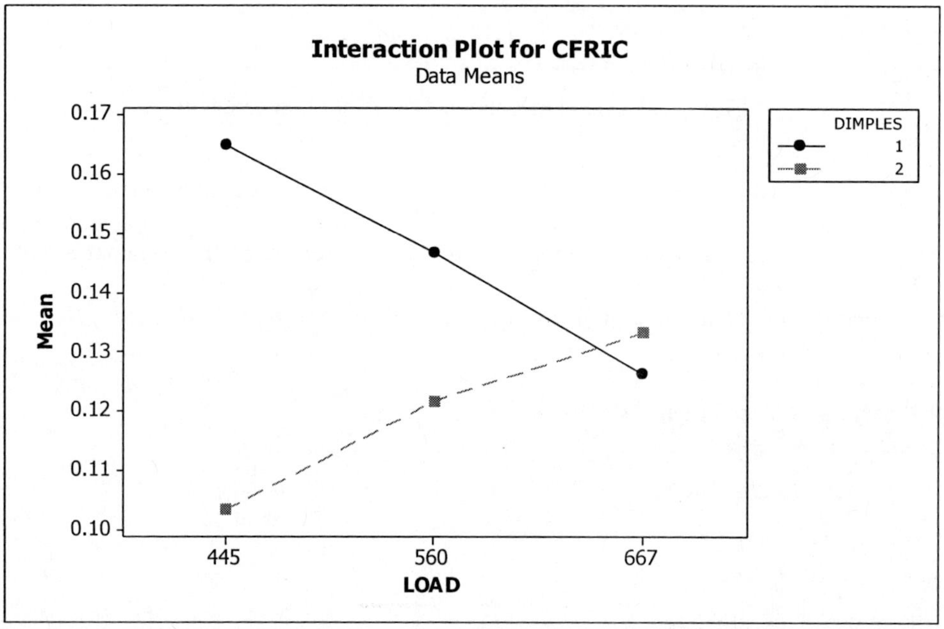

Figure 14. Interaction effect plot for coefficient of friction between dimples and load

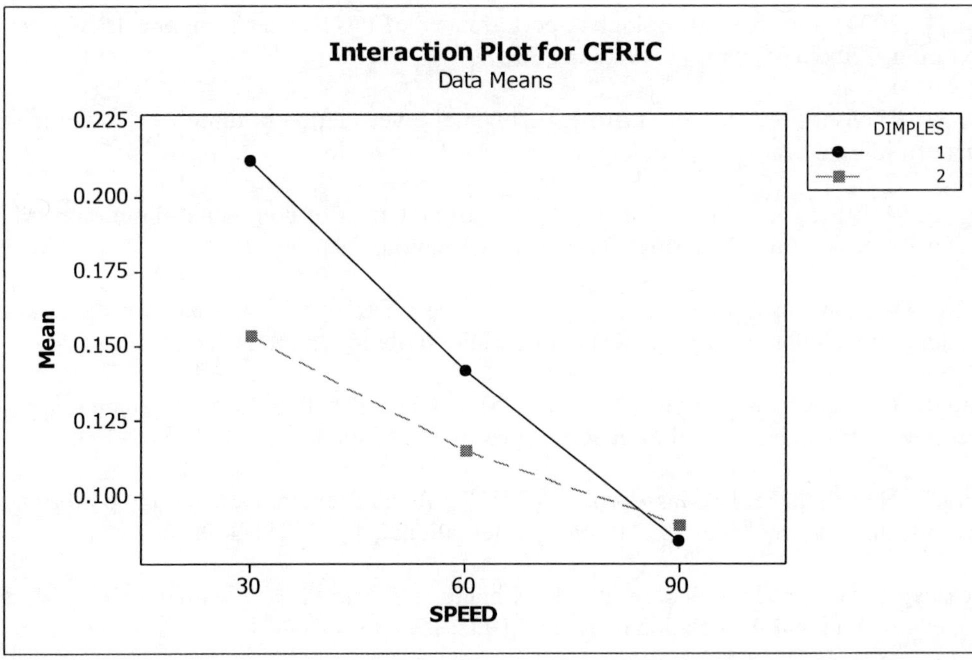

Figure 15. Interaction effect plot for coefficient of friction between dimples and speed

From figure 13 it is seen that dimples of type 2 i.e. elliptical dimples etched type results in lower value of friction coefficient than compared to the type 1 i.e., plain bearing bushing. The lower P-Value for speed indicate that this factor has a greater significance on the coefficient of friction. Figure 14 and figure 15 shows that there is interaction effect between load and dimple also between speed and dimples.

Conclusions

- Producing dimples on bearing bushing has significant effect in reducing the friction coefficient in the mixed lubrication regime.
- The elliptical etched dimples has lower value of friction coefficient than that of conventional plain bearing which can be easily distinguished with lighter oil like GE32 oil.
- Taguchi method gives accurate results as that obtained by experimentation in limited number of runs. Thus if this analysis is carried out before experimentation then it would save the time.

References

[1] Fogg, A. (1946). Fluid film lubrication of parallel thrust surfaces. Proceedings of the Institution of Mechanical Engineers, 155(1), 49-67.

[2] Sinanoglu, C., Nair, F., & Karamış, M. B. (2005). Effects of shaft surface texture on journal bearing pressure distribution. Journal of materials processing technology, 168(2), 344-353.

[3] Cupillard, S., Cervantes, M. J., & Glavatskih, S. (2008). Pressure buildup mechanism in a textured inlet of a hydrodynamic contact. Journal of Tribology, 130(2), 021701.

[4] Etsion, I. (2004). Improving tribological performance of mechanical components by laser surface texturing. Tribology letters, 17(4), 733-737.

[5] Lu, X., & Khonsari, M. M. (2007). An experimental investigation of dimple effect on the Stribeck curve of journal bearings. Tribology Letters, 27(2), 169-176.

[6] Meng, F. M., Zhang, L., Liu, Y., & Li, T. T. (2015). Effect of compound dimple on tribological performances of journal bearing. Tribology International, 91, 99-110.

[7] Qiu, M., Delic, A., & Raeymaekers, B. (2012). The effect of texture shape on the load-carrying capacity of gas-lubricated parallel slider bearings. Tribology Letters, 48(3), 315-327.

[8] Sinanoglu, C., Nair, F., & Karamış, M. B. (2005). Effects of shaft surface texture on journal bearing pressure distribution. Journal of materials processing technology, 168(2), 344-353.

[9] Tala-Ighil, N., Fillon, M., & Maspeyrot, P. (2011). Effect of textured area on the performances of a hydrodynamic journal bearing. Tribology International, 44(3), 211-219.

[10] Buckley, J. W., M. H. Buckley and H. Chiang. (1976). *Research Methodology & Business Decisions*. National Association of Accountants. 15.

[11] Yushan, L., Yueming, L., Jun, W., & Hongpeng, L. (2016). Tribological performance of dimpled thrust bearings with phyllotactic patterns. Wear, 346, 108-115.

[12] Bogy, D. B. (2002). Air film dynamics for micro-textured flying head slider bearings in magnetic hard disk drives.

[13] Tanaka, H., Gomi, K., & Miyake, Y. (1993). Micro-tribology of carbon-coated thin-film media with well-defined surface texture. IEEE transactions on magnetics, 29(1), 270-275.

[14] Ryk, G., Kligerman, Y., & Etsion, I. (2002). Experimental investigation of laser surface texturing for reciprocating automotive components. Tribology Transactions, 45(4), 444-449.

IOP Conf. Series: Materials Science and Engineering **314** (2018) 012015 doi:10.1088/1757-899X/314/1/012015

Characterisation of microstructure, mechanical and corrosion properties of pulsed MIG welded modified P91 steel weld metal

P. Sundararaj[1], M. Muthukumar[1]

[1] Department of Metallurgical Engineering, Government College of Engineering, Salem – 636011, Tamilnadu, India.

Abstract: Varying the shielding gas composition with argon and carbon dioxide influences the properties of the weldments which are evaluated using microstructural, micro-hardness and corrosion studies. The modified P91 steel samples are welded by Pulsed Metal Inert Gas welding process with different shielding gas mixture, i.e., 95%Ar-5%CO2, 80%Ar-20%CO2 and 60%Ar-40%CO2. The welded steels are studied metallographically by observing microstructures at three different regions namely at the base metal, Heat Affected Zone (HAZ) and the Weld zone. Hardness measurements are also done using Vicker's micro-hardness tester. Corrosion studies in acidic media (sulphuric and nitric acid media of four different normalities 0.5, 1.0,1.5 and 2.0) are done in the welded region and the parent metal to compare the corrosion resistance. Of all the welded samples, welds made with the shielding gas composition of 95%Ar-5%CO2 exhibits good corrosion resistance over the other two welds while the weld made with the shielding gas composition of 60%Ar -40%, shows very poor corrosion resistance.

1.Introduction

To reduce the environmental pollution from fossil fuels used in power generating plant, plants should run at high temperature and pressure. It calls for a material having improved temperature properties. Among the existing Cr-Mo ferritic steels, modified 9%Cr-1%Mo (P91) steel developed by the Oak Ridge National Laboratory in USA by adding the small amount of Vanadium and Niobium is being used worldwide today for high temperature application. These added Nb and V forms fine carbo nitrides [1,2] and strengthens the material by preventing the movement of subgrain boundaries, by impeding knitting reactions between free dislocations and subgrain boundaries by pinning sub boundary dislocation.

Application of P91 steel necessitates welding that affects the micro structures, mechanical and corrosion properties [3-7] of the metal. Arivazhagan et al [3] studied the influence of weld metal composition on mechanical and micro structure properties of welded metal and found that the presence of Nb & V greatly affects the toughness of the material. Effect of shielding gas composition on mechanical properties of welded modified P91 steel was studied by Kamaraj et al [4] and found that increasing of purity of argon content reduces inclusion percentage and increases toughness of the weld metal. During welding, alloying elements present in the electrode favors retaining of δ ferrite and greatly influences mechanical properties [5]. Even welding process has significant effect on mechanical properties and investigated by Magudeeswaran et al [6]. Murugan et al [7] used pulsed Metal Inert Gas welding process and found that it improves stable spray mode metal transfer, high deposition rate and produces refined structure due to pulsed current.

The present study deals with the effect of shielding gas composition on corrosion behavior of pulsed MIG welded modified 9%Cr-1%Mo steel and along with micro structure and micro hardness distribution studies were carried out.

2. Experimental

Modified 9Cr-1Mo steel (P91) plates were normalized at 1040 °C for 1 hour and tempered at 760 °C for 1 hour. This heat treated base metals were cut into the size of 200*60*12 mm and preheated to the temperature of 200°C, subsequently joined by single "V" groove butt joint in flat (1G) position (with included angle of 60° and the root gap of 1.5 mm) using Pulsed MIG Welding process Table 1 and 2 shows the composition of the base metal and filler metal respectively and welding parameters are shown in the table 3. Welding was performed using 1.2 mm diameter electrode of 9% Cr-1% Mo steel and with varying shielding gas composition of CO_2 and Argon mixture. Three different shielding gas compositions were used namely 95%Ar-5%CO_2, 80%Ar-20%CO_2 and 60%Ar-40%CO_2.After completion of welding, post weld heat treatment is done by heating to 760°C for an hour and then cooled down to 400°C in a controlled manner and followed by air cooling.

Microstructural characterization was carried out by Optical Microscope using the etchant called Viella's reagent (1 g picric acid, 5 ml HCl and 90 ml distilled water). Micro hardness distribution across the parent metal, heat affected zone and weld metal was determined by Vicker's micro hardness tester (MATSUZAWA-MMT-X3) with the application of 500 g of load for 15 s. The corrosion studies were conducted (Potensiostat/ Galvanostat, CH instruments, USA) using Tafel extrapolation method in two different acids (Sulphuric and Nitric acid) and at four different concentrations (0.5 N,1 N,1.5 N and 2 N).

Table 1: Composition of the base metal.

Elements	C	Mn	Si	Cr	Mo	V	Nb	Ni	Fe
Wt. %	0.08	0.39	0.50	9.4	1.0	0.25	0.09	0.13	Bal.

Table 2: Composition of the filler metal.

Elements	C	Mn	Si	Cr	Mo	V	Nb	Ni	Fe
Wt. %	0.1	0.5	0.3	9.0	1.0	0.2	0.06	0.7	Bal.

Table 3: Welding Parameters.

Parameter	95%Ar-5 % CO_2	80%Ar-20 % CO_2	60%Ar-40 % CO_2
Voltage(v)	17.3	17.3	17.3
Current(amp)	72	72	72
Wire feeding rate(m/min)	1.8	1.8	1.8
Welding speed(mm/sec)	5.06	5.14	5.03
Pulsed current(amp)	125	125	125
Background current(amp)	18	18	18

Heat input(J/mm)	245.77	242.09	247.24
No.of pass	5	5	5

3. Results and discussion

3.1 Microstructure

Microstructure of base metal, heat affected zone and weld metal captured using optical microscope are shown at Magnification of 100X in Figure 1a-g.

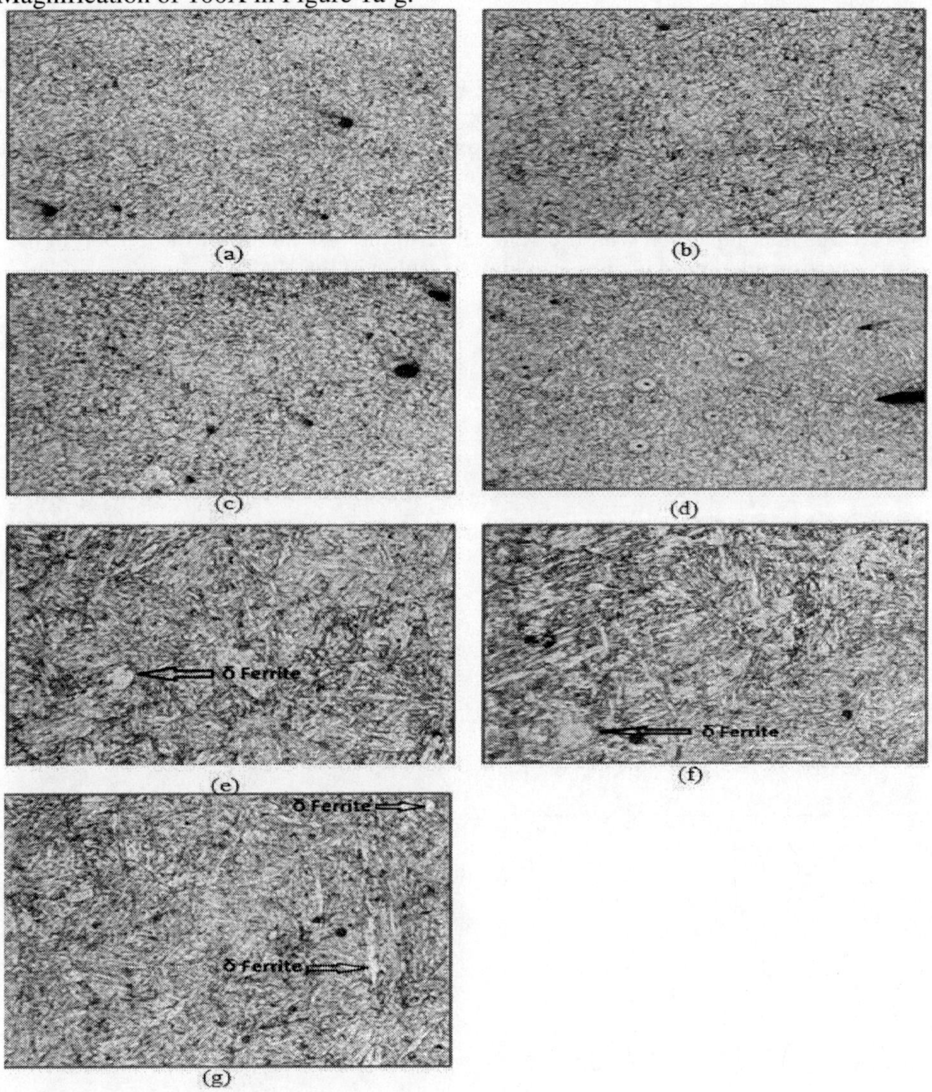

Figure 1: a) Micro structure of the base metal. b-d) Micro structure of HAZ of samples with 95%Ar-5%CO_2,80%Ar-20%CO_2 and 60%Ar-40%CO_2 respectively. e-f) Micro structure of weld zone of samples with 95%Ar-5%CO_2,80%Ar-20%CO_2 and 60%Ar-40%CO_2 respectively.

In as heat treated condition, base metal has tempered martensite.In the Heat Affected Zone of the material welded with the shielding gas composition of 60%Ar-40% CO_2, there is an initial growth of carbides accompanied by depletion of carbon from the surrounding region which is evident by the plain white structure surrounding the carbide pits.It can be observed the area fraction of δ (delta) ferrite (indicate by plain white unetched regions), which is a soft phase, is little higher than normally observed in other microstructures.The microstructure of the parent metal is the same irrespective of the welding conditions. The microstructure of the parent metal consists of tempered martensite with carbo-nitride precipitates mainly due to the addition of modifiers viz Niobium and Vanadium.

3.2Microhardness

Microhardness of the welded sample was taken with the aid of Vicker's microhardness tester and are shown in figure 2a-c. Indentations are made at the pattern of 1mm distance between each indentation from the central axis on either side.The profile takes a valley pattern with lower hardness value at the weld zone and increasing hardness as we progress towards the Heat Affected Zone and again decreases on transitioning to the base metal region. Due to the presence of soft δ ferrite at the weld zone, the hardness records a lower value in comparison with the HAZ. Due to the dissolution of the carbon in the martensite it leads to oversaturation and hence we observe high hardness in the HAZ.

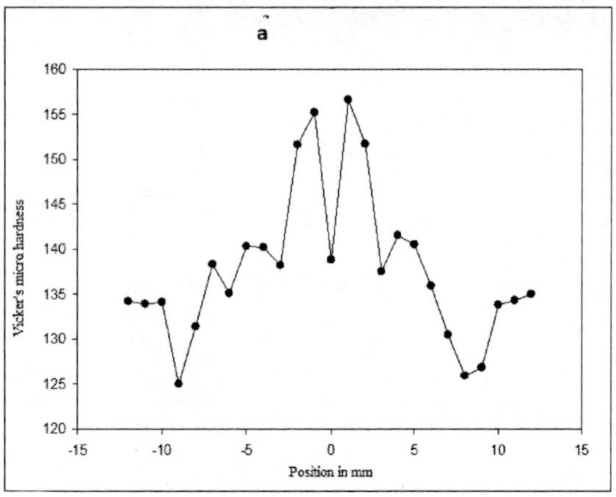

(a)

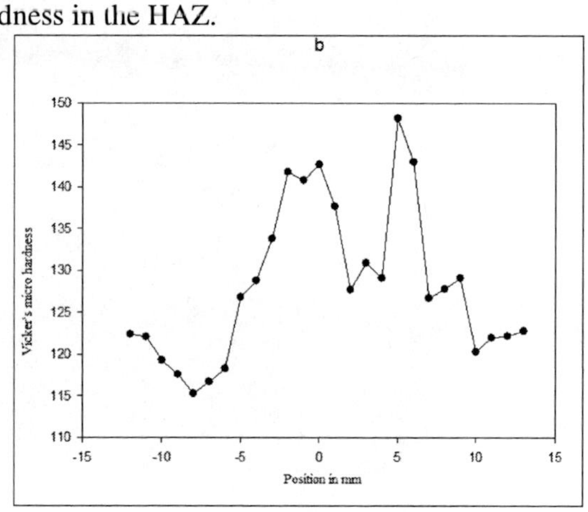

(c)

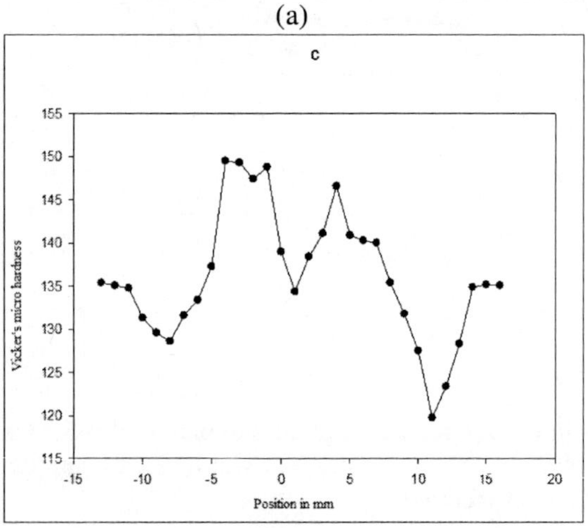

(b)

Figure 2: a-c) Micro Hardness profile of welded sample with95%Ar-5%CO_2 ,80%Ar-20%CO_2 and 60%Ar-40%CO_2 respectively.

3.3 Corrosion Behavior

Corrosion rate obtained from Tafel extrapolation method is tabulated for$H_2SO_4$and $HNO_3$for four different normalities in table 4 and 5 respectively. To compare the effect of shielding gas composition on corrosion behavior, graph is plotted for Corrosion rate vs Normality from the values given in the tables and are shown in figure 3 and 4.

Table 4: Corrosion rate at weld and base metal in Sulfuric acid.

	0.5N	**1.0N**	**1.5N**	**2.0N**
	Weld metal			
95%Ar- 5%CO_2(e+004)	1.163	1.238	1.317	1.420
80%Ar – 20%CO_2(e+004)	1.094	1.152	1.358	1.349
60%Ar – 40%CO_2(e+004)	1.203	1.255	1.345	1.539
	Parent metal			
(e+004)	1.043	1.131	1.221	1.269

Table 5: Corrosion rate at weld and base metal in Nitric acid.

	0.5N	**1.0N**	**1.5N**	**2.0N**
	Weld metal			
95%Ar- 5%CO_2(e+004)	2.289	3.296	3.949	5.711
80%Ar – 20%CO_2(e+004)	2.473	3.278	3.480	5.986
60%Ar – 40%CO_2(e+004)	3.002	4.732	5.874	7.393
	Parent metal			
(e+004)	2.200	3,263	3.747	3.894

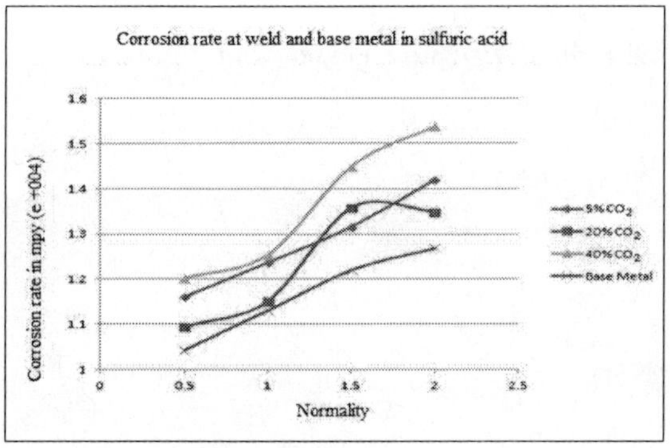

Figure 3: Corrosion rate at weld and base metal in sulfuric acid

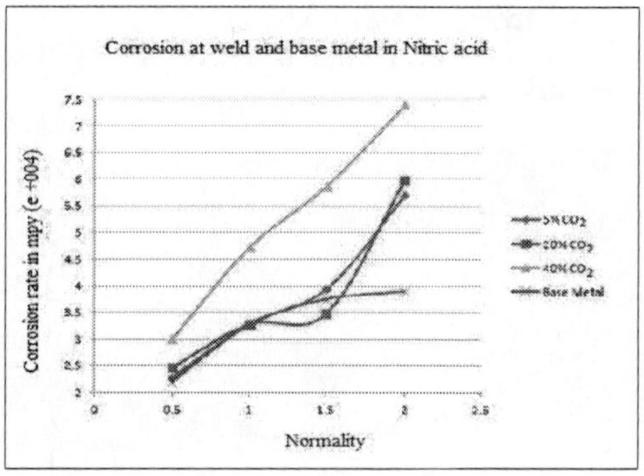

Figure 4: Corrosion at weld and base metal in Nitric acid

From figure 3, of all the joint welded with the shielding gas composition of 60%Ar-40% CO_2 is very high while the base metal records low corrosion resistance. Corrosion rate of the 80%Ar -20% CO_2 has a fluctuating rate of corrosion. From figure 4, the corrosion rate is higher in nitric acid as in comparison with the observed rate with sulphuric acid. The joint welded with the shielding gas composition of 60%Ar-40% CO_2 is very high while the base metal records low corrosion resistance. Corrosion rate of the 80%Ar -20% CO_2 has a fluctuating rate of corrosion, it reaches a lower corrosion rate than base metal for two different normalities and surges above the 95%Ar-5% CO_2 for higher concentration of corrosive medium.The sample which is welded by using 60%Ar-40%CO2has the poor corrosion resistance due to presence of more oxygen in the weldment formed during the disintegration of carbon dioxide at arc plasma

4. Conclusions
The following are the conclusions drawn from the present study.

- 80%Ar-20%CO2weld metal contains high amount of delta ferrite due to that it contain low hardness values..
- The corrosion resistance of the weldments is low in comparison with the corrosion rates of the base metal.
- The sample which is welded by using 60%Ar-40%CO2has the poor corrosion resistance due to presence of more oxygen in the weldment formed during the disintegration of carbon dioxide at arc plasma.
- Sample welded with 95%Ar-5%CO2shielding gas exhibits favourable microstructural and corrosion behavior which is better than the other two samples.

5. References
[1] Ajaja O, Ardell A J 1979 *Philos Mag* **39** 75
[2] Kassner M E 1990*J. Mater. Sci.* **25** 1997
[3] Arivazhagan B,Ranganath Prabhu,Albert S K, Kamaraj M and Sundaresan S 2009 *ASM Int***18** 999-1004.
[4] Arivazhagan B, Sundaresan S and Kamaraj M 2009 *J Mater Processing Tech* **209** 5245–53
[5] Arivazhagan B and Kamaraj M 2011 *J Mater Processing Tech* **211** 1023–29.
[6] Magudeeswaran G, Balasubramanian V and Reddy G M 2008 *Int J Hydrogen Energy* **33** 1897 – 1908.
[7] Murugan S, Kumar P V, Raj B and Bose M S C 1998 *Int. J. Pressure Vessels and Piping* **75** 891- 905.

IOP Conf. Series: Materials Science and Engineering **314** (2018) 012016 doi:10.1088/1757-899X/314/1/012016

Challenges in Special Steel Making

G Balachandran

Head (R&D), JSW Steel Ltd., Salem Works, Potaneri, Salem – 636453.;g.balachandran@jsw.in

Abstract : Special bar quality [SBQ] is a long steel product where an assured quality is delivered by the steel mill to its customer. The bars have enhanced tolerance to higher stress application and it is demanded for specialised component making. The SBQ bars are sought for component making processing units such as closed die hot forging, hot extrusion, cold forging, machining, heat treatment, welding operations. The final component quality of the secondary processing units depends on the quality maintained at the steel maker end along with quality maintained at the fabricator end. Thus, quality control is ensured at every unit process stages. The various market segments catered to by SBQ steel segment is ever growing and is reviewed. Steel mills need adequate infrastructure and technological capability to make these higher quality steels. Some of the critical stages of processing SBQ and the critical quality maintenance parameters at the steel mill in the manufacture has been brought out.

1. INTRODUCTION:

Steel long product produced, may be divided into Merchant Bar Quality [MBQ] and Special Bar Quality [SBQ]. MBQ steels, is manufactured to specified sizes with appropriate chemical limits to meet a set of properties, where the end use is non-critical. MBQ bars, is made from unconditioned billets. The bars may have liberal tolerances controls. The surface and core defects are wide and not well quantified. The bar manufacture may involve mild bending, hot forming, punching and welding. The quality norms in terms of internal porosity, surface seams are also liberal.

Special Bar Quality is a term used in long product industry, where an assured stress level in the application can be met by the steel [1,2]. They are long steel products manufactured to meet tough applications. Stress tolerance of the steel is enhanced which provide higher level of consistency and integrity. To achieve the desired level of performance chemical composition and cleanliness of the steel are critical to achieve the desired mechanical properties. While the chemistry achieved is as per the alloy design principles, the achievement of cleanliness levels is critical to achieve the special bar quality and they have a direct impact on the dynamic properties such as fatigue life of a component. The SBQ steel bars, are produced using processes that enhance quality parameters required for end use and customer specifications. The quality in the bars produced may be required to meet with the lowest levels of chemical segregation, inclusion rating, internal defects and surface defects.

Heterogeneity in a bulk steel is inherent in the manufacturing processes of the steel and SBQ has minimal heterogeneity acceptable to a customer. The heterogeneities in a steel include non-metallic inclusions [3,4], macro-segregation [5], micro segregation [6], micro-void or discontinuities that was generated in the casting [7] and that did not heal during subsequent hot deformation processes [8,9]. There is a quantified and assured acceptance level of the heterogeneity in the steel in terms of the quality of the steel bars. In terms of application, the SBQ steel performs in a highly loaded condition due to lower level of defect initiation sites. Hence, several safety critical high warranty prone components, where strength, fatigue life and durability requirements are there, fall in this category. The components are subjected to rotation, twisting and bending type of load application calls for SBQ steels.

IOP Conf. Series: Materials Science and Engineering **314** (2018) 012016 doi:10.1088/1757-899X/314/1/012016

SBQ steels are at the higher end of the quality spectrum in terms of both metallurgical consistency and dimensional accuracy. Some end uses of these steels may require superior surface quality, or special chemical restrictions, metallurgical characteristics, heat treatment or surface finishes. The internal quality of these steels is very important since machining and bending activities are involved during the processing of these steels in subsequent stages. In order to achieve the desired metallurgical quality, the steel has to be particularly clean. This makes the production route of these steels, a bit complex. Steel with high cleanliness is, achieved through the integrated steel making route or through an additional secondary steel refining process. The inputs have cleaner iron bearing inputs into an EAF in a mini mill. Internal metallurgical quality and consistency is, achieved by a meaningful reduction during the rolling process. These steels are, produced in fixed cut lengths or as wire rod coils.

2. Market Segments for SBQ:

The market segments for SBQ involves carbon steel, microalloyed steel, Low alloy heat treated steels, tool and die steels, high alloy steels, free cutting steels, case hardening steels, and stainless steels. These steels are functional grades, that is supplied, in different forms that include round bars, round corner squares, flats, round corner flats, hexagons, tubes, and special shapes. The products may also include hot rolled steel bars or cold finished steel bars. The SBQ steels have guaranteed with precise dimensions, metallurgical properties and chemistry. These steels have superior mechanical properties, defect free surface and internal quality. The products from the special bar quality steel line adapt to the customer's specific requirements, including those related to chemical composition, malleability, hardness, ductility and steel surface condition. SBQ steels are generally machined, forged or cold drawn during subsequent processing. The steel grades, is subjected to heat treatment such as direct hardening, carburizing, induction hardening, and/or nitriding. The SBQ bars are supplied as input raw material for closed die forging or open die forging, machining directly, cold working operations, wire drawing, extruded or further heat treatment etc.

The SBQ alloy steels meet critical levels of hardness and/or hardenability, strength, ductility, toughness, fatigue resistance, fracture resistance, wear resistance, machinability, and formability. Few grades are, applied to elevated temperature applications that demand high temperature strength, thermal fatigue, creep, oxidation etc. The market segments involved for use of SBQ steel for typical applications is, shown in Table1. The various types of components made from SBQ steels are, listed. The steels include carbon steel, alloy steel and stainless steels as well. The bars are used in transportation sector [automotive, railways, marine etc.]; energy sector [thermal power, hydro power, nuclear power, wind power etc.]; strategic defence, nuclear and aerospace sectors; Oil and Gas industry, tool and die industry, mining industry and Industrial Machinery etc.

3. Attributes Demanded in a Special Steel:

SBQ steels are required when certain input steel quality is, demanded for a fabrication process such as hot forging, cold forging, machining, heat treatment etc. Use of merchant bar quality may lead to failure to achieve target properties in application or excessive defect in the processing that may lead to component failure. The process of manufacturing SBQ removes excessive pipe formation, chemical in homogeneity, surface defects and internal core defects. The bars are, inspected and conditioned at every stage of the production process. The frequency and the degree of surface defects is influenced by chemical composition, type of steel, and bar size. Resulphurized grades, certain low C killed steels, and boron (B) treated steels are most susceptible to surface defects. Some end uses or fabricating procedures can necessitate one or more extra requirements. These requirements include special hardenability [11], internal soundness, non-metallic inclusion rating, and surface condition and microstructure and properties described in the standards. The quality specification for bars to which only one of these special requirements is applied is the 'restrictive requirements' of the quality. The restrictive requirements can be a single special restriction or more. Multiple restrictive requirement SBQ bars are those to which two or more restrictive requirements are applied.

Table 1 Market segments of SBQ steels and components made from long bar products

Automotive LCV, MCV & HCV
Engine Components: Crankshaft, conrod, rocker arm shaft, cam shaft, piston pin, ring gear, monotherm Piston, Cam followers, Cam rollers, Valve rocker arms, Valve rocker arm shafts, Valve springs, Push rods, Diesel injection systems components, Gudgeon pins, timing gears, fuel injection pumps forging for diesel engines, bearings Transmission: Mission gear, differential gear, output shaft, shift fork, CVT Pulley, forged parts going into drive shafts, Gear Counter shaft, gear boxes Coupling sleeves Carriers Gear shift shafts Shafts Planetary gears Sun gears Ring gears, Differential gears Crown wheels Planetary gears Sun gears Ring gears, Gears Levers, Constant velocity joint, Transmission shafts, Joint Yoke, bearings Chassis Components: Front axle, steering knuckle, Rack for Pinion gear, shock absorber spring, steering, suspension and hydraulic components, Ball joint, blades, eyes for hydraulic cylinders, Gears, Starter rings , stub axle, yokes, rings, rear axles, bearings, seamless tubes, fastners, steering suspension, hydraulic component Wheel Assembly: Wheel Hub unit, rear axles Couplings, Carriers Drive shafts, Flanges for universal driving shaft, Pinions Thrust washers Break arms, Wheel nuts and bolts; cable wires, tyre beads
Railways
Axles, rails, wheel forgings, fastners, piston rods, flange, housing pieces, wagon parts like screw couplings and other for wagons, cushioning device components, pin and steel liners, traction gears; wheels, axles, brake disks, drive unit mechanisms, and joints, for high speed rail transport, fastners, crankshafts, Portal axles for rails, connecting rod for diesel engine, claw lock for rails.
Marine / Ship Building
Propeller Shafts, Rudder Stock, Intermediate Shaft, Pinion Shaft, Forged Plates, Forged Rounds and others. Crank shaft, connecting rod, Marine engine Bolts , crank shaft, connecting rod
BioMedical
Surgical stainless steel [316, 440, 420]
Mining Equipment
Shafts, crusher forgings, bearing, drill rigs to drill bits, Rock drilling equipment, Crusher Shafts, Excavator Shaft, Transmission Shafts, Grinding Mill Rolls, Pinions, front spindle, etc.
Agriculture
parts for loaders, farm equipment, land diggers, ploughs, gears, shafts, levers and spindles to tie-rod ends, spike harrow teeth and cultivator shafts.
Steel Mill Components
Rolls & Roll Blanks used in rolling mills like, Table Rolls, Coiling Rolls, Leveller Rolls, Descaler Rolls, Work Rolls, Backup Rolls, chain steels, hooks
Oil & Gas
Mud pump forgings, Body Block Forgings, Cross Forgings, Forged Valves, X - mas tree Forgings, Well Testing Forgings. oil string Surface: Gate valve Body surface, casing head surface, shell surface, blow out preventor, Annular BOP for surface and sub-surface drilling Sub Sea : Spool body, sub sea Christmas tree, connector for sub sea drilling, Mandrel for sub sea Christmas tree, Shale Gas: crankshaft for fraction pump, fluid end fraction pump, machined fluid end
Thermal Power Plant
High temperature fastner, valve forgings, turbine forgings, valves, pipes and fittings, hydraulic shafts, flanges, valve bodies and stems, tees, elbow reducers, saddles, rock cutter bits, drilling hardware, and high-pressure valves and fittings, shafts, Rotor, Turbine Shafts and others. valves, pipes and fittings, hydraulic shafts, steam turbine rotors [High Pressure, Intermediate pressure], steam turbine excitor shaft, heat exchanger tube sheet
Hydro turbine
Generator Shafts, Rotor, Turbine Shafts, fastners

IOP Conf. Series: Materials Science and Engineering **314** (2018) 012016 doi:10.1088/1757-899X/314/1/012016

.....Contd. Table 1 Market segments of SBQ steels and components made from long bar products

Wind turbine
Components for Windmills, Windmill Main Shafts, wind turbine components, bearings and seamless tubes, transmission gear blanks, Transmission rings, crane wheel drives, gear box hollow shafts
Chemical & Petrochemical
Pressure vessel for crude refining, Shafts (Round and Square) in Fertiliser Plants, Refinery, Petrochemical, Cement, Sugar, Chemicals, valves and pipeline couplings for oil extraction
Defence
Barrels, shafts, shell forgings, gears, bearing, Tie rod, missile components, rifle triggers to nuclear submarine drive shafts. Heavy tanks, missiles, armoured personnel carriers, shells and other heavy artillery
Nuclear
Nuclear flask component, heavy forgings for reactor pressure vessels, steam turbines and generators, but extends to other engineered component, turbine rotor, generator shaft, retainer rings, reactor pressure vessel, nozzle, shells, shell flanges, Forgings for both Nuclear Island and Turbine Island
Aerospace
Landing gears, Missile parts, forgings are widely used in commercial jets, helicopters, piston-engine planes, military aircraft and spacecraft, bulkheads, wing roots and spars, hinges, engine mounts, brackets, beams, shafts, landing gear cylinders and struts, wheels, brake carriers and discs and arresting hooks, shafts, couplings, manifolds, wheel lever forging, fan forging, landing gear main leg forging
Industrial
Hook, Chains, Mooring chains for offshore platform, Input/output shafts, Pump Bodies, Wheel hubs and spindles, Ring gears and pinions, Valve bodies, Balls, Tees, Seats, Hammer union nuts, Check valves, crane wheels, and machine parts. , pins, studs, bolts, gears, shafts, axles, bolts, studs, crane wheels, and machine parts.
General Engineering
engine and transmission parts, variety of gears, sprockets, levers, shafts, spindles, ball joints, wheel hubs, rollers, yokes, axle beams, bearing holders and links. blanks, blocks, connecting rods, cylinders, discs, elbows, rings, T's, shafts and sleeves.
Tool & Die steel
Hand tools and hardware: Pliers, hammers, sledges, wrenches and garden tools, as well as wire-rope clips, sockets, hooks, turnbuckles and eye bolts, Dies for plastic moulds, die casting, forging tools and dies, hammers, screwdrivers and wrenches

SBQ quality steel bars can be produced using rimmed, capped, semi-killed, or fully killed deoxidation practice. The appropriate type is dependent on chemical composition, quality, and customer specifications. In addition, specification may demand quantified austenitic grain size, microstructure and mechanical properties and dimension. The tensile strength ranges in SBQ is similar to those applicable for the MBQ steels, while there could be enhanced fracture resistance properties such as impact or fracture toughness. There no specified size ranges established for SBQ steels. The bars can be squares, rounds, ovals, half-ovals, half-rounds, octagons, and other special special shapes such as beam blanks. A customer needed to give specific quality parameters for his application in unambiguous numerical value of acceptance depending on his end use application. Some of the typical parameters sought for SBQ steel quality is, given in Table 2. The Table brings out the quality parameters, typical equipment needed, critical parameters that need to be controlled and the requirements to qualify the steel under SBQ category. The SBQ steels have to pass through narrow chemistry bands and higher levels of customer specification. More than cost performance is the basis a customer evaluates a SBQ producer. Grade wise even simple carbon steels are enabled to perform at higher performance levels [life or load enhancement], when it is subjected to SBQ processing.

IOP Conf. Series: Materials Science and Engineering **314** (2018) 012016 doi:10.1088/1757-899X/314/1/012016

Table 2 Typical quality and control parameters demanded in a SBQ steel.

Quality Parameters	Equipment/ Processing	Critical Parameters	Requirements
Compositional	Ladle furnace	Steel making	Meeting very narrow band of chemistry
Size range	Ingot, Caster & Rolling Mill	Roll pass; Kocks or PSM mill	Strict dimensional control
Tolerances on dimension	Rolling Mill	Roll pass	European standards, for round bars EN 10060. Better than DIN/3
Manufacturing lengths	Rolling Mill	Decide by the Rolling mill	6,000 mm (+200 mm) 10% of bar below specs
Roundness / squareness	Rolling Mill	Roll pass	Out of roundness = [max dia- min dia]. Desired at most 2/3 of the diameter tolerance.
Straightness	Rolling Mill & Bar straightening machines	Rolling Mill; Cooling condition post hot rolling	Straightness is measured as the maximum height of arch, i.e. the largest deviation from the straight line. Normally the test length is 1.0 metre. Standard deviation of straightness is 2 mm/meter maximum.
Surface quality	Hot eye during rolling; Magna flux; MPI; Dye penetrant at times; ECT for peeled bar	Caster; Rolling; Cooling bed	Standard surface crack depth is max. 1% of dia.[mutually agreed]. Bars are inspected by magna flux leak detection or magnetic particle method. Normally, surface quality class D can be achieved for diameters up to Ø 80 mm and C for diameters up to Ø 120 mm complying with EN 10221.
End executions	Hot saw ; Hot shear; Chamfer machine	Steel hardness	Smaller sizes rolled products up to Ø 75-90 mm are cold sheared. Larger sizes are hot sawn or abrasive cut. Deburred at end
Decarburisation& excessive scaling	Reheating furnace; Rolling Mill	Temperature, time, Air:Fuel mixture; coating	Good surface quality and Yield losses to be reduced
Banding	Steel Grade; EMS, super heat; Caster; Rolling Mill	Columnar to equiaxed zone in concast; Degree of deformation;	Banding in alloy steels is minimized. The hardness difference between adjacent layers <5BHN
Non-Metallic Inclusion Rating	Ladle furnace; Tundish; Caster; Rolling Mill	LF melting; Casting temperature, super heat, mould powder.	Devoid of oxide inclusion rating Type B and Type C Thin and Thick series to be <0.5 ; sulphide as grade demands; Silicates Nil values
Macro inclusion	Ladle furnace; Casting	Exogeneous inclusion	Demand on extreme value statistics; Flaw size determination on Immersion UT testing at high frequency
Microstructure	Steel Grade; Caster; rolling; heat treatment	Processing& Heat treatment [Grain size; phases]	ASTM Grain size No.8 and higher to improve strength, toughness & FATT
Hardness	Steel grade; rolling; heat treatment	Microstructure	Depends on fabrication input. Mostly around 260BHN for eased of machining
Heat treatment supply condition	Heat treatment furnace	Microstructure	Required heat treatment such as annealing, normalizing or hardening & tempering

Contd..... Table 2 Typical quality and control parameters demanded in a SBQ steel.

Quality Parameters	Equipment/ Processing	Critical Parameters	Requirements
Hardenability [Jominy specs]	Steel chemistry; Hot Rolling	Chemistry; grain size	As per application it is specified
Mechanical Properties	Rolling/ heat treatment	Phase constitution Microstructure, grain size & inclusion	Strength, toughness, ductility at room temperature or high temperature, Impact toughness, FATT, Fracture toughness, Fatigue properties, creep properties etc.
Internal Soundness	Caster; Hot rolling Mill	Casting & rolling parameters	Internal defect size measured by Ultra sonic testing as per 2 to 8 mm FBH based on product thickness.

4. Processing of SBQ

In the manufacture of SBQ, one has to assess the type of infrastructure a plant has towards manufacture of the same. It is necessary to have infrastructure that has quality control at every stage of its operation. The infrastructure should address to some of the enhanced quality requirement demanded by the steel. JSW Salem works is a leading SBQ steel producer in India. The steel plant should have a means to produce steels with lowest levels of impurities within a narrow chemistry range. The infrastructure the plant has towards making SBQ has to be, examined by a customer. The typical processing route of JSW Salem Works is, given in Fig.1. The steel in the plant is made through, the blast furnace route with controlled input charge material, with quality control at every unit process with adequate equipment. The steel in JSW is made from the blast furnace – EOF route with least addition of external scrap, which is a source of certain defect causing tramp impurities such as Cu, Sn, As. Excess residual Cu can lead to surface defects that defeats the purpose of SBQ [12]. The steel made from blast furnace route is devoid of harmful residual elements unlike a recycled EAF steel.

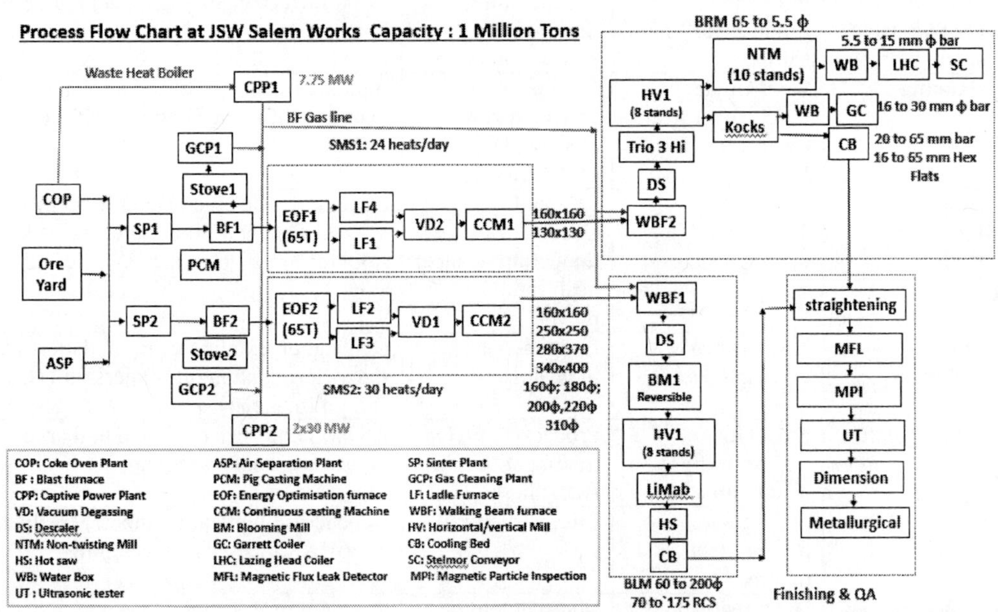

Fig.1 Process Flow chart of JSW Salem Works

4.1 Steel Making

The blast furnace hot metal pig iron is, extracted with the desired level of Si content [0.5 to 1.2%] and carbon content [~ 4.3%]. Lower levels of P, S and Ti, is desired for SBQ steels. The exothermic oxidation of the carbon and Si with oxygen in an EOF produces heat sufficient for steel making. The oxidation of all solute elements in hot metal, is carried out in the EOF. The elements with higher oxide stability [lower standard free energy of oxide] preferentially get oxidised first. While all the alloying elements are oxidized, the iron purity increases to >99.5% pure iron with dissolved oxygen. It is, desired to ensure that the tap carbon level is controlled, to achieve the lowest possible equilibrium oxygen in the bath. It is desirable to maintain EOF melt oxygen<600 ppm for SBQ type clean steel. The low phosphorous levels required by the SBQ steels is, generated by promoting P mass transfer from molten steel to slag at this stage. Some of the stringent mechanical property requirement in power plant, aerospace and defence sectors, demand extremely low P level as low P content enhances the fracture toughness and DBTT of the of the heat treated alloy steel [13]. The levels may be as low as 0.007% in power plant rotor steel forgings. The slag basicity and FeO content is controlled, to ensure dephosphorisation at the primary melting stages such as EOF furnace [14].The FeO content in the slag has to be maintained so that the equilibrium oxygen in the pure iron tapped does not increase. If the FeO content in slag is large then, the equilibrium oxygen in steel bath is higher during tap. This promotes more number of inclusions in the steel at the secondary steel making stage. Low inclusion levels are desired by SBQ steels in ladle furnace process ensure management of easily reducible oxides (FeO+MnO) towards lowest levels.

The steel from EOF is, transferred to a ladle with deoxidisers to bring down the initial dissolved oxygen in steel. Here, there are several strategies to control the inclusions. Addition of carbon along with the tapping stream, prevent ingress of air to molten steel bath due to carbon oxidation. In some of the experimental heats made by the author, observed, a high Al recovery along with high carbon pick up. This implied that the carbon is able to go into solution in molten steel and contributes to the lowering of oxide inclusions in the bath. The residual oxygen in the tapped steel is killed with a combined Si-Mn deoxidation or with large amounts of Al additions in the ladle. This initial deoxidation during tapping is called as blocking the steel heat, which brings down the oxygen level in molten steel, low enough, so that alloying elements added subsequently are fully recovered without oxidation. The initial Al killing gives lowest oxygen levels during blocking [15]. Alloying using quality controlled ferroalloys is carried out in ladle furnace, depending on the steel grade. In some practices, initial slag skimming followed by synthetic slag addition is carried out [16, 17]. The slag in LF has to have adequate lime content >50% to ensure a white slag practice, the reducing condition along with good basicity for refining reaction. Vacuum Degassing with a pressure <1mbar is carried out, where excellent degassing is promoted by intense gas stirring under reduced pressure. Nitrogen levels fall down along with hydrogen to low values and in SBQ steel [18]. The FeO content of the slag falls to very low levels <0.5% [reducing condition], which along with high slag basicity, promote excellent desulphurisation. Some of the steels, where transverse toughness properties are critical sulphide inclusions have to be reduced and hence very low levels of S <0.005% is targeted [19]. Desulphurisation takes place intensely at vacuum degassing stage where again the FeO content is monitored to the lowest level possible to achieve low dissolve oxygen content in the bath [20]. In the case of resulphurised steel, sulphur wire is subsequently added to meet the required sulphur levels[21].Usually, deoxidant addition, especially Al is carried out within vacuum as this is the element most reactive and may fluctuate in Ladle furnace. After Vacuum degassing treatment, a period of soft rinsing, at low Ar flow rates has to be carried out. Here, the flow rate of Ar has to be at low rates, which ensure formation of large plume of fine bubbles. Each of this bubbles as they ascend through the melt provide heterogeneous interfaces for inclusions to interact with each other and float to melt surface. Often, stirring efficiency is a parameter optimised during alloying and soft rinsing [22]. It must be ensured, that the duration of soft rinsing does not result in temperature fall that warrants re-arcing. Appropriate initial ladle temperature should be maintained to, account for this loss in temperature.

If there is a requirement of Ca treatment, the LF processing window in adjusted with suitable slag tailoring and Ca addition maintaining a Ca/Al ratio suitable enough to promote $CaO.Al_2O_3$inclusion which is liquid at the temperature of steel making [23]. If the required liquid phase is not achieved high temperature calcium aluminate inclusions populate the steel bath as inclusions. Usually, for high quality steel, after VD arcing is not permitted as, it exposes the melt to atmospheric oxygen during turbulent arcing. After VD the steel is very clean with least amount of gaseous elements. Sometimes, deoxidant addition such as Al content is specified and this has to be made up preferably in the VD stage only. The residual Al is important depending on the grade as this combines with N to form fine AlN inclusions that ensures fine grain sizes desired in an SBQ [24] . Fine grain sizes enhance the strength and toughness and optimum level is required.

4.2 Casting of Steel

SBQ involves both ingot and continuous casting processes and the later process predominates due to higher yield, productivity and quality in certain areas. Ingot casting is preferred where the product requires highly recrystallized grain size that warrants high reduction ratio.

Once the liquid steel is made, it is important to maintain its quality during subsequent processing. The tundish has to have a good refractory that is stable and least reactive to the melt chemical and physical interaction and should have flow behaviour where macro inclusions float out [25]. The refractory with high MgO content is preferred. It is subjected to pre-determined prolonged heating to ensures that the lining is free of moisture. The molten bath, when transferred to tundish may react with moisture to introduce hydrogen and oxygen [26]. Hydrogen measurement at this stage is a critical parameter. Appropriate post processing treatments such as stack cooling or anti flaking treatments depend on the hydrogen levels at this stage [27]. Also the nitrogen pick-up at this stage is monitored as there is a correlation between N pick up and oxygen pick up from atmosphere [28]. Casting is carried out with proper tundish flux through an SEN. Free opening of the ladle is desired and the SEN is properly shrouded to prevent ingress of air to oxidise the bath. The SEN with upward angle nozzle is preferred as the flow is smoothened and the flow pattern enables flotation of inclusion towards the top mould surface layers, where the mould powders have the ability to absorb inclusion [29]. The mould powder choice is critical to the grade. For example, the peritectic grade demand a different condition of solidification compared to a high carbon steel grade [30,31]. The wrong choice of mould powder can lead to sub-surface seams that may escape the notice during surface grinding yet show up during hot rolling. The super heat needs to be optimised [32]. High super heat favours inclusion flotation during casting, but macro segregation and centre line porosity tends to enhance. Too low a super heat tends to clog the nozzle and prevent inclusion from floating. Appropriate parameters of casting speed, primary and secondary cooling are to be ensured [33, 34, 35]. The casting mould is subjected to proper oscillation frequency to ensure smooth stripping of the billet and improper choice of condition may lead to severe ripples in the steels in surface, especially at the corners. It may also lead to transverse crack and hooke defect [36]. Another important phenomenon that will ensure proper solidification with freedom from surface and sub-surface inclusions is, obtained by the electromagnetic stirring in the mould or strand. The stirring ensures better heat transfer that gives better solidification macro structure and the lighter inclusions are, pushed to the core due to the centripetal force acting on the particles [37, 38]. Different grades have different characteristics in secondary cooling. In-appropriate choice of this parameter again results in seam cracks. The temperature during unbending in certain grades leads to embrittlement associated with AlN precipitation [39]. Automatic mould level controller is another control parameter that ensure proper turbulence free meniscus solidification [40] that results in superior surface quality with mould powder smoothly filled between the solidifying steel shell and water cooled mould.

The bloom or billet solidification structure is usually heterogeneous with a chill layer, columnar zone and core equiaxed zone [41] . The columnar zone is minimised and efforts to maximize equiaxed grains. The deformation of columnar zone product results in banded microstructure with anisotropic mechanical properties especially transverse toughness and ductility. Efforts to maximize the equiaxed zone is maximised by proper choice of superheat, casting speed, primary and secondary cooling in a continuous casting process. The structural banding is an important aspect of alloy steel and it is seldom possible to completely eliminate the banded microstructure caused by the interdendritic region rich in solute and dendrite core devoid of alloying elements [42, 43]. The acceptable level of banding is specified to be about 5 BHN difference in

IOP Conf. Series: Materials Science and Engineering **314** (2018) 012016 doi:10.1088/1757-899X/314/1/012016

some of the standards. Depending on the grade, one may encounter core central porosity or macro segregation. Efforts to minimize the same by optimising super heat of the grade being cast, casting speed, and EMS current.

Secondary cooling characteristics govern the initiation of surface cracks and one needs to optimise the parameters of the same for a given grade. There could be formation of embrittling phases during unbending in continuous casting that can initiate surface flaws [44]. Although much lower in volume, special steels produced by Electro-slag remelting, vacuum arc remelting etc. Are, classified under SBQ type steels. These steels may be demanded for use in high end applications in nuclear, defence and aerospace.

4.3 Hot Rolling of Steel

The cast blooms or billets made are hot scarfed, in some plant, while they are subject to cold grinding before being hot rolled [45]. In some plants hot charging is carried out which, enables conservation of energy apart from preventing phase transformation during complete cooling that involves volume changes and associated stresses in the bloom or billet [46, 47]. The blooms have to be heated appropriately before hot deformation in reheating furnace as a temperature gradient will result in non-uniform flow strength and inhomogeneous deformation. The thermal cycling is such that the core achieves the required temperature [48] . During the bloom re-heating, apart from choice of heating parameters, oxidation of the bloom and the surface decarburisation are important aspects that are crucial to SBQ product [49]. Scales generated in the reheating operation reduces the steel yield, which needs to be curtailed, and in some grades can be potential sites for crack initiation. The choice of the burning fuel and fuel:air ratio are factor optimised in a furnace. The soaked bloom is, subjected to initial deformation in a blooming mill where the cast structure is heavily broken down. The deformation at this stage is large enough to convert the cast dendritic microstructure to recrystallized grain structure. Phenomenon such as static, dynamic and metadynamic recrystallization refines the grain structure [50]. The reduction ratio is a monitored measure of the extent of deformation that has enabled the cast product to become a wrought product. The fine grain structure obtained has to be retained especially after the finish rolling. The deformation strain, pass schedule and rpm are critical parameters apart from the surface and core quality of the input bloom. Post rolling, the blooms are cooled in cooling table or slow cooling boxes or pits or vermiculite, to ensure achievement of suitable hardness and also diffusing out hydrogen in steel in some grades. The final hardness achievement in the bars, suitable for machining, may be specified by some of the customer, where slow cooling needs to get a softer product. Certain high alloy steel prone for hydrogen flaking needs to be given a proper anti-flaking heat treatment below A_1 temperature [51]. Hydrogen level in excess of 1.2 ppm in high hardenability steel such as C-Mn, Cr-Mo and Ni-Cr-Mo steels are prone for hydrogen flaking problem.

4.4 Heat Treatment Testing & Quality Control:

Many of the SBQ steels should show favourable response to heat treatment. They have specified hardenability, fine grains, flawless surfaces etc. This has to be consistently maintained by the steel mill to ensure consistent heat treatment response and mechanical property requirements in production at customer end. Some of the SBQ steels are inclusion engineered, where inclusion modification is carried out using Ca, Te, Pb or re-sulfurised [52 to 54]. They form fine evenly distributed secondary phases, which can enhance machinability. SBQ may also have non-heat treated steel grades such as microalloyed steel [55]. These steels, post forging achieves high strength properties.

In terms of quality control, the steels have to conform to chemistry, inclusion rating, grain size, microstructure, macrostructure, surface flaw by Magnaflux or MPI, core defect by ultra-sonic testing, decarburisation, banding level and mechanical properties at the customer specified conditions. There are other physical attributes such as the dimension, multiple length, straightness, ovality in round product or rhombodity in a square product [56]. Steel mills, which make SBQ has the specialised infrastructure to ensure dimension and defect free surfaces. The machines include straightening machines, surface grinding machines for blooms, annealing, peeling units etc.

Often SBQ steels undergo direct machining or further processing such as closed die forging, extrusion operation or cold drawing operation in a spherodize annealed condition [57] or a heat treatment after fabrication or surface hardening heat treatment post fabrication. Quality steel mill understands the requirement of subsequent processing industry and tailors the SBQ steel bar quality to suit the end use applications so that the products can be realized with consistency and least failures.

SBQ steels are, subjected to annealing, normalising, hardening and tempering heat treatments. Minimising defect initiation sites in the steel making operation enable achievement of consistent properties. Properties such as transverse impact or fracture toughness are superior in SBQ steels. The embrittlement generating impurities are least in the steel. There is a good microstructural homogeneity due to control over parameters such as hardenability. In inclusion engineered steels, there is a consistency in terms of machinability. Surface heat treatment response of the SBQ steel bars show consistency in the hardness and depth of layers and grain size in processes such as, carburising, nitriding, induction hardening etc, [58-59]. The surface quality of the bars influence the flaws generated in heat treatment and these steels and that is one reason SBQ is preferred for heat treated products.

In spite of additional costs associated with additional infrastructural necessity in manufacture and testing, productivity losses associated with higher refining times, and yield losses due to higher discards, the customer prefers SBQ to ensure assured performance. The SBQ steel market is growing. Customers generally assesses, the capability of the product by its performance in final applications.

5 Conclusion:

A review of various processing conditions in a steel mill in the manufacture of SBQ steels has been brought out. Special bar quality requires assured mechanical and quality parameters which needs to be adhered to by the steel mills. Proper infrastructure and processing technological capability and scientific understanding is needed for the manufacture of SBQ. JSW is a leading producer of SBQ with excellent infrastructure for the manufacture and quality assurance of the products.

6. References:

[1] Timken steel http://www.timkensteel.com/what-we-make/high-performance-steel/sbq-steel-bars-and-billets#sthash.K0FnkHql.dpuf

[2] Newquist Don E, Watson Peter S., Rohr David B., Brunsdale Anne E.,,Crawford Carol T. and Nuzum Janet A., "Certain special quality hot-rolled and semi-finished Carbon and Alloy steel products from Brazil", Determination of the Commission in, Investigation No. 731-TA-572, USITC Publication 1992, July, I6-I8, 2537.

[3] Zhang Lifeng and Thomas Brian, "State of the art in the control of inclusions during steel ingot casting", Met Trans B., 2006, V.37B, Oct., p.733-800.

[4] Zhang Lifeng, and Thomas Brian, XXIV National Steel making symposium, Morelia, Mich, Mexico, 26-28, Nov.2003, p.138-183.

[5] Howe A A, "Segregation and phase distribution during solidification of carbon alloy and stainless steel", Catalogue No. :CD-NA-13302-EN-C, Pub by the Commission of the European Communities, Luxemborg.

[6] Won Young-Mok, and Thomas Brian, " Simple model of microsegregation during solidification of steel", Met Trans A, 2001, V.32A, July, p.1755.

[7] Pfengfei U, "Numerical modelling of porosity and macrosegregation in continuous casting of steel", Doctoral Thesis, 2013, University of Iowa, https://ir.uiowa.edu/etd/2482;ir.uiowa.edu/viewcontent.cgi?article=4610&context=etd

[8] Liu Yazheng, Jiang Bo, Liu Guanglei, Wang Zhilin, Zhang Chaolei, Zhao Zhigangand Sun Jianlin, "Void Closure and the Sensitivity Analysis of the Process Parameters during Forging of Large Steel Ingot",

8nd International Conference on Physical and Numerical Simulation of Materials Processing, ICPNS'16, Seattle Marriott Waterfront, Seattle, Washington, USA, 2016, October, 14-17.

[9] Yu, H., Tieu, A. K., Lu, C. &Godbole, A. R. (2014). Investigation of closure of internal cracks during rolling by FE model considering crack surface roughness. International Journal of Advanced Manufacturing Technology, 75 (9-12), 1633-1640.

[10] Kor G. J. W., and Glaws P. C., Ladle Refining and VacuumDegassing", Chapter 11, Steelmaking and Refining Volume, Copyright © 1998, The AISE Steel Foundation, Pittsburgh, PA. http://studylib.net/doc/8246142/chapter-11---ladle-refining-and-vacuum-degassing

[11] DoyuglasV. Doane, " Application of hardenability concepts in heat treatment of steel ", J. Heat Treating, Vol. 1, No.1, p.5. http://www1.diccism.unipi.it/Valentini_Renzo/slides%20lezione%20met.%20meccanica/hardenability.pdf

[12] V. Leroy, R. D'Haeyer, J. Defourny, T. Hoogendoom,J. P. Birat, H. J. Grabke, W. B. Morrison, N. G. Henderson, R. D. Longbottom, T. Laux, I. Les "Effects of tramp elements in flatand long products", European Commission technical steel research, EUR Report No. CGNA16672ENC_0

[13] Cheruvu N S and Seth B B, " The influences of impurity content, tensile strength, and grain size on in-service temper embrittlement of CrMoV steels", Met Trans., A, Nov, 1989, V.20 no.11, pp.2345-2354

[14] Somnath Basu, "Studies on dephosphorisation during steel making", Doctorla Thesis, School of Industrial Engineering and Management, Dept of Material Sci. &Engg., Material Process Science, Royal Institute of Technology, SE-10044, Stockholm, Sweden, 2007. P.7-19 & 32-39.

[15] Socha Ladislav, Bazan J, Styrnal Pwetr, Pilka Vaclav, Piegza Zbygnev, Melecky Jan, Michalek Karel, Tkadleckova Marketa,"Evaluation of steel desulphurisation in the ladke during the utilisation of briquetting fluxing agent for slag" Procd of Conf Metal-2012,.5.2012 Brno, Czech Republic, EU, 2013, p. 23.-25.

[16] Gavanescu Adrian, Aannals of Faculty Engineering Hunedoara – International Journal of Engineering, IX, 2011, p.177. http://annals.fih.upt.ro/pdf-full/2011/ANNALS-2011-4-37.pdf

[17] Socha L, Michalek K., Bazan J., Gryc K., Machovc P. AK´, Opler A., Styrnal P., Archives of Metallurgy and Materials, Volume 59, 2014, Issue 2, p.809-813

[18] Turkdogan E T, "Fundamentals of Steel making", Pub by Institute of Materials, UK, 1996, p.277-284.

[19] Bul'ko B, Kijac J and Domovec M, "Optimization slag composition in ladle furnace Considering to effective steel desulfurisation, ActaMetalurgica Slovaca, 2009, 15, 2 , p.93-99

[20] Stephen Famurewa Mayowa, " Improvement of the desulphurization process by slag composition control in ladle furnace", Master Thesis, Lulea University of Technology, Dept of Chem Eng& Geo Sciences, Division of Process Met., 2009. P.9-14 & p.27-44

[21] Crawford G. P., " Wire injection or metallurgical powders into molten metal", INFACON 6, Proceedings of the 6th International Ferroalloy Congress, Cape Town, Vol.1. Johannesburg, SAIMM, 1992. pp, 271-277

[22] Shen Jiandong, Zhao Yong, Han Guijin, " Design on Optimization of Argon Bottom Blowing of Molten Steel ladle", Procedia Engineering, Volume 16, 2011, p. 284-290.

[23] Lind Minna, " Mechanism and kinetics of transformation of Alumina inclusions in steel by calcium treatment", Doctoral thesis, Helsinki University of Technology Publications in Materials Science and Engineering, 8., Sept., 2006. P. 1-22 and p. 58-60.

[24] ChengLeon M., Hawbolt E. Bruce, Ray T and CroftMeadow, " Dissolution and coarsening of aluminum nitride precipitates in low carbon steel — distribution, size and morphology", Canadian Metallurgical Quarterly, 2000, Vol 39, No 1, pp 73-86.

[25] SahaiYogeshwar and EmiToshihiko, "Tundish Technology for Clean Steel Production", World Scientific Publishing Co. Pte. Ltd., 2008, p.17-73.

[26] Dekkers, R., Blanpain, B., Plessers, J., and Wollants, P., "Steel cleanliness and hydrogen in liquid steel". VII International Conference on Molten Slags Fluxes and Salts, The South African Institute of Mining and Metallurgy, 2004, p.753

[27] VoronenkoB. I., "Hydrogen and flakes in steel", Metal Science and Heat Treatment, 1997, Vol. 39, Nos. 11 - 12, p.462-470.

[28] Doostmohammadi M, Andersson M, Karasev A and Jonsson P G, " Use of computational thermodynamic calculations in studying the slag/steel equilibrium during vacuum degassing" ,Steel Research Int., 2010, 81, V.1, p.31 – 39.

[29] Reala Cesar, Hoyos Luis, Cervantes Francisco, Mirandac Raul, Palomar Pardavéa Manuel, and Gonzalez Jesus, , "Sensitivity analysis of the three dimensional flow dynamics in the continuous casting submerged entry nozzle ", Mechanica Computational V.XXVI, 2007, p.1292-1310. www.amcaonline.org.ar

[30] Görnerup, M., Hayashi, M., Däcker, C-.Å., And Seetharaman, S," Mould fluxes in continuous casting of steel—characterization and performance tuning", VII International Conference on Molten Slags Fluxes and Salts, The South African Institute of Mining and Metallurgy, 2004, p.745.

[31] Majumdar B, " A process for developing continuous casting mould powder for steel industries" , Journal of Scientific and Industrial Research,1999, V.58, Oct., p.773-780.

[32] ZhangJian, ChenDengfu, WangShuigen, and LongMujun, "Compensation Control Model of Superheat and Cooling Water Temperature for Secondary Cooling of Continuous Casting", steel research int., 2011, 82, No. 3, p.213-221.

[33] FalkusJ, Miłkowska-Piszczek K, "Strategy of cooling parameters selection in the continuous casting of steel", Arch. Metall. Mater., 2016, Vol. 61, No.1, p.329-334.

[34] Štetina Josef, Kavicka František, Mauder Tomaš, Klimes Lubomír, Masarik Miloš, Šanab Zdeněk, Optimization of secondary cooling In order to achieve higher surface temperatures at the slab unbending point", Metal 2012, Brno, Czech Republic, EU , 5, 23- 25.

[35] SpuyD. deV. van der , Craig I K and Pistorius P C, "An optimization procedure for the secondary cooling zone of a continuous billet caster", The Journal of The South African Institute of Mining and Metallurgy JANUARY/FEBRUARY , 1999, p.49-54.

[36] CibulkaJ., KrzokR., HermannR., BocekD., Cupek J. and MichalekK., "Impact of oscillation parameters on surface quality of cast billets", Arch. Metall. Mater., 2016, Vol. 61, No 1, p.283–288.

[37] Stránský Karel, Kavička František, Sekanina Bohumil, Štětina Josef, Dobrovská Jana, Stransky, Lubomir, Brně V, and oceli Výroba, Hutnické listy č. 3/2009, p.26.

[38] ChaoXiao, Jiong-Ming Zhang, Yan-zhaoLuo, GdongWei Xiao, LianW U, Shun-XiWang, "Control of macrosegregation behavior by Applying Final Electromagnetic Stirring for Continuously Cast High Carbon Steel Billet", Journal of Iron and Steel Research, Iinternational, 2013, 20(11),p.13-20.

[39] DarsouniA, Bouzabata B and MontheilletF, "Hot Ductility of a Microalloyed Steel in the Intermediate Temperature Range", Journal De Physique IV Colloque C7, supplement au Journal de Physique 1995, 111, Volume 5, November C7, p.347.

[40] Smutný Lubomír, Farana Radim, Víteček Antonín, Kačmář Dalibor, " Mould level control for the continuous steel casting", https://pdfs.semanticscholar.org/24f7/2089ed3cd4f29dea6707f5d3ed1e83634628.pdf

[41] White C V, Krauss G and Matlock D K, " Solidification structure and the effect of hot reduction in continuously cast steel fro bars and forgings", https://www.forging.org/uploaded/content/media/143-white.pdf

[42] Grange R A, " Effect of microstructural banding in steel", Met Trans., 1971, V.2, p.471.

[43] Herring Daniel H, "Segregation and Banding in Carbon and Alloy Steel", http://www.industrialheating.com/ext/resources/Issues/October2013/Treat/ih1013-htdr-fig1-615.jpg

[44] Mišičko R., Embrittlement of continuously cast steel products and its microstructural effects , Acta Metallurgica Slovaca, 2008, 14, 1, p.50 – 58.

[45] Zhou Xiaoxu and ThomasBrian G., "Modeling Steel Slab Heat Transfer during Scarfing Processing",Continuous Casting Consortium Report to POSCO, Department of Mechanical Science and Engineering University of Illinois at Urbana-Champaign, 2010, p.1-8 & p.19-31.

[46] Abernethy K and CindereyR J, " Effect of hot charging on steel properties", European Commission Technical steel research, Report No. EUR 14943 EN , 1995, p.23-74.

[47] John Guerard, "Considerations for the hot charging of billets " , Proceedings of the International Symposium on direct rolling and hot charging of strand cast billets,Montreal, Canada Edited by J.J. Jonas, R.W. Pugh S.Yue, 1988, August 29-30, p.53

[48] Rahul S Nalawade, Prafull P P, . Balachandran G and Balasubramanian V, "Void Closure in a Large Cross Section Bars Hot Rolled from a Low Alloy Steel Ingot Casting", Trans Indian Inst. Met, 2016, V.69(9), p.831.

[49] Vander Voort George F., "Understanding and measuring decarburization", Advanced Material Processes, Feb, 2015, p.22.

[50] Bianchi J H and KarjalainenL P, "Modelling of dynamic and meta dynamic recrystallisationduring bar rolling of a medium carbon spring steel", Journal of Material Processing Technology, 2005, 160, p.267-277.

[51] Hodge J M, Orehoski M A, and Steiner J E, "Effect of Hydrogen content on susceptibility of flaking", Trans Met Soc., AIME, 1964, V.230, Aug., p.1182-1193

[52] Wilson Alexander D, " Clean Steel Technology – Fundamental to the development of high performing steels" , Advances in the production and use of steels with improved cleanliness,, ASTM STP 1361, J K Ed by Mahaney Jr , p. 73.

[53] MapelliC., Vedani M. and Zambon A., "On microstructure development and inclusion generation in a continuously cast resulphurised steel", Lat. Am. appl. res. v.32 n.3 Bahía Blanca jul./sept. 2002, p.9.

[54] Popova L. V., Lebedev D. V., and Nasibov A C, "Properties of steel 16G2AF with microadditions of tellurium", I. P. Bardin Central Scientific-Research Institute of Ferrous Metallurgy. Translated from Metallovedenie i Termicheskaya Obrabotka Metallov, 1983, No. 3, pp. 21-22.

[55] Van Tyne C J, Matlock D K and Speer J G, "Microalloyed forging steels", Colarado School of Mines, https://www.forging.org/uploaded/content/media/_376VanTyne_0_0.pdf

[56] JSW Brochure Down loads in http://www.jsw.in/steel/special-steel

[57] Nandita Gupta and Sen S. K.," Spheroidisation Treatment for Steels", Defence Science Journal, 2006, Vol. 56, No. 4, October, pp. 665-676

[58] Boyle I E, NorthwoodD.O., Bowers R, Sun X, Bauerle P, Materials Forum V.32 – 2008, Edited by J.M. Cairney, S.P. Ringer and R. Wuhrer Pub. by © Institute of Materials Engineering Australasia Ltd., p.44

[59] Otto Fredrick J and Herring Daniell H, "Gear Heat Treatment", Heat Treating Process, July/Aug, 2002, p.1-5.

Acknowledgement

The author wish to thank Mr D Ravichandar, CEO, JSW Steel Ltd., Salem Works, for the encouragement and permission granted to publish the paper.

IOP Conf. Series: Materials Science and Engineering **314** (2018) 012017 doi:10.1088/1757-899X/314/1/012017

Numerical and experimental investigation of bi-annulus heat exchanger for different alternative materials

M Zeeshan[1,2], R Duggal[1], M K Tated[1], M Singh[1]

School of Mechanical Engineering, Lovely Professional University, Jalandhar, Punjab, India 144411.

Corresponding author: er.mdzeshan@gmail.com

Abstract. Heat exchangers are widely used in various energy-recovery applications. However, for specific applications where metallic tubes are subjected to various drawbacks i.e. cost, weight, corrosion etc. polymer materials are promising alternatives. In present study, various conventional as well as promising alternatives materials are chosen for investigation computationally. Experimentally, bi-annulus heat exchanger configuration is investigated for metallic materials. The simulations carried out conclude that the dimensionless temperature parameter for Cross-linked polypropylethylene (PEX) is greater than other polymers. It increases with increasing axial length of tube. The value for dimensionless temperature is higher for copper which is used as conventional tube material. Among different polymers highest temperature is observed for PEX followed by Low density polypropylene (LDPE), Polypropylene (PP) and Polyvinylidene fluoride (PVDF). For axial length up to 70mm approx. the temperature rises for PEX, LDPE is 28.3% and 26.4% respectively. However, temperature variation is same for PP and PVDF for same axial distance. This temperature variation is increased to 72.4%, 67.2%, 58.62% and 56.89% for PEX, LDPE, PP and PVDF respectively as axial distance variation reaches the end of pipe. The inner annulus temperature for PEX material at 10% length of tube is 28.3% of temperature achieved in copper tube which increases to 72.4% for full length of tube.

1. Introduction

In almost every heat transfer or energy-recovery application, heat exchangers are prominent to overall efficiency, cost and size of system. Presently, these applications (air-conditioning, refrigeration etc.) are based on fin-tube, plate-fin heat exchanger designs constructed using copper, aluminium. There is vast literature present whether how to enhance the heat transfer. However, the operating temperature is a limitation for metallic heat exchangers. Especially for applications like heat recovery where solvent is corrosive and exotic alloys are required. Hence, polymers solve the purpose being a) flexible in manufacturing b) chemically stable c) reduced cost as well as weight. These find application in chemical industry as well as liquid to liquid, gas to gas along with liquid to gas application [1][2]. Various applications for liquid to liquid configuration have been explored and investigated [3][4] citing the advantages with respect to corrosion resistance and weight cost factor. Moreover, triple tube or bi-annulus heat exchanger basically suited for pasteurization, drying, sterilizing, cooling, freezing, refrigeration and evaporation. Theoretical analysis had been performed relating the performance of heat exchanger with size of triple pipe [5][6]. Effectiveness equation as well as mathematical formulation for heat transfer coefficient was also developed [7]. Correlations for partial coefficient was also developed [8]. The effectiveness as well as advantages of triple pipe heat exchanger is given

in review [9] as well as method of calculating heat transfer coefficient and mean temperature difference has been provided [10,11].
The application of polymers specific to triple pipe heat exchanger or bi-annulus heat transfer device is rarely discussed. This present paper deals with the material of heat exchangers particular to polymers. In order to investigate an experimental study is also carried out and numerically different parameters are varied to obtain temperature, pressure profiles. Ansys WorkBench[12] module is chosen to investigate the problem numerically. Different materials are chosen for thermal analysis namely Copper (Cu), Aluminium (Al), Steel, PP, PEX, Low density polypropylene LDPE, PVDF for numerical analysis.

2. Brief experimental details

The experimental setup consists of several components depending on the required functioning for the practical work. The components used in this experiment are Rotameter, DC water pump, Temperature Sensors, Gate valve. The temperature sensor used to calibrate the inlet and outlet temperatures of the heat exchanger in this experiment is LM35.It is a precision integrated circuit temperature device. These temperature sensors calibrate directly in Celsius with +10mv/°C scale factor and 0.5°C ensured accuracy at 25°C. The range of the temperature sensor varies between -55°C to 150°C. These sensors operates from 4v to 30v with current drain less than 60μA. LM35 temperature sensors are easily available at the markets at affordable cost and very suitable for remote applications. The setup consists of three pipes which are in concentric arrangement made up of copper. These pipes are namely center pipe, intermediate pipe and outer pipe. Single inlet is provided to both the center pipe and the outer pipe through with cold fluid flows with an equal mass flow rate in both the pipes. Hot fluid is made to flow in the intermediate pipe. Panel Mount Rotameter is placed before both the inlets to control the flow rate of the fluid. A ball valve is placed for the inlet of the outer pipe where this valve plays a major role in showing the dual mode compatibility of the heat exchanger. If this valve is closed it acts as a Double Pipe Heat Exchanger and when it is open it acts as a Bi-annulus Heat exchanger. the results one can depict that in comparison to conventional double pipe heat exchanger heat transfer per unit length, thermal efficiency and heat flux have improved. Heat transfer length has decreased from 1.369m to 0.447m in comparison of Double pipe heat exchanger to Bi-annulus heat exchanger. The flow observed in the pipes are found to be laminar for provided volumetric flow rates. Achievement of turbulent flow can enhance the heat transfer rate and can increase thermal efficiency. The results obtained from experimental result is presented in table 1 where U being overall heat transfer coefficient and L being length of heat exchanger. The outer average static temperature obtained from experimental result (49°C) as well as numerical simulation (48°C) is well in conformation.

Table 1. Experimental results.

Material	U (W/m^2K)	L (m)
Copper	197.28	0.44
GI (Stainless Steel)	193.41	0.45
GI (Mild Steel)	195.98	0.44

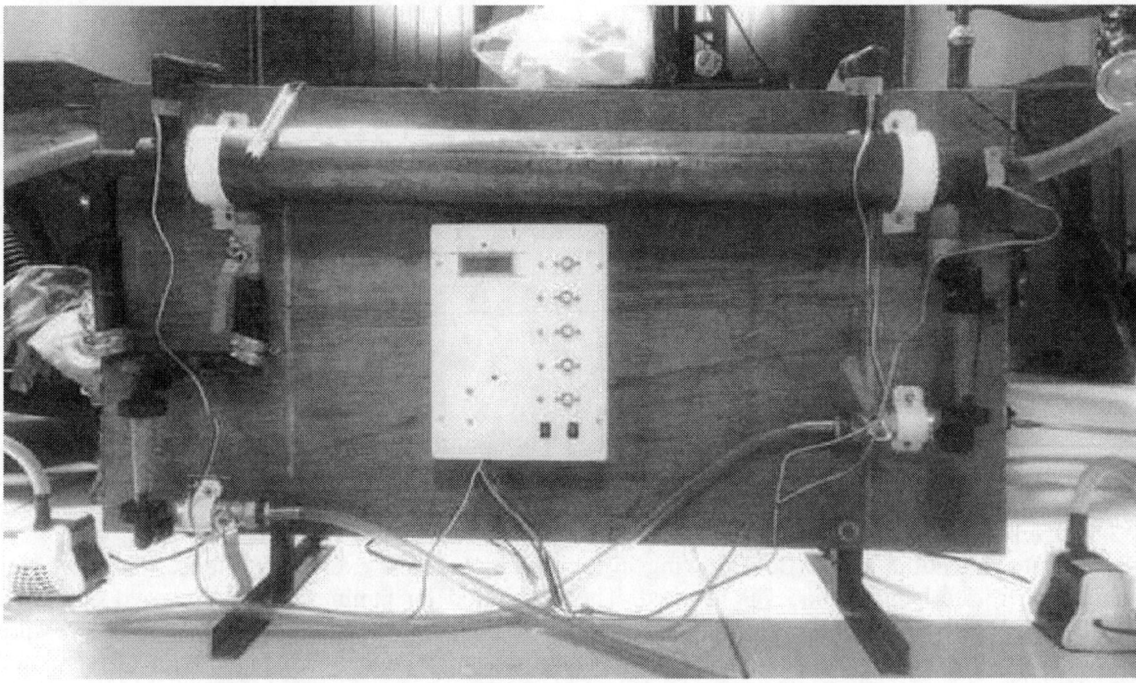

Figure 1. Experimental setup for bi-annulus heat exchanger.

3. Numerical Procedure

3.1. Mathematical model

The problem is formulated in two dimensions considering axis (700mm) as symmetric in order to reduce time and computational cost. The sketch is shown in figure 2. The problem is thermally analysed taking into consideration some of the assumptions as a) steady state operation b) no heat sink or heat source c) mass flow rate remains constant d) all the properties and parameters along with variables are considered constant. Since, the flow for all the three fluids is in same direction hence it is a co-current flow arrangement of heat exchanger. The governing equation for different fluid streams are

$$\left(\dot{m}C_p\right)_h \frac{dT_h}{dx} = (U_2P_2 + U_1P_1)(T_h - T_c) \quad (1)$$

$$\left(\dot{m}C_p\right)_c \frac{dT_c}{dx} = U_2P_2(T_h - T_c) + U_0P_0(T_\infty - T_c) \quad (2)$$

Further different non-dimensional parameters are used to plot temperature variation as follows:

$$\theta = \frac{T - T_{c,in}}{T_{h,in} - T_{c,in}} \quad (3)$$

$$X = \frac{x}{L_e} \quad (4)$$

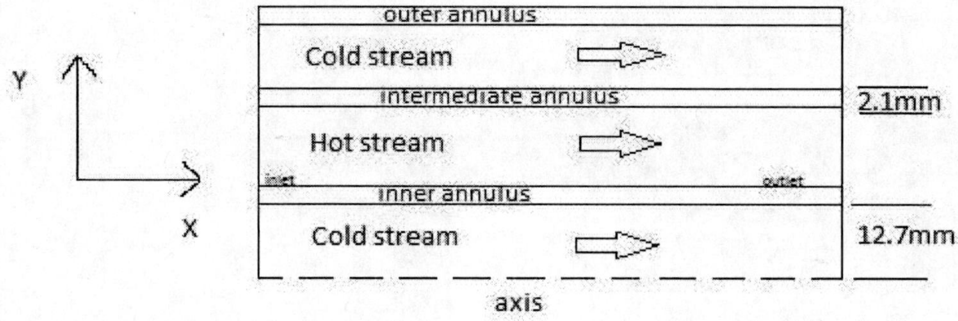

Figure 2. Two dimensional sketch of bi-annulus heat exchanger

3.2. Numerical scheme and boundary conditions

Since, water is chosen as the fluid flowing in the bi-annulus heat exchanger. Hence, the flow was laminar, therefore laminar model is invoked. The simulation for laminar flow are carried out using FLUENT 15 software package. Second order upwind scheme is used for discretization of pressure velocity coupling equations. For residuals of continuity and momentum equations, a convergence criterion of 10^{-3} was imposed whereas 10^{-6} as energy equation residuals. The convergence is determined by the residual levels and also by monitoring relevant integrated quantity like heat transfer coefficient. The solutions are obtained only after satisfying convergence criterion. The boundary conditions to which are inlet temperature of fluid for cold fluid the inlet temperature being 298K and for hot stream of fluid being 328K with 0.01388 kg/s and 0.0277 kg/s as flow rates. The outer wall of heat exchanger is considered to be insulated that is no heat is transferred from the outermost wall.

3.3. Mesh description

Two dimensional model is meshed with structured mesh. Since, structured mesh reduces errors considerably. Four different grid sizes are chosen in order to check the independency of grid, with the solution of numerical procedure. The grid chosen consist of 81,650 elements out of the numerically analyzed grid consisting elements ranging from 14,353 to 81,650. The grid independence study is given in table 2. Since, percentage deviation is less than or equal to unity for 81,650 element grid, it is chosen for further investigation. The grid size is further validated according to criteria present in literature [13]. The maximum skewness value for the generated mesh being 0.052 which should not exceed 0.85 as well as the orthogonal quality 0.99 which is close to unity providing the quality mesh for numerical procedure. The obtained mesh is shown in figure 3.

Table 2. Grid independence study.

Elements	Nusselt Number (Nu)	% deviation
14353.00	7.50	
26562.00	8.08	7.73
57750.00	8.40	3.95
81650.00	8.49	1.06

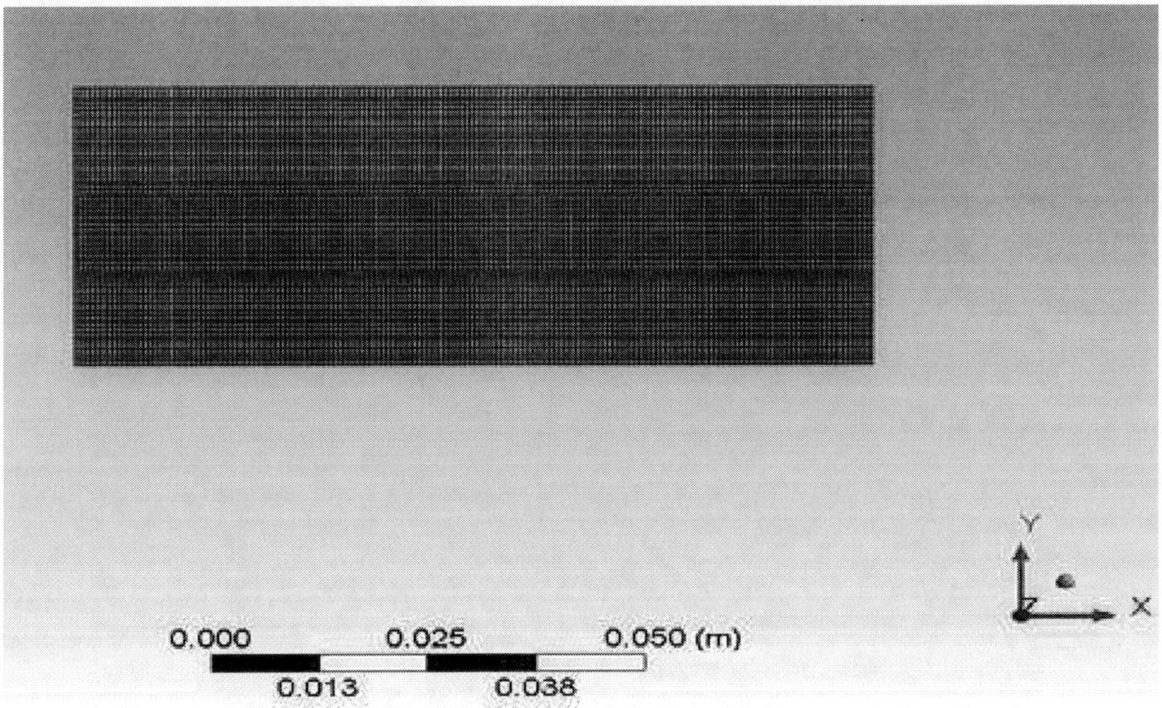

Figure 3. Two dimensional sketch of bi-annulus heat exchanger

4. Results

4.1. Isotherm

Bi-annulus heat exchanger temperature profile can be seen as isotherms considering different polymer materials as well as different metals that are conventionally used in heat exchanger application. It can be easily differentiated that as compared to metal counterpart's polymer pipes are at lower temperature. However, the difference in profile is almost negligible. Hence, it poses a promising material for heat exchanger application in chemical industry.

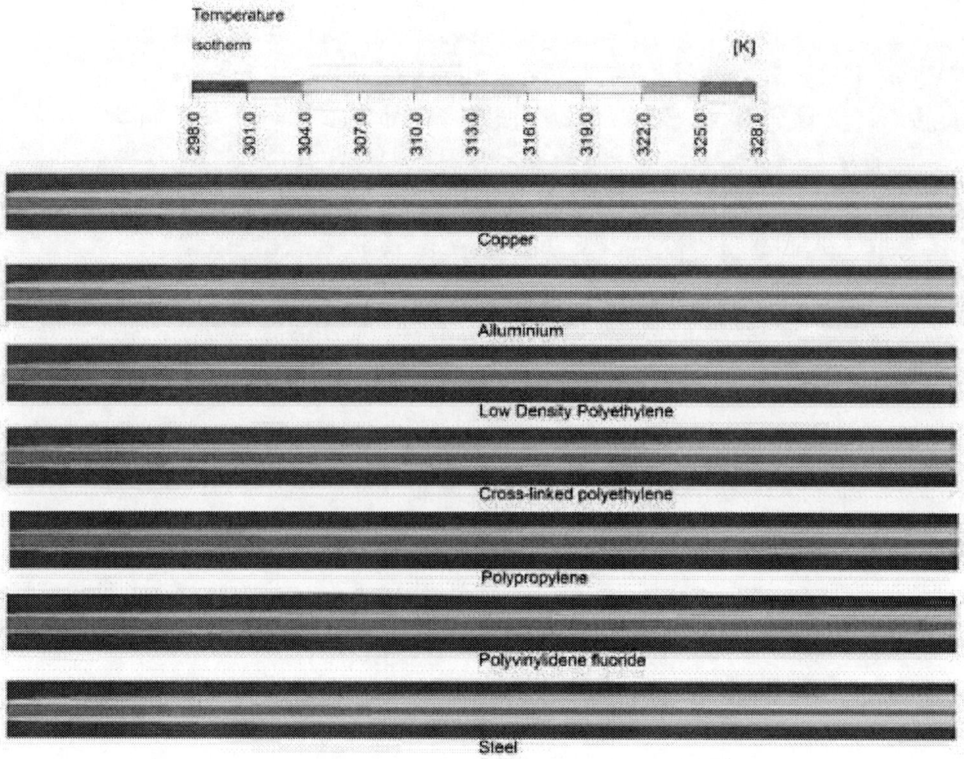

Figure 4. Isotherm contours for different material of bi-annulus heat exchanger pipes.

4.2. Heat Transfer

The temperature variation along the length of heat exchanger tube is shown in figure 5. The dimensionless temperature for PEX is greater than other polymers. It increases with increasing axial length of tube. The value for dimensionless temperature is higher for copper which is used as conventional tube material. Among different polymers highest temperature is observed for PEX followed by LDPE, PP and PVDF. For axial length up to 70mm approx. the temperature rises for PEX, LDPE is 28.3% and 26.4% respectively. However, temperature variation is same for PP and PVDF for same axial distance. This temperature variation is increased to 72.4%, 67.2%, 58.62% and 56.89% for PEX, LDPE, PP and PVDF respectively as axial distance variation reaches the end of pipe. From figure 6, it can be concluded that PEX having highest temperature rise among all polymers is compared with conventional copper material. The inner annulus temperature for PEX material at 10% length of tube is 28.3% of temperature achieved in copper tube which increases to 72.4% for full length of tube. Similarly, this variation for intermediate annulus of copper increases from 70.9% to 88.05% of PEX material. For outer annulus the variation is almost negligible for both copper as well as PEX polymer material. However, this is increased to 75% of copper temperature rise for full length of tube.

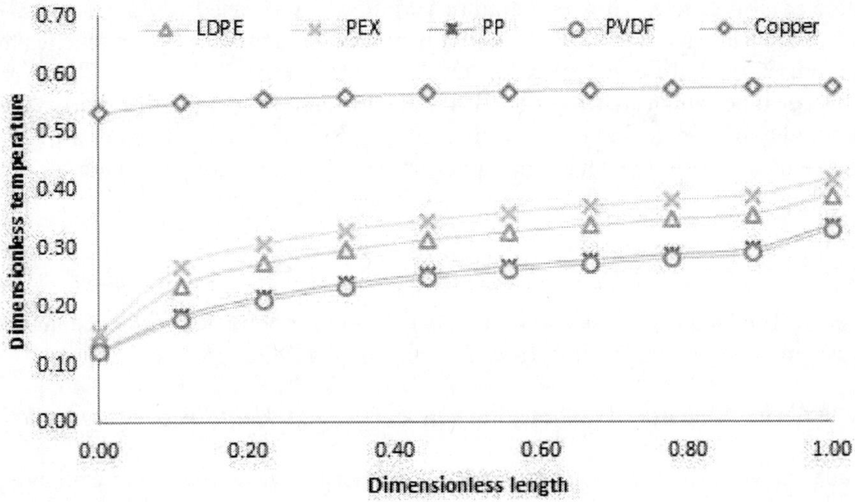

Figure 5. Dimensionless temperature variation along axial direction.

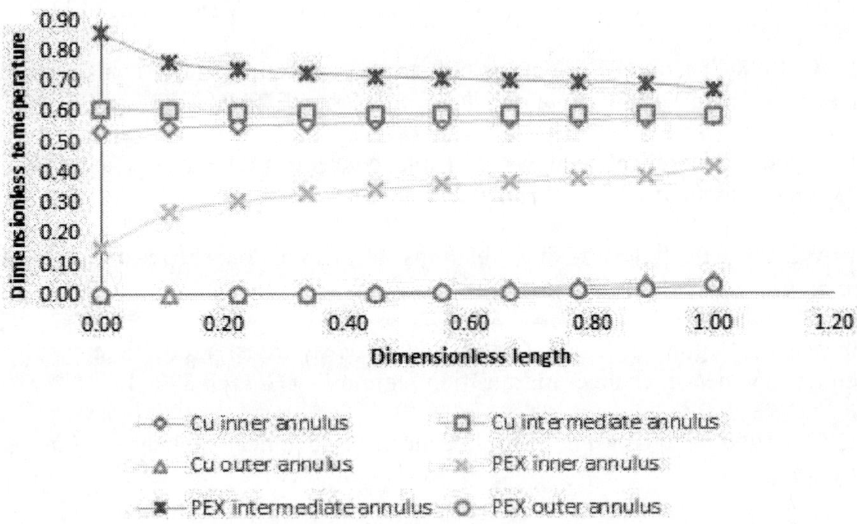

Figure 6. Comparison of different annulus walls for Copper and PEX.

5. Conclusions

Bi-annulus heat exchanger is investigated experimentally as well as numerically. The simulations are carried out for different polymer materials along with conventional tube materials mainly metal. It is concluded that the dimensionless temperature for PEX is greater than other polymers. It increases with increasing axial length of tube. The value for dimensionless temperature is higher for copper which is used as conventional tube material. Among different polymers highest temperature is observed for PEX followed by LDPE, PP and PVDF. For axial length up to 70mm approx. the temperature rises for

PEX, LDPE is 28.3% and 26.4% respectively. However, temperature variation is same for PP and PVDF for same axial distance. This temperature variation is increased to 72.4%, 67.2%, 58.62% and 56.89% for PEX, LDPE, PP and PVDF respectively as axial distance variation reaches the end of pipe. The inner annulus temperature for PEX material at 10% length of tube is 28.3% of temperature achieved in copper tube which increases to 72.4% for full length of tube. Similarly, this variation for intermediate annulus of copper increases from 70.9% to 88.05% of PEX material. For outer annulus the variation is almost negligible for both copper as well as PEX polymer material. However, this is increased to 75% of copper temperature rise for full length of tube.

6. References

[1] Zaheed L and Jachuck R J J 2004 Review of polymer compact heat exchangers, with special emphasis on a polymer film unit *Appl. Therm. Eng.* **24** 2323–58

[2] Reay D A 1989 The use of polymers in heat exchangers *Heat Recover. Syst. CHP* **9** 209–16

[3] Zarkadas D M, Li B and Sirkar K K 2005 Polymeric hollow fiber heat exchangers (PHFHEs): A new type of compact heat exchanger for lower temperature applications *ASME 2005 Summer Heat Transfer Conference collocated with the ASME 2005 Pacific Rim Technical Conference and Exhibition on Integration and Packaging of MEMS, NEMS, and Electronic Systems* (American Society of Mechanical Engineers) pp 429–38

[4] Liu W, Davidson J and Mantell S 2000 Thermal analysis of polymer heat exchangers for solar water heating: a case study *Trans. Soc. Mech. Eng. J. Sol. ENERGY Eng.* **122** 84–91

[5] Ünal A 1998 Theoretical analysis of triple concentric-tube heat exchangers Part 1: Mathematical modelling *Int. Commun. heat mass Transf.* **25** 949–58

[6] Ünal A 2001 Theoretical analysis of triple concentric-tube heat exchangers Part 2: Case studies *Int. Commun. heat mass Transf.* **28** 243–56

[7] Ünal A 2003 Effectiveness-NTU relations for triple concentric-tube heat exchanges *Int. Commun. heat mass Transf.* **30** 261–72

[8] Radulescu S, Negoita I L and Onutu I 2012 Heat transfer coefficient solver for a triple concentric-tube heat exchanger in transition regime *Rev Chim* **8** 820–4

[9] Zuritz C A 1990 On the design of triple concentric-tube heat exchangers *J. Food Process Eng.* **12** 113–30

[10] Batmaz E 2003 Overall heat transfer coefficients and axial temperature distribution of fluids in a triple tube heat exchanger

[11] Batmaz E and Sandeep K P 2008 Overall heat transfer coefficients and axial temperature distribution in a triple tube heat exchanger *J. Food Process Eng.* **31** 260–79

[12] ANSYS 2013a 2013 Ansys Fluent Help

[13] Bakker A 2013 Lecture 7 – Meshing *Applied Computational Fluid Dynamics*

IOP Conf. Series: Materials Science and Engineering **314** (2018) 012018 doi:10.1088/1757-899X/314/1/012018

An experimental approach to free vibration analysis of smart composite beam

G A Yashavantha Kumar[1] and K M Sathish Kumar[2]
[1]BMS Institute of Technology &Mgmt, Dept of Mechanical Engineering, Bengaluru-64 India
[2] BMS Institute of Technology &Mgmt, Dept of Mechanical Engineering, Bengaluru-64 India

Abstract. Experimental vibration analysis is a main concern of this study. In designing any structural component the important parameter that has to be considered is vibration. The present work involves the experimental investigation of free vibration analysis of a smart beam. Smart beam consists of glass/epoxy composite as a main substrate and two PZT patches. The PZT patches are glued above and below the main beam. By experimentation the natural frequencies and mode shapes are obtained for both with and without PZT patches of a beam. Finally through experimentation the response of the smart beam is recorded.

1.Introduction

Different types of composite material analysis are focused in literature review. By combining two or more materials a composite can be developed in order to take the advantage of desirable properties of the components. In present study a glass/epoxy composite is considered for its better strength and stiffness properties. For these composites more precisely we can control the strength and stiffness properties within the plane. Now a days the development of piezoelectric materials in the field of smart structures motivated many researchers. A structure which can sense the external disturbance and responds to it in a controlled manner is termed as smart structures. The operational efficiency of these structures due to mechanical vibrations is affected to a greater extent. It necessitates the knowledge of dynamic characteristics of a system in the field of demanding areas namely design of vibration control devices and structural health monitoring. In this paper an attempt has been made to determine the free vibrational characteristics of a smart beam experimentally.

2.Methodology

In the present work first a fixture is developed for holding the beam in a cantilever configuration. . In this study a Glass/Epoxy laminated composite beam is used. It consists of 3 layers of thickness 0.5mm in which fibers are oriented at an angle of 0^0 in all layers as shown in figure 1. The configuration of the beam considered in this study is fixed-free condition. The dimensions of the beam and the piezo patches are as shown in table 1.

Table 1. Dimensions of composite beam and PZT patches.

Particulars	'L' in mm	'B' in mm	'H' in mm
Composite beam	250	25	1.5
PZT Patches	10	10	0.5

The properties of PZT patches and laminated composite beam are listed in below table 2.

Table2. Properties of composite beam and PZT patches.

Particulars	Properties
Composite beam	E_x = 40.59GPa E_y = 13.96GPa E_z = 13.96GPa μ_{xy} = 0.22 μ_{yz} = 0.11 μ_{xz} = 0.11 G_{xy} = 3.1GPa G_{yz} = 1.55GPa G_{xz} = 3.1GPa ρ = 1830 kg/m^3
PZT-5H	E =139GPa μ =0.3 ρ =7350 kg/m^3

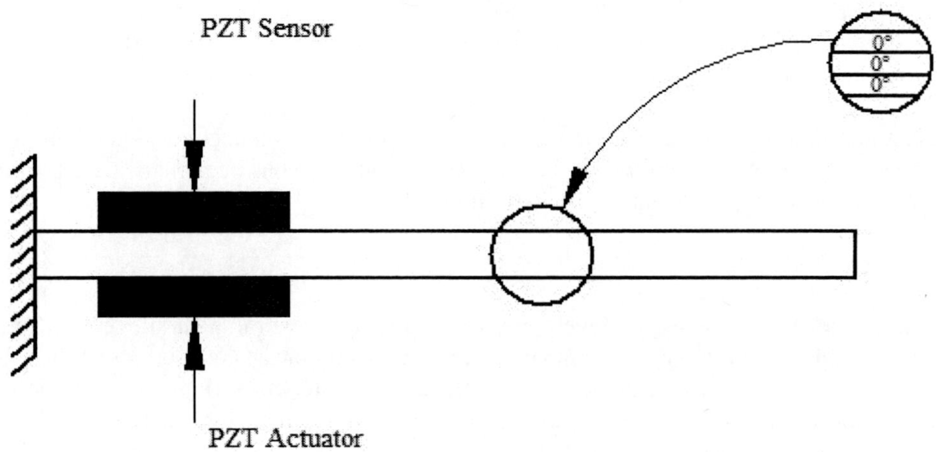

Figure1. Physical model of cantilever beam with PZT patches

The beam is marked with 20 nodes for the proper place for mounting the accelerometer and also to get the mode shapes. The connections of data acquisition system, laptop, accelerometer (model 3032A),

FFT analyzer (Data Physics, model-QUATRO), modal hammer (ENDEVCO model-2031) and cables to the system are done as per the guidance manual. Below figure 2 shows an experimental setup.

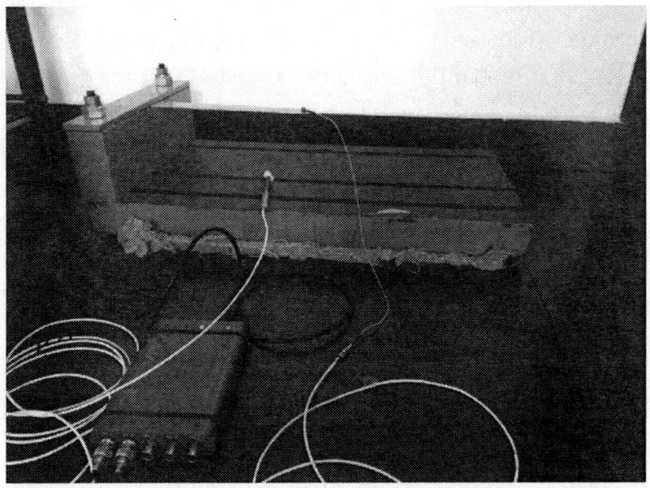

Figure2. Vibration testing of a smart beam with cantilever configuration

The beam was excited in a selected point by means of a small impact hammer (Model 2031), preferably at the free end. The input signals captured by a force transducer, fixed on the hammer. The resulting vibrations of the plate in selected points are measured by an accelerometer. The accelerometer (DYTRAN, TYPE 3032A) was mounted on the plate to the free end by using bees wax. The signal was then subsequently input to the second channel of the analyzer, where its frequency spectrum was also obtained. The response point was kept fixed at a particular point and the location of excitation was varied throughout the beam. Both input and output signals are investigated by means of spectrum-analyzer and resulting frequency response functions are transmitted to a computer for modal parameter extraction. The output from the analyzer was displayed on the analyzer screen by using ME-scope software. The present work represented the natural frequencies and mode shape of beam. The spectrum analyzer provided facilities to record all the data displayed on the screen to a personal computer's hard disk and the necessary software.

3. Results and discussions

A free vibration analysis is carried out for the smart beam and it can be taken as a basis for other dynamic analyses namely transient analysis, harmonic analysis or even spectrum analysis based on the modal superposition technique. The natural frequencies for a beam with and without PZT patches are tabulated in below table 3.

Table 3. Natural frequencies of first three modes.

Modes	Frequencies (Hz) without PZT	Frequencies (Hz) with PZT	Damping
1st	14.8	15.6	0.498
2nd	93	94.5	2.82
3rd	271	274	2.53

4. Conclusion

Modal testing is done for a smart beam with a cantilever configuration using data acquisition system. Through FFT frequency response functions are obtained. Quantitative results are presented in order to show the effect of PZT patches on modal parameters. It is found that the natural frequency of a smart beam changes slightly mainly because of the smaller size of the PZT patches compared to the size of the main beam. It has been observed that the natural frequency increases with increase in mode number. Also, the damping ratio change from mode to mode.

Acknowledgement

Authors thankfully acknowledge the Air Labs Pvt Ltd for providing the facility, constant encouragement and support in carrying out this work.

References

[1] Y.X. Zhang, C.H. Yang "Recent developments in finite element analysis for laminated composite plates." *Composite Structures* 88 (2009) pp 147 157.

[2] P.Cortes & Cantwell, "The tensile and fatigue properties of Carbon Fiber-reinforced PEEK Titanium Fiber metal Laminates", *Journal of Reinforced Plastics and composites*, **Vol 23**, 2004.

[3] Valery V, Vasiliev & Evgeny V Morozov, "Mechanics and Analysis of Composite Materials" Elsevier, 2001.

[4] Ashbee, K.H.G, "Fundamentals Principles of fibers Reinforced Composites", 2n edn. *Technomic Publ. Co. Inc.*, Lancaster, USA (1993).

[5] A.W. Leissa, Recent research in plate vibrations: Classical Theory, *Shock Vibration Digest 9*, (1977) 13–24.

[6] Kanak Kalita and Abir Dutta "Free vibration Analysis of Isotropic and Composite Rectangular Plates", *International Journal of Mechanical Engineering and Research.* ISSN No. 2249-0019, **Vol 3**, Number 4 (2013), pp. 301-308

[7] E. F. Crawley, J. Louis, 'Use of Piezoelectric Actuators as Elements of Intelligent Structures', *AIAA Journal*, October 1989.

[8] J. M. M. Silva, M. M. Maia ' Theoretical and Experimental Modal Analysis', Research Studies Press Ltd., 1998

[9] Waghulde, K. B. & Kumar, Bimlesh "Vibration analysis of cantilever smart structure by using piezoelectric smart material". *Int. J. Smart Sens. Intell. Sys.*, 2011, **Vol 4**(3), 353-375.

[10] Pourboghrat, F.; Pongpairoj, H. & Aazhang, B. "Vibration control of flexible beams using self-sensing actuators". In the Proceedings of the 5th Biannual World Automation Congress, (2002), pp. 133-139.

IOP Conf. Series: Materials Science and Engineering **314** (2018) 012019 doi:10.1088/1757-899X/314/1/012019

Improving the machinability of leaded free cutting steel through process optimization

P Sathyamurthy[a], Vetrivelmurugan R[a], Sooryaprakash J[a]

[a]-, JSW Steel limited, Salem works

Abstract

Free cutting steel grades are high sulphur grades which can be classified under two categories as Leaded and Non-Leaded. These grades are used for manufacturing components like Nuts, bolts, studs, hydraulic fittings, brake pistons where higher machining is required to get intricate shape. Machinability of these grades are affected by hard oxide inclusions and highly deformed manganese sulphide inclusions. At JSW, machinability of leaded free cutting steel is improved by various process modifications namely deoxidation through carbon and manganese, Tellurium (Rare earth element) addition and maintaining the oxygen level at 80-120ppm. Former one avoids the formation of hard SiO2 and Al2O3 compounds, Tellurium addition forms PbTe compound at the tail of MnS inclusions which resists the deformation of MnS inclusions and increased oxygen level favours the formation of less deformable oxy-sulphide inclusions. Above process modifications have resulted in achieving the low silicate content, better aspect ratio of MnS inclusions in the final rolled product. They are assessed by the characteristics of chip formation and surface roughness of machined part.

Keywords: Ladle metallurgy, MnS inclusions, Tellurium and Vacuum carbon de-oxidation

1.Introduction:

Factors influencing machinability of leaded and non-leaded free cutting steels like Surface finish, Tool life, cutting force and formation of small, stable built up edges on cutting tool [1-3] has been improved through process modification through Tellurium addition. Unless the Te/S ratio in free cutting leaded steels exceeds 0.07, Te exists in soluble form in MnS Inclusions [5]. Compounds of Te forms when Te/S ratio exceeds 0.07. Te is present in steels up-to 3% in MnS inclusions [6-11] or as compounds such as FeTe, MnTe and PbTe [12-13]. As-cast structure of leaded free cutting steels reveals mostly Fe-Te compound [6,12,13] with Pb-Te formation occurring during subsequent soaking and rolling [6,13]. Additionally, Te additions have made MnS inclusions more resistant to deformation during hot rolling and more globular [15-19] by forming a liquid envelope that could accommodate strains thereby improving the aspect ratio of MnS inclusion [11] that plays a prominent role for chip breakage.

Te being surface active in nature forms a thin film on the grain boundaries or at interface between steel and other non-metallic phases like carbides, MnS and tellurides [3]. Such distribution favours decrease in internal energy which reduces the resistance to shear. This leads to initiation of micro cracks that in turn reduces effective shear strength at tool metal interface during machining. Liquid metal embrittlement of Te and its compounds around their respective melting points generates micro cracks thereby reducing shear strength and ductility [20-21]. Globularisation of MnS inclusions were found to be by Tellurium enveloping the MnS inclusions in form of compounds like Pb-Te and Mn-Te. Both compounds Pb-Te and MnTe - MnS eutectic have melting points 1356 deg C and 1469 deg C [15,24]. During rolling process, these compounds melt and form liquid phases which is found to accommodate higher strains during rolling and restrict deformations of sulphide inclusions [14]. Also presence of Tellurium being hard phase weakens the Inclusion-matrix interphase. This enhances the possibilities of void formation during rolling deformation by de-cohesion from matrix [25]. Strain during deformations are accommodated at these voids eliminating the deformation of MnS inclusions.

2. Process Optimisation

Chemical composition of free cutting steels produced at JSW is shown in Table 1. Heat No Z001, steel made through existing practise of deoxidation through Si and Mn. Heat No Z003, steel with process modification through Tellurium addition with Mn deoxidation. Heat No Z002, steel with process modification through Mn, and Vacuum Carbon deoxidation.

Table 1. Chemical composition of Free cutting steels produced at JSW

Heat No	Grade	C %	Si %	Mn %	P %	S %	Ti %	Pb %	Te %	O2 PPM
Z001	En1A Pb	0.07	0.06	1.13	0.061	0.28	0.0010	0.23	0	60
Z003	En1A Pb_Te	0.07	0.02	1.15	0.062	0.28	0.0019	0.25	0.025	104
Z002	En1A Pb	0.07	0.01	1.05	0.062	0.25	0.0014	0.25	0	110

Process optimization is carried out with the various changes to the existing secondary steel making techniques to enhance the machinability of free cutting steel. It has been discussed in the following section,

2.1 Existing practice of deoxidation:

Free cutting steels are produced through the process route of primary steel making with energy optimizing furnace and secondary steel making followed by continuous casting. During secondary steel making the steel tapped from energy optimizing furnace is de-oxidized with Si-Mn which enables the formation of hard silicate inclusions (Refer Figure 2). These inclusions reduce the tool life drastically and sometimes cause the premature failure of tools. So a new way of making this kind of steel has been explored at JSWSL to make this steel suit for its intended application.

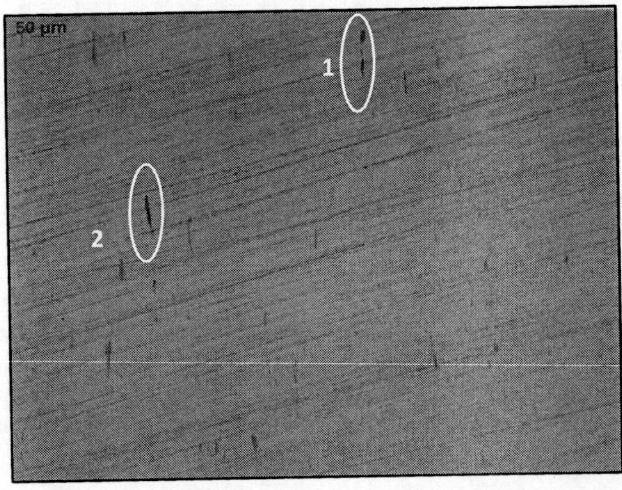

Figure 1. Micrographs showing silicate inclusions along with MnS

2.2 Process modification through Tellurium addition:

Heats are taken with Tellurium addition with modified VCD practise resulted with an Oxygen level of 104 PPM. Morphology of inclusions were studied through SEM in Backscattered mode. Pb-Te compounds were found to cling around the tail of MnS inclusions (Refer Figure 2). Pb-Te compounds resists deformations of MnS inclusions and volume fraction of Pb-Te is found to be higher than the Mn-Te.

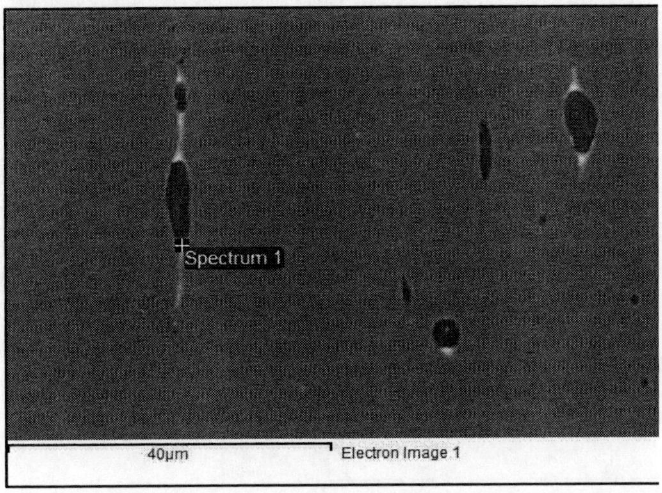

Element	Weight%	Atomic%
Mn K	1.3	1.95
Fe K	53.96	79.89
Te L	1.2	0.78
Pb M	43.55	17.38

Figure 2. SEM Micrograph of Process modified heat - En1APb with Te addition

2.3. Process modification through Vacuum Carbon Deoxidation Route:

Free cutting steels require higher oxygen content for better machinability. So, Si and Al levels are restricted to as low as possible to maintain higher oxygen in steel bath. At JSWSL, to achieve the above requirements steel refining is carried out through Mn and C deoxidation practise at low vacuum levels.

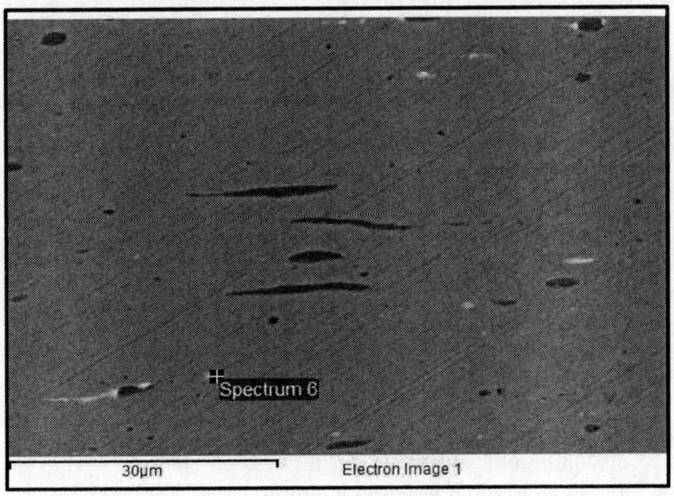

Element	Weight%	Atomic%
S K	6.67	11.8
Mn K	14.05	14.51
Fe K	70.08	71.18
Pb M	9.2	2.52
Totals	100	

Figure 3. SEM Micrograph of Process modified heat - En1APb VCD Heat

IOP Conf. Series: Materials Science and Engineering **314** (2018) 012019 doi:10.1088/1757-899X/314/1/012019

3.Results and Discussions:

Experiments were done to study the effect of process modifications on the machinability of rolled products.

After primary steel making, primary deoxidation of steel is carried out through Fe Mn alloy additions followed by other alloys to meet the grade chemistry. This is followed by optimisation of Oxygen content in steel bath through de-oxidation practise by Carbon addition under Vacuum, as efficiency of O2 removal is higher at low vacuum levels. Heats treated by Vacuum Carbon Decarburisation practise was found to finish with higher levels of O2 PPM compared with heats with Si, Mn deoxidation (Refer Fig 4). Heats were analysed for nature of inclusions formed before Pb addition, after Pb addition.

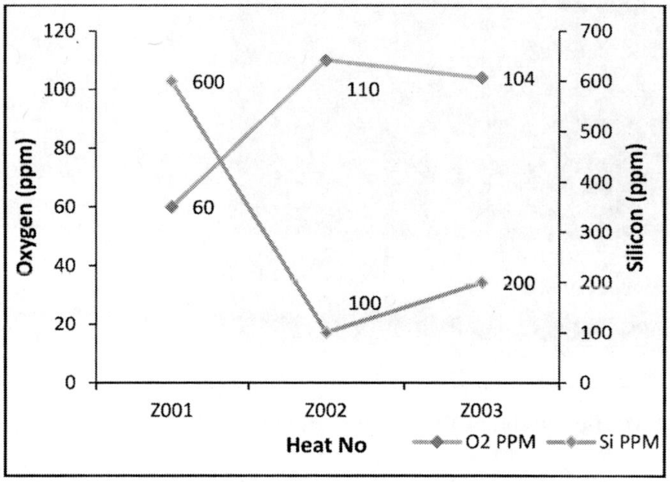

Figure 4. Comparison of Oxygen and Silicon levels in Free Cutting Steels at JSWSL

Heats with Process modifications and existing heat were subjected to plunge tests at similar cutting speeds. The chips formed after machining were analysed (Refer Figure 6) to study the chip nature along with roughness of machined samples.

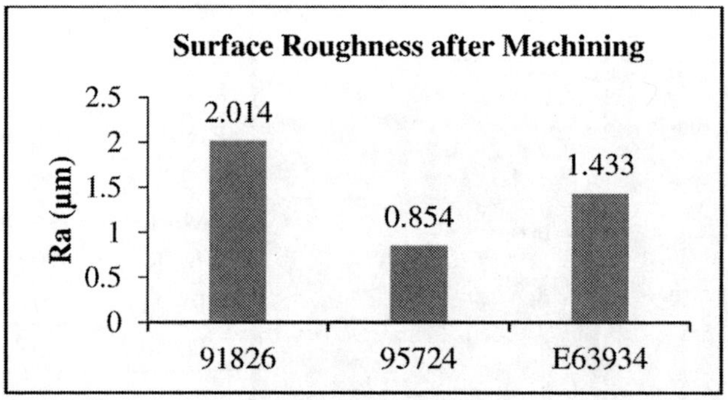

Figure 5. Comparison of Roughness values of rolled samples after machining

Heat Z001 processed through usual Si-Mn deoxidation practice resulted in continuous chips (Refer Figure 6). Chips from Z003 were found to be discontinuous with good curling nature. Heat Z002 also exhibited semi-continuous chip formation with comparatively high chip length to Heat Z003 without prominent curling chips (Refer Figure 6).

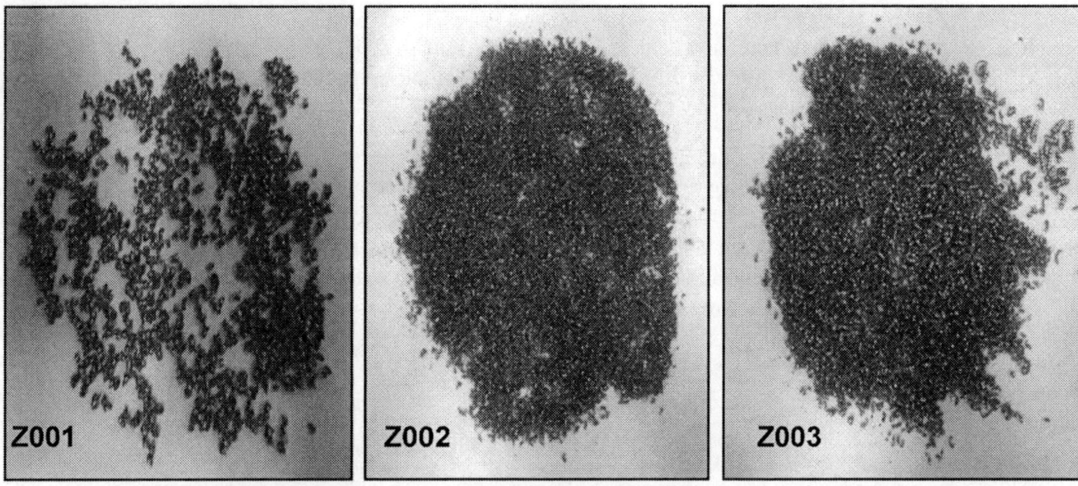

Figure 6. Comparison of Chips from different heats after machining

The inclusion samples from the rolled bars manufactured from the JSWSL heats were analysed and the observations are listed. (Refer Table 2)

Table 2 Observation of inclusions in rolled product from JSWSL heats

Heat Number	Nature of Process Modification	Inclusion Observation
Z001	Deoxidation primarily by Si	Pb tailed MnS inclusions along with Silicate inclusions (Refer Fig 1)
Z002	Te addition with deoxidation by VCD	PbTe compound at tail of MnS inclusions (Refer Fig 2)
Z003	Deoxidation primarily by VCD	Pb tailed MnS inclusions (Refer Fig 3)

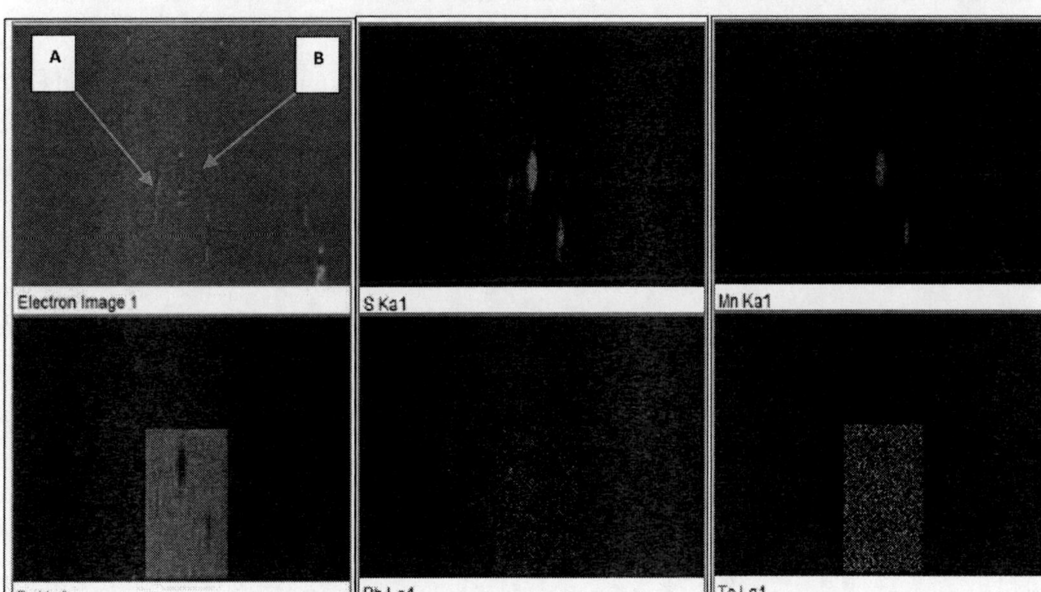

Figure 7. Elemental Mapping of Te added Heat - Z003

IOP Conf. Series: Materials Science and Engineering **314** (2018) 012019 doi:10.1088/1757-899X/314/1/012019

Elemental maps of Te added heat shows presence of PbS inclusions (Refer Figure 7 - Site A) along with MnS inclusions (Refer Figure 7 - Site B) with Te forming along the tail of MnS inclusions as an envelope.

Aspect ratio studies were done on the three heats to study the morphology of inclusions. Typical MnS inclusions were analysed through optical microscopy at 1mm from surface at 100X. Microstructural evaluation was also carried out in longitudinal direction to assess for the morphology of inclusions formed at core, mid radius and surface. Inclusions pertaining to Heat No: Z001 was found to have elongated inclusions with less globularity. Heat Nos: Z002 and Z003 were found to have globular inclusions. No significant morphology differences could be found between heats Z002 and Z003.

Table 3. Aspect Ratio of Inclusions of different heats

Heat No	Grade	Site Of Interest	No Of Inclusions (On Aspect Ratio Basis)			Aspect Ratio %		
			1-6	7-10	>11	1-6	7-10	>11
Z003	En1APb_Te	Surface	22	1	0	96	4	0
		Core	19	0	0	100	0	0
Z002	En1APb	Surface	22	3	1	85	12	3
		Core	14	3	1	78	17	5
Z001	En1APb	Surface	17	5	1	74	22	4
		Core	13	5	1	68	26	5

4. Conclusions:

JSW has taken various trials through process optimisation for manufacturing free cutting steels. Three kind of processes were analysed in detail. In normal deoxidation practise heats, silicate inclusions were observed. In VCD heat, aspect ratio of inclusions was found to be better than normal deoxidation practise heat. In Te added heat, aspect ratio of inclusions is found to be better than other two heats with 96% of inclusions at surface and 100% inclusions at core are falling within an aspect ratio 1-6. Surface roughness values were also found better for Tellurium processed heat followed by VCD and normal deoxidation heat. Chips formation was found to be discontinuous with good curling tendency in heat with Te compared to VCD heat where chips were semi discontinuous with comparatively low curling to Te added heat. Both VCD and Te added heat exhibited low level of Silicate type inclusions thereby enhancing good machinability.

IOP Conf. Series: Materials Science and Engineering **314** (2018) 012019 doi:10.1088/1757-899X/314/1/012019

References

[1] S. Nakamura: *Development of Production Process and Quality Improvement Free-Machining Steels*, ISIJ, Tokyo, (1984), **187**.

[2] R. A. Joseph and V. A. Tipnis: *Influence of Metallurgy on Machinability*, ASM, Chicago, (1975), **U**.

[3] R. H. Aborn: *The Role of Metallurgy, Particularly Bismuth, Selenium and Tellurium in the Machinability of Steels*, American Smelting and Refining Co., New York, (1968). – Compounds formed by Te, MnTe PbTe

[4] R. C. Spencer, C. D. Nagell and R. Richmond : *Metal Prog*., **82** (1962), Dec., 73.

[5] Li Daizhong, Gao Shuqin, Zhang Liefu, Wang Zeyu and Dong Xinquen: *Steel Research* (China), **56** (1985), **537**.

[6] T. Malmberg, G. Runnsjo and B. Aronsson: Scand. J.Metall., **3** (1974), **169**.

[7] T. B. Smith and D. B. Clayton: *Nature*, **198** (1963), **380**.

[8] M. Hugo, J. Bellot, J. Frey and M. Gantois: Rev. *Metall.*,**68** (1971), **397**.

[9] D. Bhattacharya and D. T. Quinto : *Metall. Trans*. A, 11A (1980), **919**.

[10] Li Daizhong, Gao Shuqin, Zhang Liefu, Wang Zeyu and Dong Xinquen: *Steel Research* (China), **56** (1985), **537**.

[11] A Josefsson: The 3rd Int. Symp. *Industrial Use of Selenium and Tellurium, Selenium-Tellurium Devel. Assoc.Inc., Darien, CT*, (1984), **148**.

[12] W.J.M. Salter and F. B. Pickering : J. Iron Steel Inst., **205** (1967), **973**.

[13] J. Bellot and M. Gantois: Trans. Iron Steel Inst. Jpn., **18** (1978), **536**.,

[14] E. Navara and K. Easterling: Metal Sci., **6** (1972), **211**.

[15] T. Kato, S. Abeyama, A. Kimura and S. Nakamura: *High Productivity Machining-Materials and Processes, ASM, New Orleans, (1985)*, **189**.

[16] T. Araki, S. Yamamoto and Y. Uchinaka: Tetsu-to-Hagane, 54 (1968), 444.

[17] A. Hauttmann: Stahl Eisen, **88** (1968), **62**.

[18] J. Bellot and M. Gantois: Trans. Iron Steel Inst. Jpn., **18** (1978), **546**.

[19] H. Ohtani, T. Hashimoto and Y. Kamada : Sumitomo Metals, **38** (1986),

[20] D. Bhattacharya: *Embrittlement by Liquid and Solid Metals*, ed. by M. H. Kamdar, TMS-AIMS, Warrendale, PA, (1984), **367**.

[21] J. R. Rellick, C. J. McMahon, Jr., H. L. Marcus and P. W. Palmberg : Met. Trans., **2** (1971), 1492.

[22] Ward, R. G. (1962). *An introduction to the physical chemistry of iron & steel making*. Edward Arnold.

[23] Turkdogan E T and Fruehan R J 1999 *Fundamentals of iron and steelmaking The Making, Shaping, and Treating of Steel* vol. 2, ed R J Fruehan (Pittsburgh: The AISE Steel Foundation)

[24] K. Kishi and H. Eda: Wear, **38** (1976), **29**

[25] H. F. Fischmeister, E. Navara and K. Easterling: Metal Sci., 6 (1972), 211.

Acknowledgements

The authors thank M/s JSW Steel Limited - Salem Works for their support to carry out this project.

Review of natural fiber composites

Rohan T[1], Tushar B[1] and Mahesha G T[2]

[1] *P G Student, Dept of Aero & Auto Engg, MIT-Manipal University, Karnataka India*
[2] *Dept of Aero & Auto Engg, MIT-Manipal University, Karnataka, India, mahesha.gt@manipal.edu*

Abstract. Development of new alternative materials to the existing traditional metals, alloys and synthetic materials is the new buzz in recent research activities at the academic and industrial level taking place all over the world. Earning carbon credits by minimizing the atmospheric pollution is getting an increase in attention by industries. One small step to conserve the atmosphere around us is to use natural resources in making fully bio degradable or partially bio degradable composite materials. Such prepared alternative materials can find applications in interior housing, automotive, marine, domestic, and other applications. Composites made by using appropriate natural fibers as reinforcements is a possibility that ensures such a reality as they can be well received in multiple disciplines of engineering. Results published from various research activities illustrates that natural fiber composites can successfully be adapted for non-structural, moderate load bearing indoor applications. Further, the few deficiencies in the natural fibers can be overcome by subjecting them to morphological changes by various physical or chemical treatment methods. The overall objective of this paper is to provide a comprehensive overview of the property profiles of Natural Fiber Composites.

1. Introduction

Bio based fibers as a reinforcement in the composite materials is getting more attention now a days. The benefits of using natural fibers as reinforcing media are many. Some of these are biodegradable properties which eases disposal problems, low investment, possibility of thermal recyclation, user friendly conditions, light weight structures, adequate specific strength properties[1,2,3,4] etc. Composites are basically, the sandwiched structure of two or more materials, maybe different in characteristics and composition combined with certain resins which maybe man made or synthetic based on the application, where the composites are used. Fibre properties are determined by the physical, mechanical and chemical properties of its constituents and their interfaces. Several studies have claimed that fibre's mechanical properties are determined by the secondary wall's middle layer [3, 5, 6].
There are certain disadvantages of using natural fibers in composites like moisture absorption, fast degradation, and low resistance to higher temperatures. Many of these problems can be overcome by modifying the fiber structure by physical or chemical treatments. In this study, we try to review some of the research activities carried out on natural fibers, their processing into composites, their characterization based on mechanical, thermal, acoustic, fire retardancy and shock absorption properties, etc.

2. Natural fibers as a reinforcement

Natural fibers are categorized into animal fibers and plant cellulose fibers. Plants that produce natural fibers can be divided into primary and secondary depending on the utilization. Primary plants are grown for their fibers while secondary plants are plants where the fibers are extracted from the waste product. There are six major types of fibers namely; bast fibers, leaf fibers, fruit fibers, grass fibers, straw fibers etc. Generally, much higher mechanical strength and stiffness can be obtained by using plant fibers, but before that, it has to be conditioned mechanically and should undergo certain chemical treatments. Fiber selection as reinforcement in the composites basically dependent upon the application, where it's going to be implemented.

In several research works, Banana fiber has been characterized for mechanical properties [2, 3, 4, 5] and it is concluded that usage of alkali treatment shown the improvement in tensile and flexural strength.
In a study involving pineapple leaf fiber (PALF) [6], green composites were prepared using twin screw extruder. Tensile properties of composites with pineapple leaf fiber and poly lactic acid (PLA) were increased significantly with fiber percentage of 40%.
Vakka, Sisal, Bamboo and Banana fibers were tested in a research work [7] for different mechanical properties, and the researchers claimed that extraction of Vakka fiber is simple. The processing of fibers results into an excellent quality, quantity and lengthy fiber useful for fabrication of large composite components. The extra light weight of the Vakka fiber makes it suitable for designing of lightweight materials.

3. Manufacturing processes

Suitable manufacturing process must be utilized for transformation of the basic materials to the final shape without causing any defect in the product. For the proper fabrication selection process with biodegradable polymer related composite, design and manufacturing engineer focuses on a number of criteria, including desired properties, size and shape of the resultant composites, processing characteristics of raw materials, production speed and the manufacturing cost. For small to medium sized components, injection and compression moulding are preferred because of their simplicity and fast processing cycle. However, for large structures, open moulding and autoclave process typically manufacture them.
In a study involving mechanical characteristics of unidirectional sisal / glass fiber reinforced hybrid polyester composites [1], hand lay-up technique was used for the creation of composites.
In processing natural and wood fiber composites [8], Thermoset mat compression moulding, and Thermoset injection moulding were adapted for the production of green composites. In Mat compression process, a fiber mat is used which was made of natural fibers. In the compression moulding, the mat is sprayed, with resin and compressed into its final contour in a hot tool; with the use of air permeability, the parts could be covered easily in a vacuum covering process.
In Analysis of the deformability of flax fiber non-woven fabrics [9], Resin transfer moulding process (RTM) was used for composite production. It's one of the main manufacturing process to produce composite parts for the transport industries [10] [11]. It is an increasingly common form of moulding using liquid composites. It's primarily used to mould components with large surface areas, complex shape and smooth finishing.

4. Natural resins

The process of resin secretion occurs in special cavities or passages in most of the plant species. It is formed in the specialized structures called "ducts". Resins come out from the bark of the trees and gets hardened when exposed to air. Natural resins of importance to the furniture coatings are rosin, damar, copal, sandarac, amber, and manila.
In studies of the medicinal plants, valuable resins with photochemicals and bioactive properties provided scientific information about triterpenes that will be important for further research on the use of the various photochemicals found in resins and plants to sustain human health [12].
Ivica et al [13], proved in their study that the properties of the resins can be improved by treatment of the fibers with phenolic composition and antimicrobial process. Gram-positive bacteria were more susceptible to resin samples and different combination of pure phenolic, cherry bud samples had high amounts of naringenin and showed strong activity against Bacillus subtilis and Listeria monocytogenes. Chen et al [14], Wood resins are valuable natural products with wide utilizations resin-containing woods, natural wood resins are usually complex mixtures consisting of various compounds. Wood resin samples can be measured directly by Fourier transform infrared (FTIR) spectroscopy and principal component analysis can resolve the absorption bands of different compounds with respect to their positions in the resin containing woods.

5. Mechanical properties

In recent years, the mechanical performance of natural fiber composites seemed to be improved significantly. Enhancement in the overall properties in natural fiber composites has occurred due to improved fibre selection, extraction, treatment and interfacial engineering as well as composite processing [49].
Mechanical properties of sisal and glass fiber reinforced composites has been studied by Sanjay et al [1]. They've compared the properties like tensile, impact and flexural strength of both sisal and glass fiber reinforced composites and concluded that the tensile strength of 30% glass fiber – 70% sisal composition and flexural strength of 70% glass fiber – 30% sisal composition and impact strength of 50% glass fiber – 50% sisal composition was found to be better than the remaining two compositions

Tara Sen et al [15], have studied properties of Sisal, bamboo, coir and Jute, and tried using them in the structural upgradation. In the study, strength of the material has been improved by the usage of Fiber Reinforced Polymers as a raw material. Different properties of these natural fibers like specific gravity, water absorption, and mechanical properties like tensile strength, modulus of elasticity, were studied and tested for their application into the structural upgradation.
Mechanical behaviour of sisal / glass fiber reinforced polyester hybrid composite was studied by Sanjay, et al [16]. In this study, sisal fiber was hybridized with glass fiber with the hand lay - up process. They found that the tensile strength of 30 % GF- 70% sisal composition and flexural strength of 70% GF- 30% sisal composition and impact strength of 50% GF- 50% sisal composition were found to be better than the remaining two composites.
In other study, mechanical properties of hybrid glass fiber – sisal jute reinforced epoxy composites were studied by Ramesha, et al [17]. The incorporation of sisal fiber with GFRP found to be having better properties than the jute fiber reinforced GFRP composites.
In a study involving the fracture and impact properties of short discrete jute-fiber reinforced cementitious composites (JFRCC), by Xiangming Zhou et al [18] they concluded that jute fiber had less deterioration in pulverized fly ash (PFA)/PC matrix than in ground granulated blast furnace slag (GGBS)/portland cement (PC) matrix upon impact failure.
Sisal fiber/ glass fiber hybrid nano composite were tested by Nataranjan, et al [19]. It has been found that the tensile strength of nano sisal composite is relatively more than sisal/nano sisal composite and much higher when compared with sisal composite. They've also found that the percentage elongation of single fiber in tensile testing is less than that of the hybrid composite.
In a study involving banana fiber, by Santhosh, et al [20], they've taken both treated and untreated banana fiber for the development of the hybrid composite material. In this study, the untreated banana fiber was treated by NaOH to increase wettability. The untreated banana fiber and NaOH treated banana fiber were used as reinforcing material with both Epoxy resin matrix and Vinyl ester resin matrix. Coconut shell powder was used along with both untreated and treated banana fiber as a filler material. They've concluded that the alkali treatment of banana fiber improved the mechanical properties like tensile, flexural and impact strength of both the epoxy/ vinyl ester and hybrid composite.
In a study involving the study of bamboo natural fiber composite, by Kumar [21], the researcher evaluated the tensile strength, flexural strength and surface hardness of the bamboo – epoxy composites. The effect of fiber loading on tensile strength, flexural strength and surface hardness also was studied on the bamboo epoxy composite. In this study, with the hand layup technique it was found that maximum value tensile strength, flexural strength, and micro surface hardness of bamboo–epoxy composite was achieved at 25% wt of fiber loading.
In a study done by Bakare et al, [22], on flax and flax basalt reinforcement, by introducing reinforcement, it was found that the surface weight of the flax basalt reinforcement was lesser than the flax. The researchers found a noticeable improvement in mechanical and thermal properties.
The mechanical properties of the kenaf fiber was studied by Saba et al [23]. They found that the kenaf fiber has good tensile strength and flexural strength which was proved by many mechanical tests which enables it to utilize in variety of applications, such as auto-industrial, light weight constructional applications, customary products like yarns, fabrics and ropes. He also found that kenaf/epoxy composites showed better mechanical properties in comparison with polymeric matrix.
A fiber named Hibiscus sabdariffa has been studied by Singh et al [24]. In this study, they've done scanning electron microscope testing. The UF (urea formaldehyde) was used as the matrix. The tensile and compressive strength has been studied and from that, they've concluded that the fiber has a good scope in in the fabrication of natural fiber reinforced polymer composites, having vast number of industrial applications.
A polymer composite involving sisal fiber which is reinforced with coconut spathe and ridge gourd has been studied by Girisha et al [25]. Hybridization of these natural fiber with coconut spathe and ridge gourd provided considerable improvement in tensile strength when compared to the individual reinforcement. The tensile strength was found to be increased to 65 %, which is maximum for the hybrid combination of sisal and ridge gourd.
Low density of natural fiber composites lead to better comparison for specific properties. Applications of Natural Fiber Composites has extended in load bearing and outdoor applications such as automotive exterior underfloor panelling, sports equipment, and marine structures [50].

6. Vibration and noise absorption properties

Noise pollution is the major problem now a days in metropolis. It causes not only annoyance and sleep disturbance, but also leads to severe problems like heart attack, learning disabilities and tinnitus [10, 26, 27]

Currently, in many of the buildings, automobiles and other transport vehicles, sound absorbing materials are used to reduce the noise and vibration through absorption of acoustic waves, because noise or the sound waves has to propagate through the sound absorber and it'll get absorbed into that later [11].
As most of the commonly used sound absorptive materials having their source as fossil fuel, it limits the usage of them at industrial level. To overcome this, some materials like Antiaris toxicaria barkcloth were tested for sound absorption using type 4206 impedance tube according to ISO 10534-2 standard using two quarter inch condenser microphones type 4187 and it has been found that the Antiaris toxicaria barkcloth had higher sound absorptive properties at higher frequencies [10].
Natural fibers were found to have ease in handling and good acoustic insulation properties. [28, 29, 30]
In a study of free vibration characteristics of composite beam, which was made of banana/sisal fibers[31], it's been found that the chemical treatment improves the mechanical and free vibration properties of polymer composites due to the enhancement of interfacial bond between fiber and matrix as the result of chemical treatment.
In a research involving the sunflower composite which was made up of crushed stem particles and chitosan binder, with high porosity and low density, sound absorptive properties were found to be increased as in comparison with the previous composition when the material was highly compact and had low porosity [32].
In a study of damping properties of flax fiber composites [33], it's been found that the damping properties of composites increased with the twist and crimp amount through enhanced intra-yarn and inter-yarn friction respectively.

7. Thermal properties

Rapid consumption of fossil fuel based resources is increasing and because of that, the problems like global warming and other environmental issues are also affecting the normal life. Hence, there is need for developing natural fiber composites and biodegradable polymers, as they have high specific strength and eco-friendly as well. But the main obstacle in developing and using natural fiber based composites at higher temperature is that they are less thermally stable and more flammable as compared to synthetic fibers.
In a study, thermos - mechanical properties of the bio based composites made from lactic and thermoset resin and flax and flax/ basalt fiber reinforcements [22], the mechanical and thermal properties were found better than the pure flax composite. It was established that with the PMLA, (pentaerythritol methacrylated lactic acid resin) upto 60 wt % fiber loading possible, while with flax fibers alone, composite could have upto 40 wt % fiber load.
In another study, Srinivasan et al [34] found that, hybrid composite of banana, flax with GFRP has better thermal stability and flame resistance than the flax, banana fiber composites.
In a study, Ramie fiber was applied with the thermal treatment by Jian-Min Yuan et al [35], and it has been found that, the thermal treatments are efficient in modifying the surface and inner structures of the ramie fibers, and the water absorption, oxidative decomposition and thermal degradation greatly affect the cellulose structure.

8. Bio degradability

Now a days, due to large amount of usage of synthetic fiber composites like glass, aramid, etc., it's causing a problem to the environment as it's not degradable as easily as compared to the natural fibers. The development of natural fiber reinforced biodegradable composites promotes the use of environmental friendly material. Use of green materials provide alternative way to solve the problems associated with agriculture residues. [7, 36]
Bio degradable materials, which are also bio based in nature are ideal for the sustainable development. It has the potential for use in recyclable products, which can be collected, processed and reused for animal feed or left in the environment to be degraded naturally, releasing the materials that can fertilize and condition the soil to increase crop yields. [37]
In the study of natural fiber reinforced biodegradable polymer composite, [10] Sahari et al did the analysis of polymer matrices which were derived from natural resources such as poly lactide (PLA), thermoplastic starch (TPS), cellulose and polyhydroxyalkanoates (PHAs).
In this work, they've found that, biodegradable composites are having competitive properties with the existing fossil plastic materials, but their present low level production and high cost restrict them to be applied in industrial application. In addition, it's also having hydrophilic properties, which makes it difficult to design. Bamboo fiber, its extraction and its reinforced polymer composite material has been studied by Yu ying wang et al [38]. Firstly, they did the analysis of the bamboo structure and later they've extracted it. Basically there are three types of extraction methods of bamboo fibers, viz., mechanical extraction, and chemical extraction, combination of mechanical and chemical extraction. After the extraction process, the fiber was reinforced with the polymeric materials and then different mechanical properties as well as cost analysis, CO_2 absorption, recyclability, renewability, energy of extraction, biodegradability were studied. They've found that the bamboo fiber is very much comparable with the glass fibers, as it was lighter in weight, had low cost and the main advantage, biodegradable.

There are mainly five types of biodegradable bio-based polymers commercially available: starch based polymer, polylactic acid based polymers (PLA), cellulose ester based polymers, bacterial polyhydroxyalkonate based polymers (PHAs) and protein based polymers.
In a study involving the composites of biodegradable matrix corn gluten meal material in wood fiber reinforced [39,40], and used water, glycerol, octanoic acid and polyethylene glycol and tested them as plasticizers for corn gluten meal (CGM).Also, they've found that water helped to get down the glass transition temperature. Mechanical properties were seemed to be the best for modest structural applications and has the good bio degradability.
In the study including starch based biodegradable reinforced composites with date palm and flax fibers [41], hybrid composites containing date palm and flax fibers, 25% wt. each, were examined. In that, it had been concluded that thermoplastic starch (TPS) composites are biodegradable and hydrophilic. The water uptake and rate of biodegradation was reported as inversely proportional to the fiber content.

9. Fire retarding properties

Basically, natural fibers are preferred over the synthetic fibers because it has several advantageous properties as mentioned below:
1. Low specific weight, which results in higher specific strength and stiffness than glass.
2. It is a renewable resource.
3. Producible with low investment at low cost.
4. Friendly processing.
5. CO_2 used while oxygen is given back to the environment.
But, natural fiber reinforced composites also has some disadvantages like high flammability, hydrophilic nature, etc. Because of the poor internal contact between hydrophilic natural fiber and hydrophobic PP matrix. Compatabilizers can be added to improve the compatibility between fiber and matrix which results in the enhancement of mechanical properties of the composites [42].
In studying the flame retardant, mechanical and thermal properties of sisal fiber composites, [42], it was added with some flame retardants like Ammonium polyphosphate (APP), $Mg(OH)_2$, Zb, APP/$Mg(OH)_2$ and APP/Zb. After incorporating these flame retardants into the composite, it resulted in the improved flame retardant and thermal stability of the Polypropylene (PP) composites.
In a study involving the PP/wood fiber composites, by Zhen Xiu Zhang et al [43], firstly, they produced the composite with the help of twin screw extruder and later they treated these with the Ammonium polyphosphate (APP) as a flame retardant. They found that as the percentage of APP in the composite increases, its tensile strength decreases but the tensile strength was observed to be increased upto certain limit when the composites were treated with silica as a fire retarding agent.
Natural fiber polypropylene composites containing magnesium hydroxide were analysed by M Sain et al [44]. In this study, flame retardant properties of boric acid or zinc borate and magnesium hydroxide were studied with natural fiber PP. And they found that 25% magnesium hydroxide could reduce the flammability effectively (almost 50%) of natural fiber filled polypropylene composites.
Bamboo chopsticks fiber - reinforced poly green composites were tested for their flame-retardant properties by Yu-Ying Wang et al [38, 45, 46]. In this study, recycled bamboo chopsticks fibre were used with poly lactic acid (PLA). Later, the surface of chopstick fiber and nano clay surface were modified with coupling agents. They concluded that the composites achieve a good grade according to the different composition of the flame retardants. Also, the scanning electron microscope analysis showed that the inner face adhesion of the composites was effectively enhanced after the surface modification of the fiber. Results also showed that the resulting such green composites were not only flame retardant, but also had good thermal and mechanical properties.
In a study involving kenaf / PP fiber composites, the ammonium polyphosphate as flame retardant were used [47, 48] under 3 different compositions. In the thermal analysis of compounded PP/ kenaf composites showed lesser rate of decomposition as compared to those of pure PP and PP/kenaf compounds. Also, the results showed the flame retardant property of PP/kenaf composites could be improved with increased APP content, irrespective of their type.

10. Conclusion

Innovations taking place worldwide on alternative materials are shedding light on feasibility of using natural resources such as plant fibers for reinforcing the base matrix materials. Many cellulosic fibers such as flax, sisal, Jute, bamboo have excellent specific strength and low density. As all the natural fibers are biodegradable, safe disposal of the materials at the end of their service life is assured. Natural fiber reinforced materials possess some inherent advantages such as better acoustic and dampening properties due to their porous structure. Certain disadvantages such as poor mechanical, thermal, moisture resistance and fire resistance properties can be overcome by modifying them suitably with chemical treatments or by adding suitable additives. If such naturally available resources are processed and used selectively, much burden on fossil fuels can be reduced and value addition to the locally available natural fibers can be achieved.

11. References

[1] R Sanjay, R Arpitha, G Vasundhara and B Yogesha *Study on Mechanical Characteristics of Unidirectional Sisal/Glass Fiber Reinforced Polyester Hybrid Composites* 2012 ISSN (Online): 2319-7064

[2] J Santhosh, N Balanarasimman, R Chandrasekar, S Raja *Study of properties of banana fiber reinforced composites* 2014 *IJRET* eISSN 2319-1163

[3] Chandramohan D and Marimuthu K *Tensile and Hardness Tests on Natural Fiber Reinforced Polymer Composite Material* 2011 IJAEST Vol. **6**, pp 97 – 104

[4] Herrera-Franco, P J, Valadez-Gonzalez *A study of the mechanical properties of short natural-fiber reinforced composites* 2005 Composites Part B: Engineering Vol. **36/8** pp 597-608

[5] Ishak M. A., Maleque F, Belal Y and Sapuan S M *Mechanical Properties Study of Pseudo-Stem Banana Fiber Reinforced Epoxy Composite* 2007 The Arabian Journal for Science and Engineering, Vol. **32** pp 359-364

[6] Supranee Kaewpirom, Cherdthawat Worrarat 2014 *Preparation and properties of pineapple leaf fiber reinforced poly (Lactic acid) green composites* Fibers and Polymers Vol.**15** No.7 pp 1469-1477

[7] K Murali Mohan Rao, K Mohana Rao, A V Ratna Prasad *Fabrication and testing of natural fibre composites: Vakka, sisal, bamboo and banana* 2010 Materials and Design 31 pp 508–513

[8] A K Bledzki , A Jaszkiewicz , M Murr, V E SPERBER *Processing techniques for natural- and wood-fibre composites* 2008 ThuÈringisches Institut fuÈ r Textil- und Kunststoff-Forschung, Germany

[9] Fatma Omrani, Peng Wang, Damien Soulat, Manuela Ferreira, Pierre Ouagne *Analysis of the deformability of flax-fibre nonwoven fabrics during manufacturing* 2016

[10] Samson Rwawiire, Blanka Tomkova, Jiri Militky, Lubos Hes, Band Madhukar Kale *Acoustic and thermal properties of a cellulose nonwoven natural fabric (barkcloth)* 2017 Elsevier, Applied acoustics **116** pp 177-183

[11] Igor M. De Rosa, Carlo Santulli, Fabrizio Sarasini *Acoustic emission for monitoring the mechanical behaviour of natural fibre composites: A literature review* 2009 Elsevier, Composites: Part A **40** pp 1456–1469

[12] Liliana Hernández-Vázquez, Susana Mangas, Javier Palazón, Arturo Navarro-Oca˜na *Valuable medicinal plants and resins: Commercial phytochemicals with bioactive properties* 2010 Elsevier, Industrial Crops and Products **31** pp 476–480

[13] A Pizzi , R Kueny, F Lecoanet, B Massetau, D Carpentiera, A Krebsa, F Loiseau, S Molina, M Ragoubi *High resin content natural matrix–natural fibre biocomposites* 2009 Elsevier, Industrial Crops and Products **30** pp 235–240

[14] Jian-bo Chen, Qun Zhou, Su-qin Sun *Direct chemical characterization of natural wood resins by temperature-resolved and space-resolved Fourier transform infrared spectroscopy* 2016 Elsevier, Journal of Molecular Structure **1115** pp 55-62

[15] Tara Sen, H N Jagannatha Reddy *Application of Sisal, Bamboo, Coir and Jute Natural Composites in Structural Upgradation* 2011 International Journal of Innovation Management and Technology Vol. **2** No. 3

[16] R Sanjay, R Arpitha, G Vasundhara, B Yogesha *Study on Mechanical Characteristics of Unidirectional Sisal/Glass Fiber Reinforced Polyester Hybrid Composites* International Journal of Science and Research (IJSR), ISSN (Online): 2319-7064

[17] M Ramesh, K Palanikumar, K Hemachandra Reddy *Comparative Evaluation on Properties of Hybrid Glass Fiber-Sisal/Jute Reinforced Epoxy Composites* 2013 Procedia Engineering **51** pp 745 – 750

[18] Xiangming Zhou, Seyed Hamidreza Ghaffar, Wei Dong, Olayinka Oladiran, Mizi Fan *Fracture and impact properties of short discrete jute fibre-reinforced cementitious composites* 2013 Elsevier, Materials and Design **49** pp 35–47

[19] Natarajan, Bharathidhasan, Thanigaivelan, Şuresh *Sisal Fiber / Glass Fiber Hybrid Nano Composite: The Tensile and Compressive Properties* 5th International & 26th All India Manufacturing Technology, Design and Research Conference (AIMTDR 2014) December 12th–14th, 2014, IIT Guwahati, Assam, India

[20] J Santhosh, N Balanarasimman, R Chandrasekar, S Raja *Study of properties of banana fiber reinforced composites* IJRET: International Journal of Research in Engineering and Technology eISSN: 2319-1163

[21] Dheeraj Kumar *Mechanical Characterization of Treated Bamboo Natural Fiber Composite* 2014 International Journal of Advanced Mechanical Engineering ISSN 2250-3234 Volume **4** pp 551-556

[22] Bakare, F O, Ramamoorthy, S K, Akesson, D, Skrifvars M *Thermomechanical properties of bio-based composites made from a lactic acid thermoset resin and flax and flax/basalt fibre reinforcements* 2015 Composites: Part A doi: http://dx.doi.org/*10.1016/j.compositesa.2015.09.002*

[23] N Saba, M T Paridah, M Jawaid *Mechanical properties of kenaf fibre reinforced polymer composite: A review* 2015 Elsevier, construction and building materials **76** pp 87-96

[24] Singha A, Thakur Vijay Kumar *Mechanical properties of natural fibre reinforced polymer composites* 2008 Indian academy of sciences Bull. Mater. Sci. Vol. **31**, No. 5 pp 791–799

[25] Girisha C, Sanjeevamurthy, Gunti Rangasrinivas 2012 *Tensile properties of natural fiber-reinforced epoxy-hybrid composites* International Journal of Modern Engineering Research (IJMER) Vol.**2** Issue.2 pp-471-474 ISSN: 2249-6645

[26] *Moszynski P WHO warns noise pollution is a growing hazard to health in Europe BMJ 2011:342*

[27] Berardi U, Iannace G *Acoustic characterization of natural fibers for sound absorption applications* 2015 Build Environ2015 http://dx.doi.org/10.1016/j.buildenv.2015.05.029

[28] P L Menezes, P K Rohatgi, M R Lovell *Studies on the tribological behavior of natural fiber reinforced polymer composite* 2012 Green Tribology, Springer pp 329–345

[29] D B Dittenber, H V Ganga Rao *Critical review of recent publications on use of natural composites* 2012 infrastructure Compos Part A Appl. Sci. Manuf. **43** pp 1419–1429

[30] W Brouwer *Natural fibre composites: where can flax compete with glass* 2000 Sampe J **36** pp 18–23

[31] M Rajesh, Jeyaraj Pitchaimani, N Rajini *Free Vibration Characteristics of Banana/Sisal Natural Fibers Reinforced Hybrid Polymer Composite Beam* 2016 Elsevier, Procedia Engineering **144** pp 1055 – 1059

[32] Narimane Mati-Baouche, Hélène de Baynast, Philippe Michaud, Thomas Dupont, Philippe Leclaire *Sound absorption properties of a sunflower composite made from crushed stem particles and from chitosan bio-binder* 2016 Elsevier, Applied acoustics **111** pp 179-187

[33] F Duc, P E Bourban, J –A E Manson *The role of twist and crimp on the vibration behaviour of flax fibre Composites* 2014 Elsevier, Composites Science and Technology, **102** pp 94-99

[34] V S Srinivasan, S Rajendra Boopathy, D Sangeetha, B Vijaya Ramnath *Evaluation of Mechanical and Thermal Properties of Banana-Flax based Natural Fibre composite* 2014 http://dx.doi.org/10.1016/j.matdes.2014.03.014

[35] Jian-Min Yuan, Yan-Rong Feng, Li-Ping He *Effect of thermal treatment on properties of ramie fibers* 2016 10.1016/j.polymdegradstab.2016.09.012

[36] Azwa ZN, Yousif BF, Manalo AC, Karunasena W *A review on the degradability of polymeric composites based on natural fibres* 2013 Mater Des 2013;47:424–42

[37] Y Cao, S Shibata, I Fukumoto *Mechanical properties of biodegradable composites reinforced with bagasse fibre before and after alkali treatments* 2006 Elsevier, Composites: Part A 37 pp 423–429

[38] Yu-Ying Wang, Yeng-Fong Shih *Flame-retardant recycle d bamboo chopstick fiber-reinforced poly (lactic acid) green composites via multifunctional additive system* 2016 Elsevier, Journal of the Taiwan Institute of Chemical Engineers 0 0 0 pp 1–7

[39] F X Espinach, M Delgado-Aguilar, J Puig, F Julian, S Boufi, P Mutje *Flexural properties of fully biodegradable alpha-grass fibers reinforced starch-based Thermoplastics* 2015 Composites Part B, doi: 10.1016/j.compositesb.2015.07.004

[40] Wisnom M R *The relationship between tensile and flexural strength of unidirectional composites* 1992 Journal of Composite Materials 26(8):1173-80

[41] M duho V I C, S peterson and K jayaraman *Natural-fibre ± biodegradable polymer composites for packaging*

[42] Rachasit Jeencham, Nitinat Suppakarn, Kasama Jarukumjorn *Effect of flame retardants on flame retardant, mechanical and thermal properties of sisal fiber/polypropylene composites* 2013 Composites: PartB,doi:http://dx.doi.org/10.1016/j.compositesb.2013.08.012

[43] Zhen Xiu Zhang et al *Effect of flame retardants on mechanical properties, flammability and foamability of PP/wood–fiber composites* 2012 Elsevier, Composites: Part B 43 pp 150–158

[44] M Sain, S H Park, F Suhara, S Law *Flame retardant and mechanical properties of natural fibre–PP composites containing magnesium hydroxide* 2004 Elsevier, Polymer Degradation and Stability 83 pp 363–367

[45] Reti C , Casetta M , Duquesne S , Bourbigot S , Delobel R *Flammability properties of intumescent PLA including starch and lignin* 2008 Adv Polym Technol 2008 ;19: 628–35

[46] Zhan J, Song L, Nie SB, Hu YA *Combustion properties and thermal degradation behavior of polylactide with an effective intumescent flame retardant* 2009 Polym Degrad Stab 2009; 94(3):291–6

[47] Aruna Subasinghe, Debes Bhattacharyya *Performance of different intumescent ammonium polyphosphate flame retardants in PP/kenaf fibre composites* 2014 Elsevier, Composites: Part A 65 pp 91–99

[48] Azwa ZN, Yousif BF *Characteristics of kenaf fibre/epoxy composites subjected to thermal degradation* 2013 Polym Degrad Stab; 98(12):2752–9

[49] Srinivasababu N, Murali Mohan Rao K & Suresh Kumar *Tensile properties characterization of okra woven fiber reinforced polyester composites* International Journal of Engineering, (IJE) Volume **(3)**: Issue (4).

[50] G. T. Mahesha, S. B. Satish, K. M. Vijaya, K. S. Bhat *Preparation of the unidirectional grewia serrulata fiber reinforced polyester composites and evaluation of tensile and flexural properties* 2016 Journal of natural fibers, 13: 5, 547-554

Acknowledgements

We, Rohan T, and Tushar B, owe a great many thanks to a great many people who helped and supported us during the writing of this review. Our deepest thanks to our lecturer Mr. Mahesha G T, the guide of this work for guiding and correcting various documents of us with attention and care. He has taken pain to go through the review work and make necessary correction as and when needed. We would also thank to our institution and our faculty members without whom, this work would have been a distant reality. We also extend the heartfelt thanks to the well-wishers, friends and family for their encouragement and cooperation.

IOP Conf. Series: Materials Science and Engineering **314** (2018) 012021 doi:10.1088/1757-899X/314/1/012021

Vehicle performance evaluation in side impact (MDB) using ES-II dummy

T S Ganessh, Praveen Bansode, Vidyakant Revankar, Sunil Kumar
Mahindra and Mahindra ltd., Chennai, Tamilnadu, India.

Abstract. Side impact collision is one of the leading causes of death. Protection of people during lateral collision is challenging because of relatively small space available to restraint occupant compared to front. Hence, it is imperative to protect the occupants in side collision. It is a function of vehicle type and restraints for side protection. This paper focuses on evaluation of injury parameters of the ES II dummy during the lateral collision of different vehicles with different spaces, sections and materials. Thus the comparison will enable us to understand the sensitivity of space, B-pillar section and material which affects the injury parameters. This study will help automotive engineers to design side impact crashworthy vehicles.

1. Introduction

Indian road accident statistics show that side impacts account for approximately 20% of all impacts and 35% of total fatalities. For example, side impact crashes were 26% of the total amount of fatal accidents that took place in 2012 in the US. This information has been retrieved from the road accident sampling system-India (RASSI), which has the only vehicles database in India, in relation to real-world severe crash tests in the US. A similar trend has been identified in Europe, where side impact road collisions correspond to approximately one quarter of the total amount of serious-to-fatal injuries.

It is for this reason that developing safety systems to prevent occupants from injuries during Side Impact crashes has a significant potential in terms of road fatalities reduction. Because of this, all regulations include injury criteria requirements that must be accomplished in Side Impact crash testing. These regulations intend to enhance the safety of the new vehicles to be set in the market by obliging Original Equipment Manufacturers (OEMs) to develop restraint systems based on the vehicle's behavior to Side Impact crashes. In this way, the vehicle occupants have a major protection in case of lateral accident. Side impacts also require more attention in that there is considerably less crush zone for absorbing energy in the side of passenger cars compared to the front and rear structures, and consequently the occupants sit almost within the crush zone, which often results in severe thoracic injuries. Research and development into minimizing structure intrusions during side impacts of passenger cars are done to reduce the effects of side impacts.

Most side impacts can be classified into two types they are car-to-broad-object and car-to-narrow-object. Car to narrow object are the impacts that are done with pole. In this paper car to broad object test is considered i.e. impact of car with mobile deformable barrier (MDB). Three passenger vehicles were taken for comparison and the performance of ES II dummy during the side MDB test is compared and analyzed with space, section and deformation.

2. Side Impact Test

IOP Conf. Series: Materials Science and Engineering **314** (2018) 012021 doi:10.1088/1757-899X/314/1/012021

Side impact tests are carried out on the vehicles to observe the structural integrity of the vehicle and thereby providing safety to the occupants of the vehicle. The test is done with a deformable barrier placed perpendicular to the vehicle longitudinal axis.

2.1. Regulation

The vehicles taken for comparison are M1 category. Impact is done on the driver side using mobile deformable barrier (MDB) which has the impactor and the trolley. Impactor is a crushable section mounted on the front of the MDB and the trolley is a wheeled frame which travels along its longitudinal axis at the point of impact. The test setup is as shown in the figure 1. Impact will be done at 50kmph.

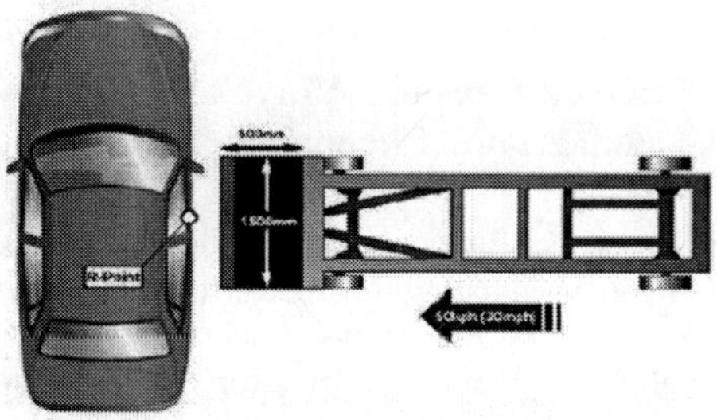

Figure 1. Test setup showing the MDB and vehicle.

2.2. Injury Criterion

2.2.1 The head performance criterion (HPC).

When head contact takes place this performance criterion is calculated for the total duration between the initial contact and the last instant of the final contact.HPC is the maximum value of the expression:

$$(t_2 - t_1)\left\{\frac{1}{t_2 - t_1}\int_{t_1}^{t_2} a.dt\right\}^{2.5}$$

Where a is the resultant acceleration at the centre of gravity of the head in meters per second per second divided by 9.81 recorded versus time and filtered at channel frequency class 1000 Hz; t1 and t2 are any two times between the initial contact and the last instant of the final contact. The value shall be less than or equal to1000.

2.2.2 The thorax performance criteria.

- Rib Deflection Criterion (RDC) is the maximum value of deflection on any rib as determined by the thorax displacement transducers, filtered at channel frequency class 180 Hz. The value shall be less than or equal to 42 mm;
- Viscous Criterion (VC) is the maximum value of VC on any rib which is calculated from the instantaneous product of the relative thorax compression related to the half thorax and the velocity of compression derived by differentiation of the compression, filtered at channel frequency class 180 Hz. The value shall be less than or equal to 1.0 m/sec.

2.2.3 The pelvis performance criterion.

Pubic Symphysis Peak Force (PSPF) is the maximum force measured by a load cell at the pubic symphysis of the pelvis, filtered at channel frequency class 600 Hz. The value shall be less than or equal to 6 kN.

2.2.4 The abdomen protection criterion.

IOP Conf. Series: Materials Science and Engineering **314** (2018) 012021 doi:10.1088/1757-899X/314/1/012021

Abdominal Peak Force (APF) is the maximum value of the sum of the three forces measured by transducers mounted 39 mm below the surface of the crash side, CFC 600 Hz. The value shall be less than or equal to 2.5 kN internal force.

2.3 EuroSID

EUROSID is the EURopean Side Impact Dummy and has been developed in four European countries under the auspices of the former EEVC WG 9. EUROSID-1 is the production version of the dummy and the successor to the prototype dummies manufactured during 1987/9. The EUROSID-1 is essentially the same as the production prototype apart from some minor design and bio fidelity enhancements. The European Directive 96/27/EC on side impact that has gone into effect on October 1st, 1998, specifies the EUROSID-1 dummy as the injury assessment device to be used in the test procedure. EuroSID-II or ES II dummies are the successor of ES I dummies. These ES II dummies are used in the side impact tests. ES II dummies have the instrumentations in head, neck, shoulder, thorax, abdomen and leg. ES II dummy is shown in figure 2.

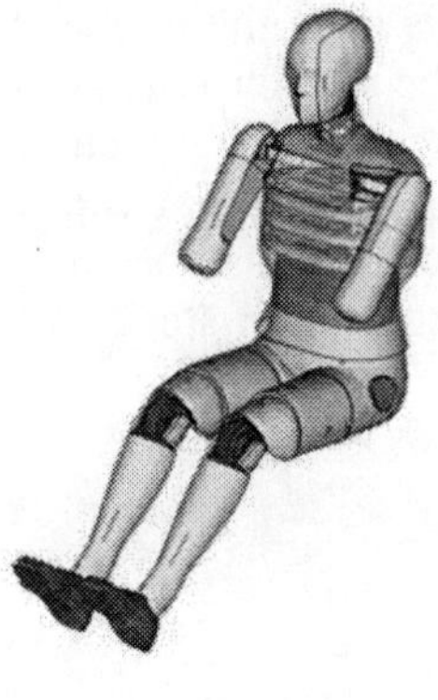

Figure 2. ES II dummy model.

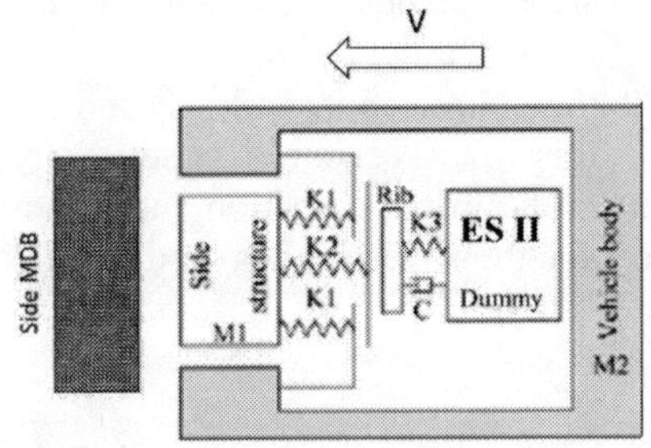

Figure 3. Dynamic model of Side Impact.

3. Dynamic model of Side Impact

The simplified dynamic model for vehicle side impact (MDB) impact is illustrated in figure.3. In this model the struck side of the vehicle is set as one part, M1 and other parts of the vehicle body are set as the rigid part, M2. M1 is connected with the vehicle body by linear elastic springs K1, representing the stiffness of the side structure of the vehicle. The spring K2 is equivalent to the stiffness of the door trim.

Let the impact force be F_{impact} and the side structure intrusion be $X_{intrusion}$. $K_{equivalent}$ is introduced to represent the combined spring constant of K1 and K2. The approximate relationship between F_{impact} and $X_{intrusion}$ can be described according to Hooke's law:

$$F_{impact} = K_{equivalent} X_{intrusion}.$$

4. Side Structure of Passenger vehicles

In general passenger car side structure consists of A-pillar, B-pillar, sill, doors, and crash beams behind the doors. During the side impact (MDB), sill, B-pillar structure and A-pillar structure will be engaged as shown in the figure 4 & figure 5.The stiffness of a vehicle's side structure is much lower than that of the front structure and there are not enough components or space to absorb the impact energy. Consequently, there will be a substantial intrusion concentrated in the impact zone.

IOP Conf. Series: Materials Science and Engineering **314** (2018) 012021 doi:10.1088/1757-899X/314/1/012021

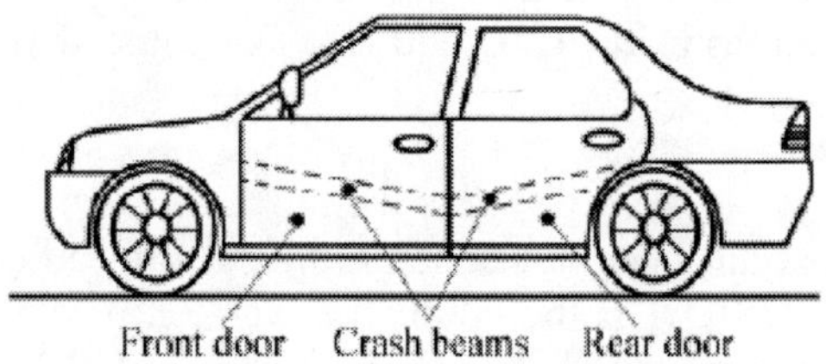

Figure 4. Vehicle side structure.

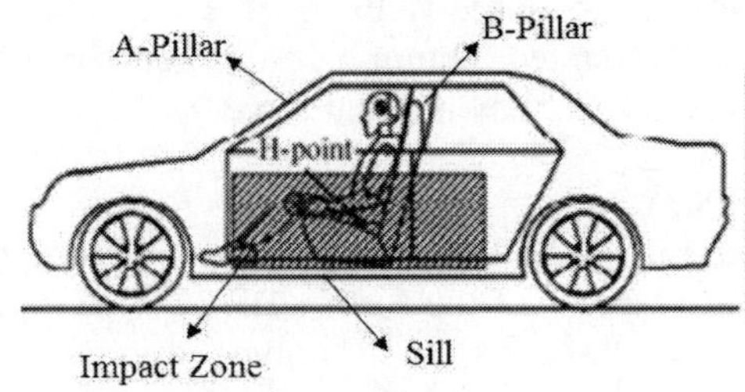

Figure 5. Impactor contact on side structure.

4.1. Side section of B-pillar, Sill, Door.

In this study three different vehicles are considered. The sections of B-pillar, door along with crash beam and sill of each vehicle are taken. These three vehicles have the nomenclature as V1, V2, and V3. This will be used in the entire paper for comparison. The section A-A is the b-pillar section and Section B-B is the door section with sill and crash beam taken in the respective vehicles as shown in figure 6 and figure 7. We could observe from the comparisons that the vehicle V2 has more reinforcements on the b-pillar section and the cross section of the crash beam is larger in vehicle V3. Materials used for these sections are cold rolled steel, high strength steel, dual phase steel and hot formed steel.

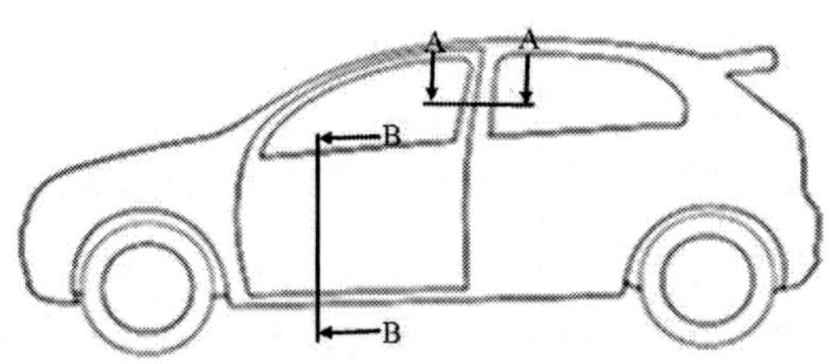

Figure 6. Section A-Aand Section B-B

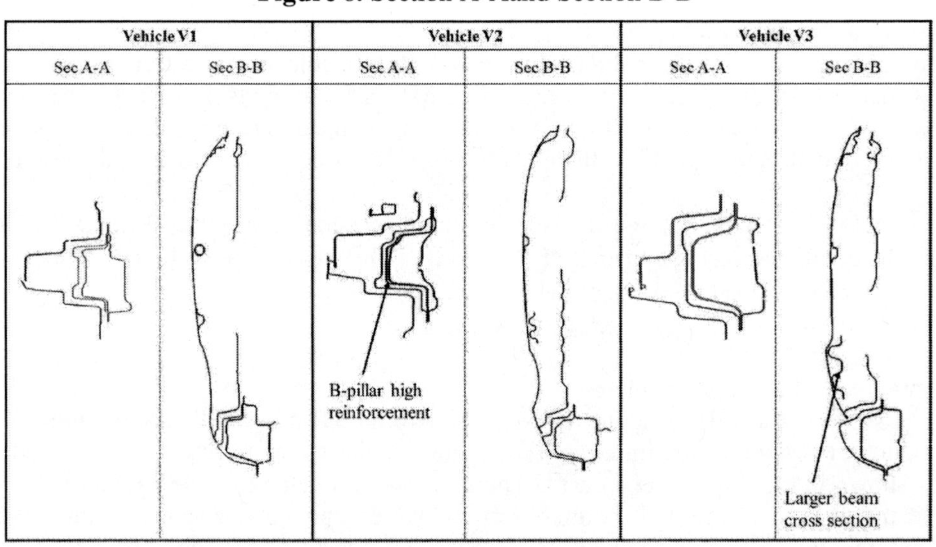

Figure 7. Sections of B-Pillar and Door

4.2. Space comparison

Space available around the ES II dummy when it is seated at intended position on the driver side is tabulated for all the three vehicles in table 1.These spaces are important for our study as we will be able to know how the vehicle performs with respect to space in the event of side collision. It can be observed from the values that the vehicle V2 has small space when compared to vehicles V1 and V2.

Table 1. Space Comparison of vehicles.

Vehicle Model	Head to window	Shoulder to window/padding	Elbow to door	Pelvis to door	Knee to door/padding	Side Restraints
V1	166	135	69	144	215	Yes
V2	220	80	62	129	112	No
V3	215	134	55	152	150	No

5. Deformation

Post-test the deformation profiles of the side door and the B-pillar of all the considered vehicles are shown in the figure 8 & figure 9. It is seen that the deformation is small in Vehicle V2 and V3 compared to V1. This can be attributed due to fact that the B-pillar of the vehicle V2 has more reinforcements and the presence of larger crush beam in the door in vehicle V3.

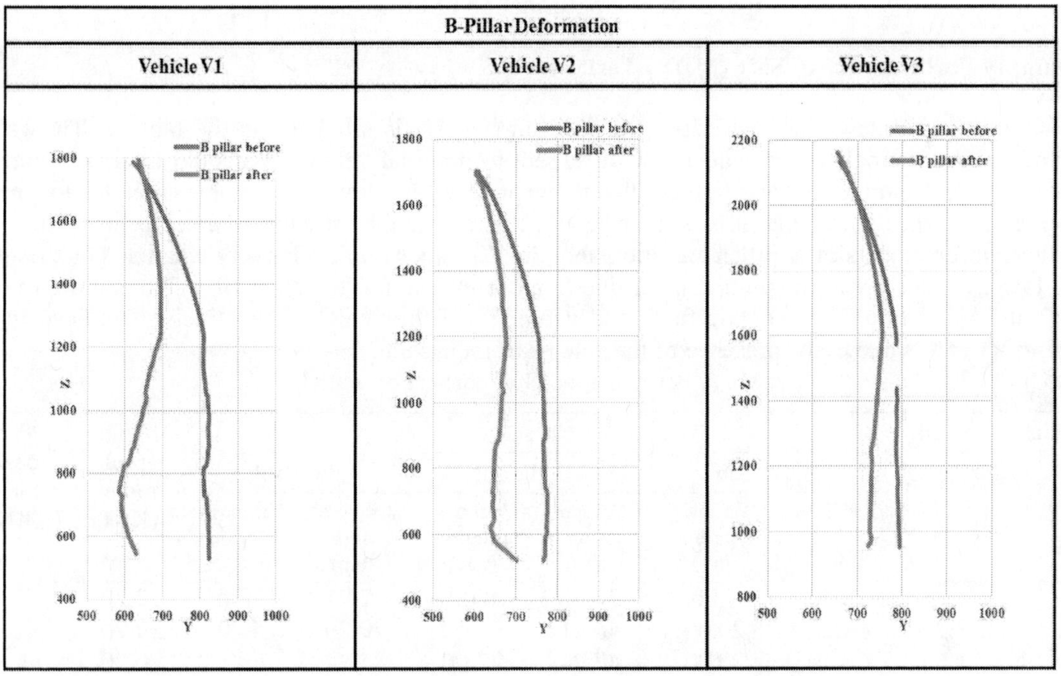

Figure 8. B-Pillar deformation of the vehicles.

IOP Conf. Series: Materials Science and Engineering **314** (2018) 012021 doi:10.1088/1757-899X/314/1/012021

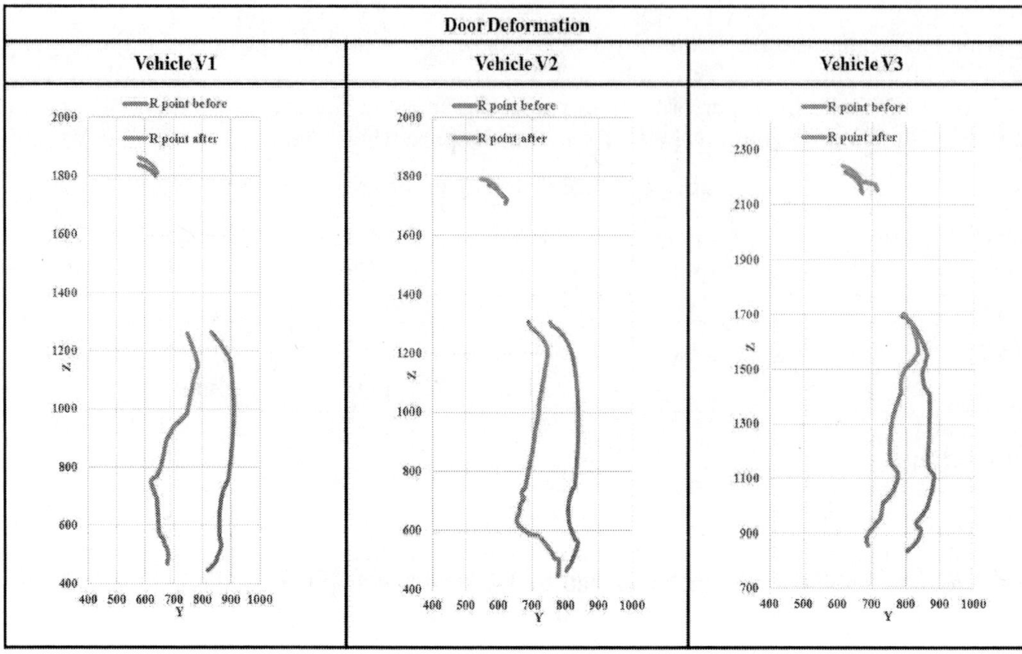

Figure 9. Door deformation of the vehicles.

6. Dummy Performance in Side (MDB) Test.

The dummy performance in each case after side impact test is tabulated in the table 2. The values shown here are normalised percentages with respect to nominal values. We could interpret from the values in table 2 and figure 8 and figure 9 that the vehicle V2 has lower rib compression due to smaller deformation of the door and B-pillar similar to V3. It is due to the presence of crash beam in the door and the reinforcements at B-pillar, the intrusions in V2 and V3 are relatively smaller. On the other hand lower rib compression is more in V1 due to more intrusion. The upper rib value is more in case of V2 and V3 due to higher pelvic force and geometry of the vehicle respectively. The head injury value in V1 is less due to the presence of the side restraint system.

Table 2. Performance Comparison of vehicles.

Vehicle	Head	Chest						Abdomen peak force (KN)	Pelvic peak force (KN)
	HPC	Upper Rib		Middle Rib		Lower Rib			
		Compression (mm)	Viscous criterion (m/s)	Compression (mm)	Viscous criterion (m/s)	Compression (mm)	Viscous criterion (m/s)		
V1	1.14	24.52	4.00	34.60	8.00	47.71	15.00	27.20	17.33
V2	8.23	64.26	20.00	48.21	17.00	29.29	15.00	39.20	36.33
V3	1.83	64.19	20.00	47.24	27.00	25.84	7.00	20.80	11.50

7. Conclusion

This paper investigates the sensitivity vehicle parameters (Interior space, B-pillar and door section deformation) which affect the injury parameters of ES-II dummy in the event of side (MDB) collision. It can be noted that, higher the intrusions higher is the injury parameters. Rib deflections are one of the important parameters to be monitored in the side impact. V1 which has relatively higher intrusion, resulted in higher lower rib deflection whereas V2 and V3 has relatively lesser lower rib deflection due to lesser intrusion.

These intrusions can be managed by having reinforcements on the B-Pillar and crash beam on the door. The reinforcements can be decided based on the space availability around the occupant in the cabin at the concept stage itself, thus reducing the injury value by reducing the intrusion level. In addition to these providing side restraints helps us in getting better performance of the occupant.

8. References

[1] Teng T L, Chang K C and Nguyen T H, *Crashworthiness Evaluation of Side-Door Beam of Vehicle* (Technische Mechanik October 2007) p 290-292
[2] WangDazhi, DongGuang, Zhang Jinhuan and Huang Shilin, *Car Side Structure Crashworthiness in Pole and Moving Deformable Barrier Side Impacts* (Tsinghua Science And Technology December 2006) p 728
[3] Automotive Indian Standards: AIS-099: Approval of Vehicles with regards to the Protection of the Occupants in the event of a Lateral Collision.
[4] Waagmeester C D, Twisk D, Philippens M M G M and Van Ratingen M R, *Status of Side Impact Dummy Developments in EuropeES-2 Prototype Development* (EVC WG12 Report August 2001) p 3-4
[5] Euro NCAP side impact test protocol.
[6] rassi.org.in/publications.html
[7] http://www.iihs.org/iihs/topics/t/general-statistics/fatalityfacts/passenger-vehicles#Crash-types, 8th March 2014.

IOP Conf. Series: Materials Science and Engineering **314** (2018) 012022 doi:10.1088/1757-899X/314/1/012022

State of the Art Control Measures for Aluminium Fade and SEN Clogging during Steelmaking Operations

Ashok Kamaraj [1#], R Saravanakumar [2], M Rajaguru [3]

[1] *Metal Extraction and Forming Division, CSIR-National Metallurgical Laboratory, Jamshedpur, India – 831007*
[2] *Alumni, Indian Institute of Technology, Kanpur, India*
[3] *Rashtrya Ispat Nigham Ltd (RINL), Vishakapatnam, India*
[#] *srhsbioboy@yahoo.co.in / ashokk@nmlindia.org*

Abstract: Crack formation, fatigue failure of components and other process interruptions in liquid steel practices such as ladle nozzle clogging, SEN clogging, break outs are mainly due to residual deoxidation products such as alumina present in steels. The present paper deals with the issues in steel processing operations and provides state-of-the-art control measures for clean steel production. Investigations regarding the residual alumina content and its consequences at integrated steel plant shows that, Al-fade of maximum 0.02% is observed, which produces equivalent of 37 Kg of alumina in the liquid steel. Furthermore, slag carry over, re-oxidation, improper argon rinsing practice, aluminium consumption at secondary steelmaking practices also influences the alumina formation during steelmaking practices. The residual alumina not only affects the quality of steel, but also results in process interruptions such as ladle choking, SEN choking, subsequent break outs etc. various steelmaking practices influences clogging and aluminium fade are discussed and possible suggestions are given to improve the cleanliness of steel.

1. Introduction

Since 1960, secondary steelmaking practices are employed in steel plants for production of quality steels. Secondary steelmaking practices primarily includes argon purging, electric arc heating, vacuum and wire treatment. Deoxidation, desulphurisation, degassing and inclusion control are the refining techniques involved in secondary steelmaking sections to enhance the quality of steels. Though the steel plants are practicing tough control in quality of steel products during steelmaking processes, there is always ample scope to improve it. Production of steel at every stage is technologically significant, as it dictates steel quality as well as the service performance of various steel components. Inclusion free steels demand careful process conditions during various stages of refining treatment.

Various steel plants are suffering from process interruptions such as ladle nozzle clogging, SEN clogging, break out etc during continuous casting of billets and blooms. It affects the day to day production in many folds. Problems related to castability mentioned above are mostly arises due to improper deoxidation and incomplete removal/modification of non-metallic inclusions in liquid steel. Steels produced for automotive and structural applications are aluminium killed, as it is a prerequisite for continuous casting of liquid steel. It is well known that, aluminium killed steels are seriously suffering from castability issues due to difficulty in removal of fine alumina inclusions formed in later stages of refining practice.

Ladle containing liquid steel is hold ideally for minimum of 10-15 minutes between argon rinsing/ladle refining station and before to start casting of liquid steel in continuous caster. During this period the liquid steel in ladle is subjected to only thermal convection and however, all possible natural convections are arrested. Slag metal interfacial reaction is expected to takes place during this period. Therefore, it is always expected that change in concentration of trace elements present in liquid steel. Elements like aluminium, calcium, titanium, boron, magnesium etc are prone to diminish during these holding periods. A little change in these solutes causes significant changes in characteristics of non-metallic inclusion as well as hampers the thermo-mechanical processing of steels in later stages. Diminishing of aluminium (aluminium fade) in liquid steel indicates the formation aluminium oxide or aluminium nitride. Precipitation of alumina in liquid steel before casting ensures poor castability in continuous caster. It encourages nozzle clogging, SEN clogging, break outs and other quality defects in cast billets/blooms. However, owing to the high oxygen affinity, aluminium in liquid steel could not be avoided from fading. Detrimental effect of oxides precipitates in the later stages of refining on castability also should be controlled. The alumina particles formed during deoxidation by aluminium at the ARS or in later stages, if not controlled, also contaminate the final steel product and decrease the cleanliness index. These inclusions pose several metallurgical problems in the final steel. Ductility is appreciably decreased with increasing amounts of oxides and sulphides in steels [1]. Larger inclusions act as stress riser, which induce cracks in the finished products [2] and also they lower the resistance to hydrogen induced cracks [3]. Hard and brittle inclusions are also act cause for fatigue failure of steel components [4].

Therefore, in the present investigation efforts are taken to understand the effect of process parameters on aluminium fade in liquid steel during casting. Another aim of this investigation is to explore the scope to control the effect of aluminium fade such that, castability of aluminium killed steels should be maintained or improved. This paper provides a key understanding about the science behind the problem, and gives improved understanding to realize inclusion free steel.

2. The science of alumina formation and clogging

When the deoxidizer enters in to the liquid bath, it reacts with the dissolved oxygen in the melt to nucleate the oxidation product. Growth of particles follows the mechanisms includes Ostwald ripening [5], Brownian motion [6,7] and Stokes collision. Rising of oxide particles is naturally due to buoyancy, particles attachment to bubbles in gas stirred ladles and fluid transport also facilitates the oxide particles to float up to the top slag. It is also expected that some of the inclusions would stick to the refractory walls of the reactors.

The particles reaching the slag-metal interface have to react, separate and dissolve in the slag phase, which is governed by the dissolution kinetics of the particles in slag. It is further controlled by the composition, basicity, viscosity and melting temperature of the slag. Classical core reaction model [8] for dissolution manifests that up to 10μm, the rate limiting step is boundary layer diffusion and below 10μm, the rate limiting step is dissolution reaction between slag and inclusions.

Alumina inclusions are usually precipitates and grow in dendritic fashion, in high oxygen environment [9], cluster type in Al-killed steels [10] and coral like structure due to ostwald ripening of dendritic clustered or clustered alumina inclusions [9]. Alumina inclusions readily form a three dimensional clusters (about 1 - 5μm) via collision and aggregation due to their high interfacial energy [11, 12]. However, small size particles are entrapped in the liquid steel as inclusions and therefore transferred to the downstream operations along with the steel matrix.

In the tundish, transport of deoxidation products to the nozzle wall is facilitated by fluid motion. Within the tundish recirculation zone, turbulent velocity fluctuations oriented in all directions. Fluctuations towards the tundish wall enable accumulation and deposition of inclusion [13]. Turbulent eddies present in the tundish, even in the absence of recirculation zone will transport deoxidation products to the nozzle wall [14]. Increased nozzle roughness also increases the probability of interception of entrained deoxidation particles present in liquid steel [13,15]. Due to various such factors that the agglomeration of solid inclusions takes place at the tundish nozzle and results in so called unwanted process interruptions during casing termed as clogging or choking of SEN. Not only alumina, but also calcium aluminates or calcium sulphides present in liquid steels have also been observed to be responsible for clogging in calcium treated heats [15,16]. Sintering of alumina particles takes place at temperatures above 1530°C [17], where the particle-wall or particle-particle bonding is relatively rapid [18] and assist for faster progress in clogging phenomena.

3. Plant data collection

A brief overview of the process route followed in an integrated steel mill from where the samples and process data are collected for the present investigation is presented here. After tapping of liquid steel from the converter (BOF), the liquid steel is transferred to argon rinsing station (ARS) where chemical and thermal homogeneities are targeted. While rinsing with argon, aluminium wire is injected from the top to deoxidize the bath as well as alloying elements are added to meet the required chemical composition in predefined time period. If needed, the molten steel is taken to chemical reheating unit, to compensate the loss in temperature incurred during rinsing and transit between different. After all the secondary metallurgical operations are completed, the liquid steel is transferred to continuous casting section and teemed to the tundish for the commencement of casting.

Steel samples are collected from various sections (after tapping, ARS and during teeming) and it is analysed for the chemical composition using OES. Slag samples are also collected from the ARS, after completion of secondary metallurgical treatment. Slag samples are also analysed using XRF for the chemical composition.

4. Results & discussion

Based on the chemical analysis of steel samples collected from ARS and tundish, aluminium fade (difference in Al between ARS and tundish) is calculated and the results are presented in this section. **Figure 1** shows the variation of various identified process parameters with aluminium fade for the studied heats

Figure 1 (a) reveals that wt% aluminium fade proportionally increases with the amount of aluminium in liquid steel at tundish, irrespective of the steel grades. It could be attributed to the amount of aluminium present in liquid steel is more, causes the driving force available for aluminium fade high. Concentration gradient exist between liquid steel and top slag plays vital role in deciding the extent of fading in liquid steel refining process. **Figure 1 (b)** clearly shows the reverse trend for amount of aluminium added into liquid steel as deoxidizer. It is expected that the amount of Al added would be high for the heats/grades which have high wt% Al in steel composition. Eventually, Al present in liquid steel is high and also driving force for Al oxidation exists, even after ladle processing for various other reasons. Therefore, Al fade is expected to be high for the heats which consumes more amount of Al during secondary processing. However, the reverse trend observed for amount of aluminium added into liquid steel suggests that complete deoxidation. Such that alumina removal at early stage of process and also fine alumina inclusions find more time for floatation, otherwise alumina forms in latter stages as a consequence of aluminium fade. Therefore, these two contradicting factors originated for same reason and responsible for Al fade and decide the extent of alloying elements fading.

Figure 1 (c) shows that wt% aluminium fade increases steeply with increasing in tapping temperature of liquid steel in BOF convertor. Tapping temperature could be controlled by BOF convertor operational practice. It depends upon various process parameter such as after blow duration, end carbon, various additions during process like iron ore, lime, dolomite etc. However, importance of tapping temperature on Al fade needs to be addressed clearly. Very high tapping temperature would result in high amount of dissolved oxygen in liquid steel along with high amount of FeO in primary steelmaking slag. The residual FeO, which ever not reduced completely after deoxidation would be high for the heats tapped at high temperature from LD convertor. Therefore, improper slag killing always have adverse impact on recovery of alloying elements added during tapping and LRF station. **Figure 1 (d)** shows that Al fade increases gradually as the lifting temperature of liquid steel at LRF increases. In the standard industrial practice, liquid steel in ladle from BOF reaching at LRF station are having far lower than the temperature required by continuous casting section. Therefore, raising the liquid steel temperature to desired temperature at LRF station is one of the primary objectives. This effort of increasing liquid steel temperature at LRF causes substantial decrease in the aluminium present in liquid steel before casting due to driving force created by temperature increment.

Figure 1 (e) depicts the dependency of total rinsing time at LRF for liquid steel on aluminium fade. It shows that prolonged purging of liquid steel with argon either from bottom or top causes minimum aluminium fade before casting. This could be explained by the following reasons. Prolonged argon purging of liquid steel encourages excellent chemical and thermal homogeneity via., mixing. Once the system approaches the thermodynamic equilibrium, the slag - metal reaction is expected to complete. Therefore, the driving force for aluminium fade would not exist further and the fading occurs before

IOP Conf. Series: Materials Science and Engineering **314** (2018) 012022 doi:10.1088/1757-899X/314/1/012022

casting would be minimum. However, long term treatment of liquid steel at LRF will not always an encouraging industry practice. This would badly affect the availability of liquid steel for continuous casting section, since LRF is an intermediate section and expected to act as buffer between BOF and caster. Therefore, always an optimal time for LRF treatment should be identified and it should be maintained to avoid Al fade as well as to ensure continuous supply of liquid steel to caster.

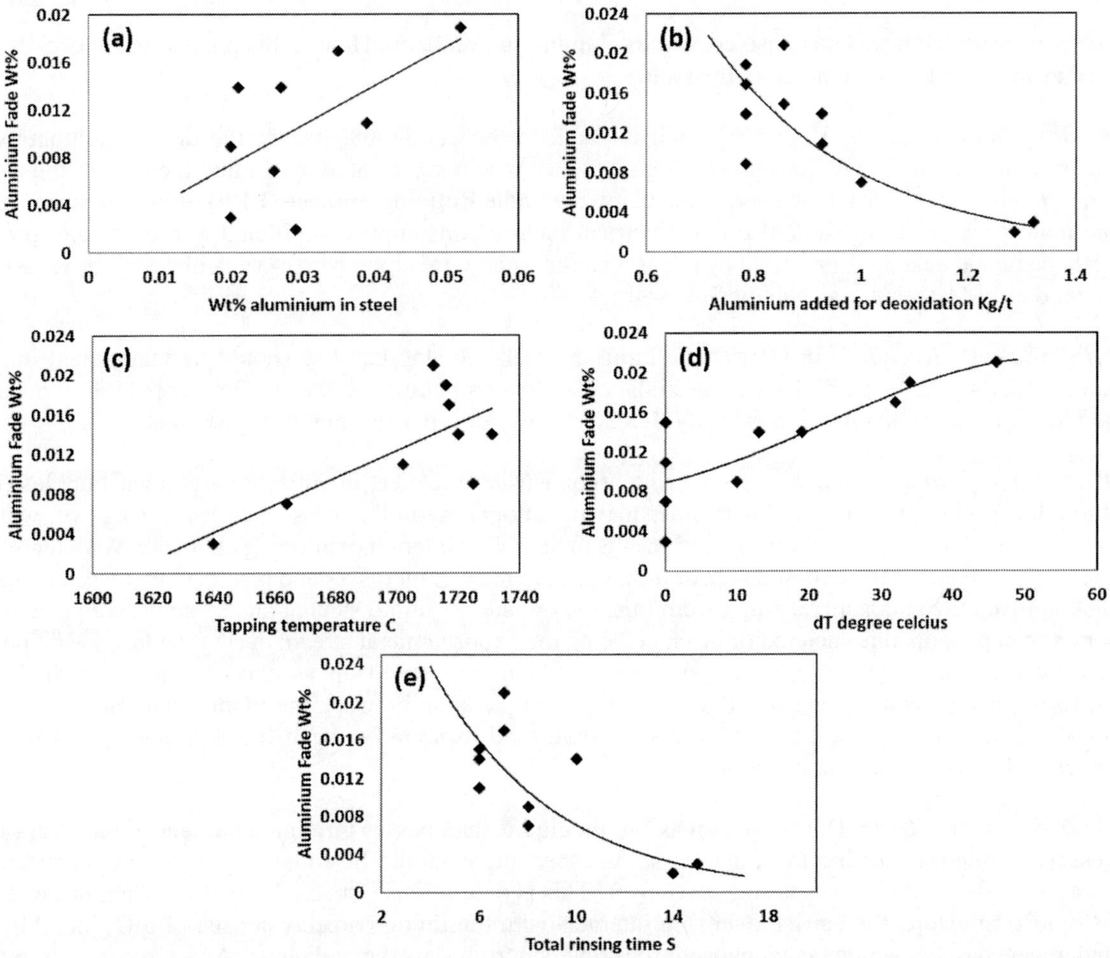

Figure 1: Variation of aluminium fade with various process parameters (a) wt% aluminium in steel, (b) weight of aluminium added as deoxidiser, (c) tapping temperature, (d) change in temperature of liquid steel between tapping and ARS and (e) total rinsing time at ARS

4.1. Reasons for formation of alumina

Though the reasons for aluminium fade exists in studied heats are discussed in previous section with various process parameters, a generic discussion on favorable conditions, prime reason for formation of alumina inclusion and the control measures in steel making practices are given in this section.

4.1.1. Deoxidation: Fe-Si, Fe-Mn are used as a deoxidizer in earlier days. Due to high residual silicon and manganese (≤ 0.5-1%) in the bath, now Aluminium is used as a deoxidizer in the form of wire/bar/shots. It results in dissolved oxygen content of below 5-20 ppm with residual Aluminium of 0.02- 0.04%. This practice leaves alumina as a de-oxidation product and the respective chemical reaction as follows,

$$2[Al] + 3[O] = (Al_2O_3) \quad (1)$$

4.1.2. Desulphurization: Sulphur transfer from the metal to slag may take place with the help of CaO added into the liquid steel. The corresponding reaction could be presented as,

$$FeS_{\ steel} + (CaO) = (CaS) + (FeO) \quad (2)$$

Oxygen liberated by this step also consumes aluminium or silicon. Hence, this process gives CaS as well as increased Al_2O_3 content in the bath.

4.1.3. Refining Practice: Before the melt is taken to the continuous caster, the drop in temperature incurred during all the previous processing steps has to be compensated by heating the melt to the right temperature (super heat). However, in the plant, the Ladle Refining Furnace (LRF) is located away from the main processing line and also due electrical energy consumption and lengthy processing time in LRF, chemical heating is preferred by injecting aluminium wire along with oxygen blowing those results in exothermic reaction via., equation (1)

4.1.4. Basicity Decline: For efficient refining of steel, the slag basicity should be maintained in the range of 2.8 – 3.2. If the basicity of secondary slag lowers below 2.8, the slag becomes highly viscous and has low inclusion retention capacity. It also causes improper refining of liquid steel.

4.1.5. Slag eye area & Reoxidation: During argon rinsing practice, the injected argon gas bubbles rise through the melt, pushing the slag to sides leaving an opening in the top slag called as slag eye, which exposes metal to the atmospheric air and results in aerobic oxidation of alloying elements. With increase in gas injection rates the exposed area of liquid steel in the eye increases and results in increased oxygen pick up from the atmosphere. Apart from this, the transfer of liquid steel at any stage provide a chance for oxygen pick up, one such major instance being the exposed metal stream during the furnace tapping. All these re-oxidation costs excess aluminium which might end up as excess alumina inclusions. N.A.Mc Pherson et al [19] reported that, the nitrogen picks up between the tundish and mold is found to be 5 ppm. Considering 250 T heat and assuming all aspirated oxygen forms alumina, this would generate 1.4 Kg of alumina in two heats.

4.1.6. Slag Carry Over: During steel making, the liquid steel passes through a number of metallurgical vessels. While transferring from one vessel to other, there should be some slag drained inadvertently along with the metal known as slag carry over. This is deleterious, especially the BOF slag, because of its oxidizing nature, the carried over slag interacts with the lining (fireclay consist of SiO_2 and Al_2O_3) and lowers basicity which favors phosphorous reversal from slag to metal phase. Apart from the problem of phosphorous reversion, this also necessitates increased addition of Al which results in increased alumina formation.

4.1.7. Inclusion Size: Stokes rise velocity for large size particle is higher than the smaller size particle. Therefore, that the smaller particles remain entrapped in the liquid steel. Argon rinsing assist agglomeration and coagulation of inclusion, so that inclusions can float up to the slag phase. However, it is experienced that floatation of fine alumina inclusions are challenging task for the steelmakers to produce extra clean steels.

4.2. Challenging Remedies:

4.2.1. Calcium Treatment: Calcium addition (Ca-Si powders filled in steel tubes) transforms pure Al_2O_3 to calcium aluminate inclusions, which is expected to possess liquid state at 1873 K, when the CaO content has reached 35%. If the Ca amount is too small, solid aluminates are formed $CaO.6Al_2O_3$, $CaO.2Al_2O_3$, $CaO.Al_2O_3$), which are even more detrimental than alumina inclusions with regard to nozzle clogging [14, 20].

$$3\,[Ca] + \langle Al_2O_3\rangle inc = 2\,[Al] + 3\,(CaO) \quad (3)$$

$$3\,\langle CaO\rangle inc + 2\,[Al] + 3\,[S] = 3\,\langle CaS\rangle inc + \langle Al_2O_3\rangle inc \quad (4)$$

In high sulphur steels (S>0.005%) the injected calcium causes desulphurization rather than inclusion modification, as explained by the above mentioned chemical reactions. It forms calcium aluminate inclusions surrounded by CaS rich rim of small size (5μm) that are easily deformable and therefore, mechanical properties of the steel would not adversely have affected. Calcium treatment cannot be suggestable recommendation for the steel plants operating with high sulphur content in liquid steels (> 100ppm).

4.2.2. Slag arresting using DART: DART is a device made of refractory used to control the slag carry over in BOF converter during tapping. The material of the DART should have density value in between steel and slag, such that it floats near to the interface and prevents carry over during tapping process. Improper calibration and positioning of DART are the problems associated and it is not a reliable device to control slag carry over in day to day practice for quality steel production.

4.2.3. Ar Injection through Stopper Rod: It is also a proposed mechanism being practiced in steel plants for instant removal of alumina deposited (clogging) in the tundish outlet i.e, mouth of the SEN. The cylindrical arm with argon injection facility is inserted right upright above the tundish outlet causes flushing of the deposits. The solid particles present in the choked region will enter into the solidifying strand and remain in the product as an inclusion. Therefore, the choking problem cannot be solved without compromising the steel quality.

4.3. Control Measures:

- Sulphur control right from the blast furnace avoids reaction (2 and 4) during steel refining practices. Employment of external de-sulphurization unit as a hot metal pre-treatment, addition of lime during tapping from BOF irrespective of the grade will decrease the sulphur content in liquid steels. The latter process improves cleanliness of steel, thereby decreasing the aluminium consumption and maintaining the appropriate basicity of secondary slag.
- Once, the sulphur content of the hot metal is controlled by either way, calcium treatment helps to modify the inclusion.
- Synthetic slag formation/addition helps to trap the inclusions present in liquid steel. The favourable slag of that kind is CaO- 50 % SiO2- 7 % Al2O3:43 and should have a melting temperature lower than 1350°C [22].
- Installation of IR-Camera with slide gate mechanism to control the slag carry over during liquid steel transfer practice from one metallurgical vessel to other vessel.
- Employing ladle furnace in the on- line mode, thereby temperature drop can be adjusted by electrical heat rather than the chemical heating practices.
- Optimization of argon rinsing practice is required to control the slag eye opening and hence oxygen picks up and alloying elements fading can be minimized.
- In ARS, the heat should be finished with a gentle rinsing, which will let the smaller alumina to agglomerate and float up with much less temperature drop and re- oxidation.

These types of process modifications will assist to some extent to improve the alumina formation, thereafter reduces nozzle clogging. Apart from these, the present investigating authors suggest to re-engineering of the tundish design for enhanced floatation of inclusion.

- Re-Engineering of Tundish Design: Tundish being a reservoir for molten metal in continuous casting operation plays a major role in removal of inclusions to float out. This is facilitated by efficiently modifying the flow fields by using flow modifiers. For instance, L.Zhang et al [23] studied the inclusion removal efficiency and reported that the flow control devices are favorable for the inclusion removal. The total removal ratio is 51 wt% without flow control devices and the inclusions bigger than 72μm in radii can be totally removed. The total removal ratio is 79 wt% with flow control devices and the inclusions bigger than collision 61μm in radii can be totally removed. After a decade, Rajasekar [24] improved the tundish design with flow modifiers, his mathematical model based on population balance approach suggests that, operating the tundish at higher bath height and higher flow rate one can attain the inclusion removal efficiency of about 82.85% and the number density of inclusion larger than 35μm is significantly less in value. Zhang et al [25] summarizes from various research works towards inclusion control that, the two-layer (upper layer- rice hull; lower layer – low melting basic flux) flux practice in tundish, protection by argon gas from steel shrouding from ladle to mold, new techniques at SEN (such as swirl nozzle technique, step nozzle technique, *multiport* nozzle, and oval offset bore throttle plate etc.) are quite useful *and promising* techniques. Hence, the authors suggest to re- engineer the tundish design along with process modifications, s*uch* that *the* choking problem *may be* solved without contaminating the final product and hence the dream of clean steel *could* be realized.

5. Conclusions

Investigation carried out provided a chance to study the alumina formation at successive stages in steel making operations. This paper emphasizes the science of-alumina formation, influence of plant practice in aluminium fade and hence, formation of alumina. At the end it suggested some process modifications to reduce the population of alumina and also it is recommended to re-engineer the tundish design for tackling the choking problem towards clean steel production.

6. References

[1] P. K Trojan, 1988 *ASM Hand Book,* ASM International Metals Park **15**, pp 88-97

[2] T.J. Baker and J.A.Charles, 1977, *The Iron and Steel Institute*. Scarborough, UK, pp 88-94.

[3] A. D Wilson, 1999, Advances in the Production and Use of Steel with Improved Internal Cleanliness, J.K. Mahaney, ed ASTM Philadelphia PA, pp 73-88.

[4] E.Fuchs and P.Jonsson, 2000, *High Temp Material Proc* **19**, pp 333-343.

[5] C. Wagner, 1961, Z, *Elektrochem* **65**, pp 581-591.

[6] E. T Turkdogan, 1972, *J.Iron Steel.Inst,* **210(1),** pp 21-36

[7] Y.Miyashita and K. Nishikawa, 1968, *Trans.ISIJ* **8**, pp 181-185.

[8] R.Rastogi and A.W. Cramb, 2001, *Steel Making Conf. Proceedings ISS PA*, **84** pp 789-829.

[9] H. Ooi, T. Sekine, and G. Kasai, 1975, *Trans.ISIJ* **15**, pp 371-79

[10] L. Zhang and B.G. Thomas, 2002, *7th Eur. Electric Steelmaking Conf.* venice, Italy **11**, pp 2.77-2.86.

[11] R. Dekkers, B. Blanpain, and P. Wollants, 2003, *Metall.Trans.B* **34**, pp 161-171.

[12] S. Dawson, N.D.G. Mountford, I.D. Sommerville, and A. McLean: 1990, *Iron and Steel Maker*, pp 33-42.

[13] F.G.Wilson and T. Gladman, 1987, *Iron Making and Steel Making* **14**, pp 296-309.

[14] W. Fix, H. Jacobi and K. Wünnenberg, 1993 *Steel Research,*.**64**, pp 71-76.

[15] Y.K.Shin, 1988, *Iron Making and Steel Making* **15**, pp.143-149.

[16] T.B. Braun, J.F. Elliott, and M.C. Flemings, 1979, *Metall.Trans.B,* **10B** pp 171-184.

[17] Ken-ichiro Uemura, Masamitsu Takahashi, Shinji Koyama, Masaki Nitta, 1992 *ISIJ International*, **32**, pp 150-156.

[18] N.A.Mcpherson, 1985, *68th ISS Steel Making Conference*, Detroit , **68 ,** pp 13-25

[19] G.M.Faulring, 1980, *Iron and Steel Maker* **7**, pp 14-20.

[20] Deo. B, Boom. R, 1993 "Fundamentals of Steelmaking Metallurgy" pp 267.

[21] Deo. B, Boom. R 1993 "Fundamentals of Steelmaking Metallurgy" pp 225.

[22] L.Zhang and S. Taniguchi, 2000, *Metall.Trans.B* (2000) **31 B** pp253-266.

[23] K.Rajasekar, 2009, M.Tech Thesis, IIT- Kanpur.
[24] L. Zhang, B.G. Thomas, 2003, XXIV *National Steelmaking Symposium*, Mexico, pp 138-183

IOP Conf. Series: Materials Science and Engineering **314** (2018) 012023 doi:10.1088/1757-899X/314/1/012023

Multi-response optimization of process parameters for GTAW process in dissimilar welding of Incoloy 800HT and P91 steel by using grey relational analysis

Lakshmanan vellaichamy[1], **Sathiya Paulraj**[1,*]
[1] Department of Production Engineering, National Institute of Technology, Tiruchirappalli 620015, Tamilnadu, India
*Corresponding author: psathiya@nitt.edu; Tel.:+91 431 2503510; Fax: +91 431 2500133

Abstract- The dissimilar welding of Incoloy 800HT and P91 steel using Gas Tungsten arc welding process (GTAW) This material is being used in the Nuclear Power Plant and Aerospace Industry based application because Incoloy 800HT possess good corrosion and oxidation resistance and P91 possess high temperature strength and creep resistance .This work discusses on multi-objective optimization using gray relational analysis (GRA) using 9CrMoV-N filler materials. The experiment conducted L9 orthogonal array. The input parameter are current, voltage, speed .The output response are Tensile strength, Hardness and Toughness. To optimize the input parameter and multiple output variable by using GRA. The optimal parameter is combination was determined as A2B1C1 so given input parameter welding current at 120 A, voltage at 16 V and welding speed at 0.94 mm/s. The output of the mechanical properties for best and least grey relational grade was validated by the metallurgical characteristics.
Key words: optimization, Grey relational grade, Microstructure, Fractography

1. Introduction

In next generation of power plant and nuclear plant industry, the dissimilar welding was used in many applications. They are used in super heater, reheater and heat exchanger tube etc Earlier the ferritic and austenitic based materials were used for this dissimilar joint but to create many problem such as cyclic thermal stress between two base metal due difference in co efficient of thermal expansion and also due to stress oxidation cracking and accelerated creep which occur in ferritic side interface. To avoid this issue one such material the austentic based Incoloy 800HT and ferritic based P91 materials are used [1].The Incoloy 800HT are nickel based super alloy and oxidation and carburization resistance at high temperature and also they possess good corrosive resistance steel[2]. But in ferritic based P91 materials they have high tensile strength and creep properties compared to other ferritic steels and also it's a excellent mechanical properties [3]. This dissimilar welding is suitable for GTAW process in order to avoid the cyclic thermal stress to using nickel based filler material .Additionally GTAW process is being multi pass welding they have improve tempering of welds and also toughness of weld can be increased[4].In general the quality of the weld depends on the Input process parameter .Traditionally the skill of the operator and engineer is considered while selecting the trail and error method with suitable process parameter , also it is more time consuming and as the required specification of the weld does not meet some times. So to overcome this problem the author has used design of experiment (DOE), evolutionary algorithms and computational network to select the suitable welding parameter and required output and meet the specification.[5]. Gray relational analysis (GRA) is used to convert multiple responses into single response by assigning gray relational grade (GRG) to individual responses [6]. Shanmugarajan .B et al.[7] applied Taguchi based gray relation analysis results to optimum parameters have been derived by considering the responses such as depth of penetration, weld width and heat affected zone (HAZ) width. Hsuan-Liang Lin et al [8]. Analysis of dissimilar welding between JIS SUS 304 stainless steel and SAE1020low carbon steel. This paper presents an integrated approach using the Taguchi method (TM), grey relational analysis (GRA) and a neural network (NN) to optimize the weld bead geometry in a novel gas metal arc (GMA) welding process K.F. Tamrin et al [9]. applied the GRA technique and it was found that the weld speed has dominant effect on joint characteristics in comparison to stand-off distance at a fixed laser power. R. Adalarasan et al[10]. The G-PCA method employed for optimization combines the non-parametric approach of PCA with the uncertainty handling capabilities of GRA to predict the optimal setting of parameters in friction welded dissimilar joints involving AA 6061 and AA 6351 alloys. Deepan Bharathi Kannan et al[11] successfully implemented GRA technique for predicting optimization parameter in laser welding of Nitinol shape memory alloys.

From the above literatures, it is clear that none have been carried out in dissimilar welding of Incoloy 800HT and P91 steel using GTAW process in optimization and characterization. It was also clear that GRA can be used to optimize the GTAW process parameter to obtain the desired quality weldments. In this work, multi-objective optimization using GRA has been carried out to optimize the GTAW process parameter (welding speed, current and voltage). The output response was tensile strength, Impact Toughness and Hardness in room temperature. By analyzing the grey relational grade, the most influential factor was determined. Further, the best and least grey grade obtained experiments mechanical properties were validated by their metallurgical features like microstructure and SEM fractography.

2. Experimental Procedure

The base metal dimension is 200mmX 100mm X 4mm cutting by using wire cutting EDM machine and the chemical compositions of the as-received base metals and filler metals have been studied using wet spectroscopic method and are presented in Table 1.

Table 1. Base material and filler material Chemical compositions (in wt %)

		C	S	N	Cr	Ni	Mn	Si	Ti	Nb	Cu	P	Al	Co	Mo	V	Fe
Base Materials	Incoloy 800HT	0.06	0.002	0.013	20.63	30.56	0.69	0.46	0.34	0.01	0.10	0.008	0.87	0.08	-	-	46.177
	P91	0.11	0.0038	-	9.46	0.09	0.46	0.31	-	-	-	0.015	-	-	0.88	0.20	Bal
Filler Material	9CrMoV-N	0.12	0.003	0.04	9.10	0.42	0.72	0.26	-	0.06	0.08	0.008	0.002	-	0.94	0.19	Bal.

The filler metals such as 9CrMoV-N have been chosen as these filler metals are compatible with the base metals employed, and also thickness of filler materials is 2.4 mm. Standard V-groove butt joints (single V-groove having a root gap of 2 mm, size land of 1 mm, and included angle of 70°) were employed for welding these dissimilar metals The weld experiments were conducted by using manual GTAW with direct-current and negative electrode. The GTAW process parameters and levels are shown in table 2.

Table 2. Welding process parameters and levels

Factor	Unit	Level1	Level2	Level3
Welding current	A	90	120	150
voltage	V	16	18	20

Welding speed	mm/sec	0.94	0.98	1.2

Several trials were carried out to select the upper and lower levels of the process parameter. Taguchi L9 orthogonal array was selected, and the experiments were carried out accordingly. The process parameter and their levels are given in Table 3. Shows the process parameters used during the welding process.

Table 3. Experimental desgin using L9 orthogonal array and performance result

Experiment number	Current (amps)	Voltage (volts)	Speed (mm/s)	Tensile strength (Mpa)	Hardness (Hv)	Toughness (J)
1	90	16	0.94	598.24	181.215	46
2	90	18	0.98	603.12	180.876	47
3	90	20	1.2	583.14	187.651	43
4	120	16	0.98	634.34	165.976	52
5	120	18	1.2	586.23	193.124	43
6	120	20	0.94	644.21	169.243	51
7	150	16	1.2	638.67	175.908	50
8	150	18	0.94	642.26	161.745	51
9	150	20	0.98	588.17	183.654	44

A specially designed fixture with a copper back plate that could clamp the base metals firmly so as to avoid distortion and bending. During welding to maintain Interpass temperature at 200°C.The weldments were characterized for any defects using gamma ray radiography technique. Subsequent to the results, after welding the samples are shown in Figure.1

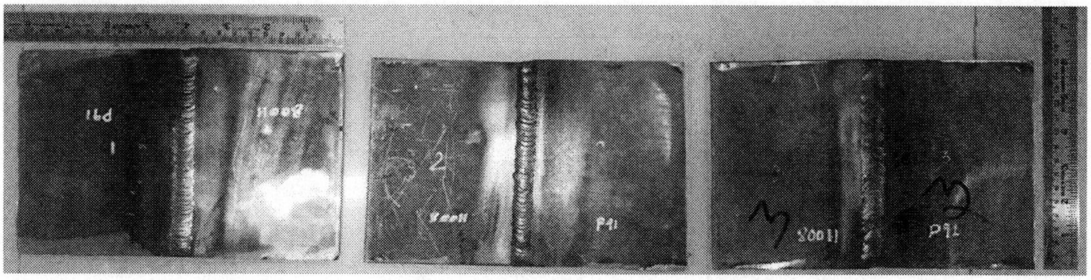

Figure 1. Welded specimens of 9CrMoV-N filler materials

The welded samples were cut into different coupons for further analysis. For metallographic analysis using Sic paper for different grade like 120 upto1200 and further automatic polishing machine the specimen was polished by mirror to provide mirror like polish , The specimens were etched in two different solutions: electrolyte etching solution (10g of CrO_3 +100ml distilled water) followed by villas reagent solution (1g of $C_6H_3N_3O_7$ + 5ml of HCL + 100ml of C_2H_6O). Was employed to examine the microstructure of Incoloy 800HT and P91. Further to estimate the mechanical properties of dissimilar joints various tests were performed. Nine samples were tested for tensile, Hardness and impact test for determining results

Tensile studies were carried out on the weldments which were fabricated as per ASTM E8/8M standards. These samples were tested at a strain rate of 2 mm/min at room temperature. Charpy V-notch impact test samples were prepared and tested according to ASTM E23-12C standards of sub-sized specimens by simple beam method. Notches were made such that fracture occurred only within the weld fusion zones. Furthermore the fractured samples were characterized for SEM analysis to determine the mode of fracture. Hardness measurement was carried out on the composite region of the weldments across weld zones by using Vickers's micro-hardness tester with a load of 500grams for a dwell time period of 10 s at regular intervals of 0.25 mm.

2.1 Grey relational analysis (GRA)

The transformation of S–N ratio values from the original response values was the initial step. For that the equation (1) of 'larger the better' was used. Subsequent analysis was carried out on the basis of these S/N ratio values. This is shown in Table 4.

Table 4. S-N ratio for the responses

Experiment Number	Tensile strength (Mpa)	Hardness(Hv)	Toughness(J)	S-N ratio for UTS	S-N ratio for Hardness	S-N ratio for Toughness
1	598.24	181.215	46	55.53751	45.16388	33.25516
2	603.12	180.876	47	55.60807	45.14762	33.44196
3	583.14	187.651	43	55.31546	45.46702	32.66937
4	634.34	165.976	52	56.04644	44.40091	34.32007
5	586.23	193.124	43	55.36136	45.71672	32.66937
6	644.21	169.243	51	56.18055	44.57021	34.1514
7	638.67	175.908	50	56.10553	44.90571	33.9794
8	642.26	161.745	51	56.15422	44.17662	34.1514
9	588.17	183.654	44	55.39006	45.28001	32.86905

$S/N_L = -10 * \log(1/y^2)$ (1)

In GRA, initially the experimental data are normalized. By using this normalized data, grey relational coefficient were evaluated, the grey relational grade was obtained by averaging the GRC values related to selected experimental results.

2.2 Grey relational generation

GRG can be categorized into three types namely Smaller the Better, Larger the Better or Nominal is a better (NB) criterion. The preferred quality characteristics for ultimate tensile strength, yield strength and impact toughness are Larger the Better criterion; then it is expressed by using equation (2):

$$y_i*(k) = \frac{y_i(k) - \min y_i(k)}{\max y_i(k) - \min y_i(k)} \quad (2)$$

Where i=1,.....m; k=1,2,3,...n; m= no of experimental data, n= number of factor, $y_i(k)$=original sequvance, $y_i*(k)$ after gray relation generation; min $y_i(k)$ and max $y_i(k)$ are the minimum and maximum value of $y_i(k)$,respectively. The normalized value are shown in Table 5

Table 5. Sequences of each performance characteristic after data processing

Experiment Number	UTS	Hardness	Toughness
1	0.257	-0.641	0.355
2	0.338	-0.630	0.468
3	0.000	-0.838	0.000
4	0.845	-0.146	**1.000**
5	0.053	**-1.000**	0.000
6	**1.000**	-0.256	0.898
7	0.913	-0.473	0.794
8	0.970	0.000	0.898
9	0.086	-0.716	0.121

2.3 Grey relational coefficient (GRC)

The calculation for grey relation coefficient was done using equation (3):

$$\varepsilon_i(k) = \frac{\Delta \min + \omega \Delta \max}{\Delta_{oi}(k) + \omega \Delta \max} \tag{3}$$

Where

$\varepsilon_i(k)$ is the grey relation coefficient; Δ_{oi} is deviation among $y_o^*(k)$ and $y_i^*(k)$; $y_o^*(k)$ = ideal (reference) sequence; $\Delta\max$ = highest value of $\Delta_{oi}(k)$, $\Delta\min$ = least value of $\Delta_{oi}(k)$.

2.4 Grey relation grade

The grey relational grades (GRG) are determined by taking average of the Grey Relational Coefficient related to every observation as presented in equation (4):

$$\Gamma_i = \frac{1}{n} \sum_{=1}^{Q} i(k) \tag{4}$$

Where, Q = total quantity of responses and n denotes the quantity of output responses. The GRG represents level of relationship among the reference or ideal sequence and the comparative sequence. If larger GRG is obtained for the equivalent set of process parameters compared to other sets, it is considered to be the most favorable optimal setting.

3. Results and Discussion

The GTAW dissimilar welding process on Incoloy 800HT and P91 was performed according to L9 orthogonal array to investigate the effect of the welding process parameters, namely, welding current, voltage and welding speed on the output responses, ultimate tensile strength, Hardness and impact toughness. An effort has been taken to determine the best possible set of welding parameters for dissimilar welding the Incoloy 800HT and P91 effectively and efficiently.

3.1 Multiple response models using GRA

By using GRA complicated optimization, problem can be solved effectively. The higher grey relational grade will have better multiresponse characteristics. Table 6 shows the grey relational grade for all experiments. Hence, it is clear that experiment 6 has the optimal parameters setting for best multi-response characteristics, such as ultimate tensile strength, yield strength and toughness.

Table 6. Grey relational coefficient, grey relational grade and their Rank

Experiment Number	Gray Relation coefficient			Gray relation Grade	Rank
	UTS	Hardness	Toughness		
1	0.402	0.234	0.437	0.357	6
2	0.430	0.235	0.485	0.383	5
3	0.333	0.214	0.333	0.294	8
4	0.763	0.304	1.000	0.689	3
5	0.346	0.200	0.333	0.293	9
6	**1.000**	**0.285**	**0.830**	**0.705**	**1**
7	0.852	0.253	0.708	0.604	4
8	0.943	0.333	0.830	0.702	2
9	0.354	0.226	0.363	0.314	7

3.2 Mean Response table for GRG using S/N ratio

Mean response for the Gray relation grade value for every level of the input parameters is shown in Table 7. Its been calculated by taking the average for each level group in all the levels of process parameters. Since it denotes the level of correlation between references

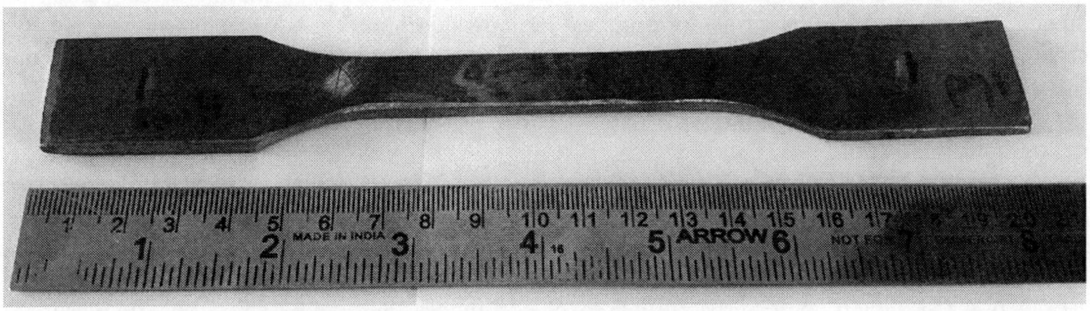

Figure 2. Tensile specimen before testing

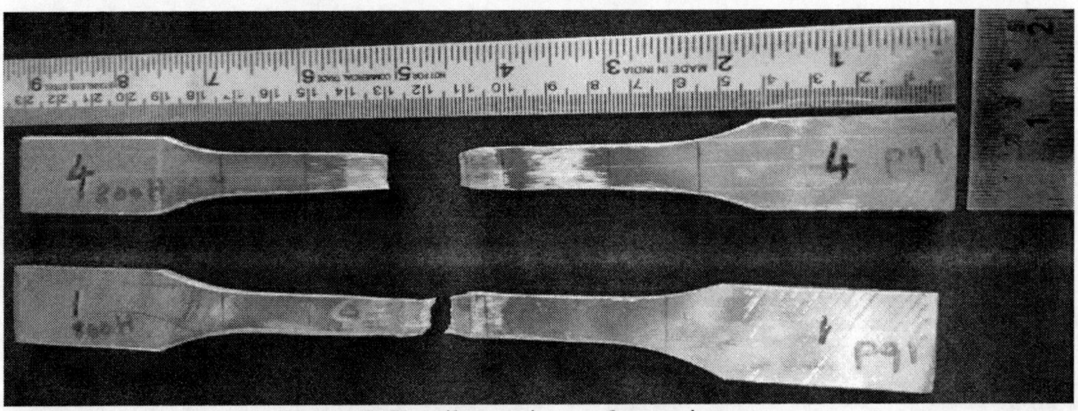

Figure 3. Tensile specimen after testing

Table 7. Mean response for the Gray relation grade

Parameters	Symbol	Level1	Level2	Level3	Main Effect (Max-Min)	Rank
Welding current	A	0.345	**0.562**	0.540113	0.218	1
Voltage	B	**0.550266**	0.459371	0.437455	0.11281	3
Welding speed	C	**0.588132**	0.462019	0.396942	0.19119	2

Sequence and obtained sequence, the higher value of average grey grade indicates stronger correlation between them. It indicates optimal level of process parameters. It depicts that the optimal set of process parameter is A2B1C1, which means the current of 120 A, voltage of 16 V and welding speed of 0.94 mm/s. Current was the main influencing factor, followed by the welding speed and voltage.

The influence of GTAW welding process parameters on the output response is shown in table.3. Tensile testing is one of the most basic types of mechanical test used to qualify a welding procedure. By performing this test, it is easy to determine how the material will react against forces when applied in tension. As the material is being pulled, the strength of the weld joint along with the elongation can be calculated. The point of failure is of much importance, and it is typically called as

IOP Conf. Series: Materials Science and Engineering **314** (2018) 012023 doi:10.1088/1757-899X/314/1/012023

ultimate tensile strength. The base material ultimate tensile strength of Incoloy 800HT in and as received condition was 530 MPa and P91 is 580Mpa. Before testing of tensile specimen as shown in Figure .2. The entire welded specimen exhibited higher strength than the base material. This section discusses the reason for the need for variation in output response for experiment 6 and 8 in relation to some of the metallurgical characteristics. Figure. 3 shows the fractured location for the tensile tested specimens.

From the Figure.3 it is seen that the fracture location is 10mm away from the weld center for experiment number 4, whereas the fracture occurred in the weld region for experiment number1. From the output result obtained during tensile testing, The tensile value 1.95 Mpa and GRC value 0.003 and more for experiment 6 as compared to experiment 8.

.

3.3 Fractography analysis

The impact tested specimen was analyzed on the fractured surface by scanning electron microscope to investigate the type of fracture that has occurred. There were two types of fracture, brittle and ductile, depending on the ability of the material to undergo plastic deformation before the fracture. The ductile mode of fracture undergoes extensive plastic deformation ahead of the crack, and the crack was stable which could resist further extension unless the applied stress was increased. In brittle mode, the crack was unstable and propagated rapidly without increase in applied stress. The ductile mode of fracture is preferred in most of the applications. Figure. 4a corresponds to the fractography images taken for experiment 6 showed deep and wide dimples which revealed ductile fracture mode, whereas Figure. 4b which corresponds to experiment 8 showed mostly cleavages micro voids which revealed brittle mode of failure.

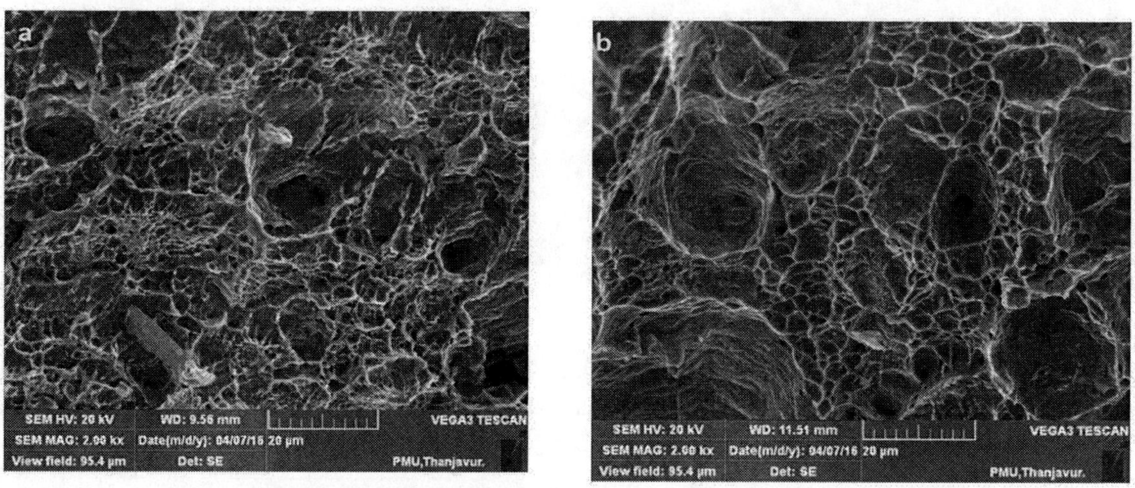

Figure 4 a &b Fractography Images of Tensile samples

3.4 Microstructure analysis

The microstructure of the weld region for experiments 6 and 8 is shown in Figure. 5a & b. The microstructure for experiment 6 shows in Figure 5a.The full austenitic structure carrying fine cellular and equiaxed grains over the welded regions, there is no hard dendritic structure during solidification period because experiment 6 low heat input process compare to experiment 8 .whereas the experiment 8 weld zone microstructure showed in Figure 5b.The weld zone Microstructure cellular and harder dendritic structures, which are prone to cracking. The harder dendritic structure may be due to the higher heat input during the welding. The structure of the weld zone clearly explains the need for the variation in mechanical strength for experiment 6 and 8.from the weld metal Incoloy 800HT side is twinning available so tensile strength is automatically increased.

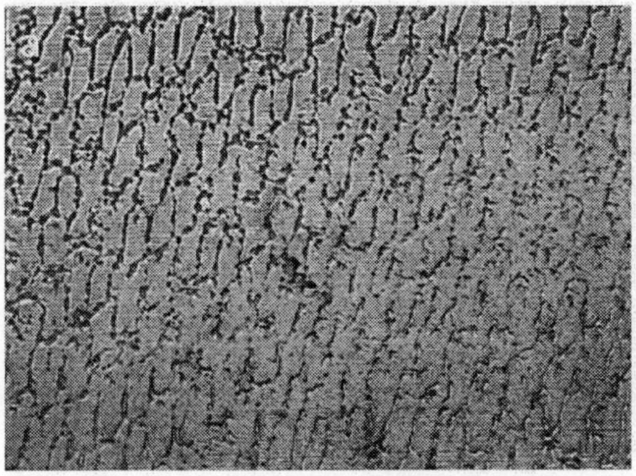

Figure.5 (a) weld zone microstructure experiment 6

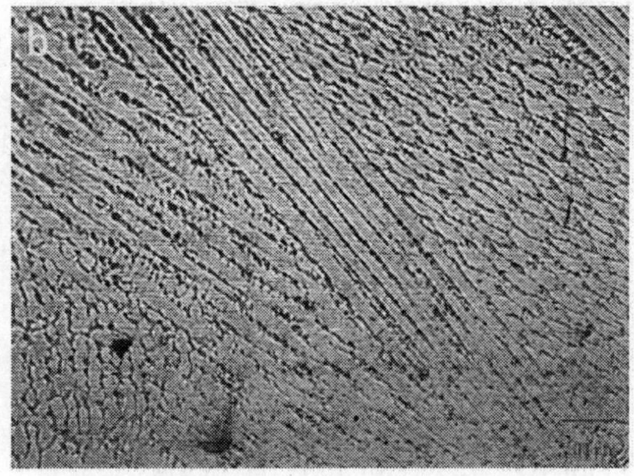

Figure.5 (b) weld zone microstructure experiment 8

3.5 Results of confirmatory test

The results of confirmatory test are shown in Table 8. It is found that prediction of optimal parameter setting is valid.

Table 8. Results of the confirmation test

		Optimal control parameters	
	Initial condition	Prediction	Experimental
Levels	**A2B3C1**	**A2B1C1**	A2B1C1
Pitting potential	0.705	0.736	0.762

Improvement of Grey relational grade: 0.057

3.6 Confirmation run

The purpose of confirmation run is to verify the results obtained. These tests are conducted at an optimum level of input and process variables with three replications. The welding was done at an optimum setting of variables for producing better tensile strength, Hardness and Toughness .Result of confirmation runs presented in Table 8 , all the deviations are less and therefore the optimum values obtained are acceptable.

4. Conclusions

- In this study, Taguchi L9 array with grey relational analysis has been used to optimize the multiple performance characteristics such as ultimate tensile strength and Hardness and impact toughness.
- An optimum combination of three set of test parameters of grey relational grade for quality weld joints was found to be welding current of 120 A, voltage of 16 V and welding speed of 0.94 mm/s.
- Based on the results of GRG, it was observed that the welding current exerted a significant influence on multiple responses followed by welding speed and voltage.
- The mechanical properties were correlated with the metallurgical characteristics. The impact fractography analysis showed ductile mode of fracture for experiment 6 and mixed mode of brittle and ductile fracture for experiment 8.
- The microstructure of the weld region revealed finer cellular structure for experiment 6, whereas harder dendrites were formed for experiment 8 due to higher heat input.

5. References

[1] M. Sireesha, S. K. Albert & S. Sundaresan,2003,Metallurgical changes and mechanical behavior during high temperature aging of welds between Alloy 800 and 316LN austenitic stainless steel, Materials Science and Technology .Pp 1411-1417, 19:10,http://dx.doi.org/10.1179/026708303225006042

[2] G. Sayiram, N. Arivazhagan , 2015,Microstructural characterization of dissimilar welds between Incoloy 800H and 321 Austenitic Stainless Steel Materials Characterization ,102 ,Pp180–188.

[3] M Sireesha,., S. Sundaresan, & S.K Albert, 2001,Microstructure and mechanical properties of weld fusion zones in modified 9Cr-1Mo steel. J. of Materi Eng and Perform 10:Pp 320. Doi:10.1361/105994901770345033

[4] B. Arivazhagan, M. Vasudevan , 2014,A comparative study on the effect of GTAW processes on the microstructure and mechanical properties of P91 steel weld joints, *Journal of Manufacturing Processes* 16,Pp 305–311

[5] Arun Kumar Srirangan, Sathiya Paulraj ,2016,Multi-response optimization of process parameters for TIG welding of Incoloy 800HT by Taguchi grey relational analysis., *Engineering Science and Technology, an International Journal* 19 Pp 811–817.

[6] Mukundraj V. Patil , 2017, Multi response simulation and optimization of gas tungsten arc welding, Applied Mathematical Modeling, 42 Pp 540–553

[7] B Shanmugarajan. , Rishabh shrivastava , P.Sathiya., G. Buvanashekaran ,2016,Optimisation of laser welding parameters for welding of P92 material using Taguchi based grey relational analysis. Defense Technology 12, Pp343–350

[8] HL. Lin, 2012,The use of the Taguchi method with grey relational analysis and a neural network to optimize a novel GMA welding process ,*Journal of Intelligent Manufacturing* 23, Pp 1671-1680. Doi:10.1007/s10845-010-0468-2

[9] K.F.Tamrin , Y.Nukman , N.A.Sheikh , M.Z.Harizam, 2014,Determination of optimum parameters using grey relational analysis for multi-performance characteristics in CO2 laser joining of dissimilar materials, Optics and Lasers in Engineering 57,Pp40–47.

[10] R. Adalarasan, M. Santhanakumar and A. Shanmuga Sundaram, 2014,Optimization of weld characteristics of friction welded AA 6061-AA 6351 joints using grey-principal component analysis (G-PCA) *Journal of Mechanical Science and Technology* 28 (1) Pp 301-307.

[11] T.Deepan Bharathi kannan, T.pavani priya, P.Sathiya, T,Ramesh, 2017,Metallurgical aspects and optimization of Yb: YAG laser welded NiTional shape memory alloy, Materials Today Proceedings,DOI:10.1016/jmat.pr.2017.01.147.

IOP Conf. Series: Materials Science and Engineering **314** (2018) 012024 doi:10.1088/1757-899X/314/1/012024

Investigation on wear and corrosion behavior of equal channel angular pressed aluminium 2014 alloy

S P Divya[123], G Yoganandan[2], J N Balaraju[2], S A Srinivasan[1], M Nagaraj[1], and B Ravisankar[1]
[1]Department of Metallurgical and Materials Engineering, National Institute of Technology, Tiruchirappalli, India
[2]Surface Engineering Division, National Aerospace Laboratories, Bangalore, India
[3]Corresponding author: perumaldivyagce@gmail.com

Abstract. Aluminium 2014 alloy solutionized at 495°C, aged at 195°C was subjected to Equal Channel Angular Pressing (ECAP). Dry sliding wear tests were conducted using pin on disc tribometer system under nominal loads of 10N and 30N with constant speed 2m/s for 2000m in order to investigate their wear behavior after ECAP. The Co-efficient of friction and loss in volume were decreased after ECAP. The dominant wear mechanism observed was adhesion, delamination in addition to these wear mechanisms, oxidation and transfer of Fe from the counter surface to the Al 2014 pin were observed at higher loading condition. The corrosion behavior was evaluated by potentiodynamic polarization (PDP) and electrochemical impedance spectroscopy (EIS) in 3.5% NaCl solution. The results obtained from PDP showed higher corrosion potential and lower corrosion density after ECAP than base. Electrochemical impedance spectroscopy (EIS) showed higher charge transfer resistance after ECAP. Surface morphology showed decreased pit size and increased oxygen content in ECAP sample than base after PDP.

1. Introduction

Equal Channel Angular Pressing is one of the emerging severe plastic deformation (SPD) techniques that improve mechanical and micro structural properties without changing the dimension of bulk unlike cold working, coating, surface treatments etc. For past decade EACP is efficiently employed for various alloys of aluminum to improve their properties by making it ultrafine grained material. Aluminum 2014 (Al-Cu alloy) is the material that is used for truck frames, aircraft structures, automotive parts, cylinders pistons, machine parts and structural applications due to its high strength imparted by $CuAl_2$ precipitates. ECAP improves hardness and tensile strength of Al 2014 alloy which has been reported [1]. Indeed, wear resistance is one of the important properties that influence the effective functioning of the material. Previous studies on the wear behavior of ECAPed materials reported that there is a significant improvement in wear resistance after ECAP, but some ended up with contradictory results i.e. ECAP can also decrease wear resistance in spite of its improved hardness due to lower ductility and lower strain hardening capability [2]. One the other hand, grain refinement plays vital role on the corrosion behavior of the material. Since ECAP tailors mechanical properties by the formation of ultrafine grains, equal importance must be given to their corrosion behavior, particularly in aluminum alloys where a protective oxide film imparts corrosion resistance. The effect of ECAP on the corrosion behavior of materials is still unclear. Corrosion current density has been shown to increase or decrease or remain unaffected in the presence of ultrafine grains. It has been reported that ECAP processing decreases the corrosion resistance of the base material due to high grain boundary area, internal stress, increased dislocation density and second phase that provide more sites to initiate corrosion [3]. However reports are also available to explain the improved corrosion resistance due to ECAP processing [4]. Nevertheless, study on the corrosion behavior of ECAPed aluminum copper alloys is scanty. Therefore, systematic study on the corrosion and wear behavior of aluminum alloys will be helpful to extend the knowledge on the frictional characteristics and electrochemical behavior after ECAP. Hence, in the present investigation, an attempt has been made to study the wear and corrosion behavior of the ECAPed Al 2014 alloy and the results where compared with the base material.

2. Experimental procedure

Al 2014 alloy rod of diameter 15mm in stock condition and with following chemical composition (in wt%): Al-92.56%, Cu-4.7%, Mn-0.831%, Mg-0.751%, Fe-0.234%, Zn-0.0946% was taken for the present investigation. The as-received rods were machined to the diameter of 11.7mm and with the length of 70mm to match with die specifications. The samples were subjected to solutionizing at 495^0 C, followed by ageing at 195^0 C for 5 hours to attain peak hardness [1]. ECAP process was conducted in room temperature using a split die with an angle of 90°between the two channels and 30° outer arc of curvature. The sample and channels of die are coated with Molybdenum disulphide (MoS_2). Load of 30 tons was applied and samples were processed using route C for three passes. Route C follows 180° rotations in same direction between every pass. The mechanical and micro structural properties were presented in our previous work [1]. Wear pins of 8mm diameter and 10mm height are made of ECAP samples for pin-on-disc wear test using DUCOM Vaccum tribometer with hardened EN 31 steel discs as counter body. The wear tests were conducted under the loads 10N and 30N for a sliding distance of 2000m at humidity of 65±5% and ambient temperature of 25±5°C. The circular samples for corrosion test were cut from the core of the ECAPed sample perpendicular to the direction of pressing and were polished using 1200 grit SiC emery sheets followed by ultrasonification in acetone for 10 minutes. Electrochemical tests such as potentiodynamic polarization and electrochemical impedance tests were carried out. Samples were immersed in all the solutions for 30 min before test to attain stable potential (open circuit potential). EIS studies were carried out in the frequency range of 100 kHz to 10 mHz. After EIS the system was allowed to reach steady state and then potentiodynamic polarization (PDP) tests were performed by scanning the electrode from -200mV to 200mV with respect to OCP. The sweep rate was 0.001V/s. The PDP plots obtained have been represented as potential vs current density and that of EIS as Nyquist plot (real vs. imaginary impedance). The surface morphology of the tested samples was analyzed by scanning electron microscopy (SEM) and energy dispersive X-ray analysis (EDAX) was used for elemental analysis.

3. Results and discussions

3.1. Wear behavior

3.1.1. Volume loss. Figure1(a) shows vickers hardness of base and ECAPed sample. From the figure it is evident that the hardness of Al 2014 improved after 3passes of ECAP from 95HVN to 165HVN. Figure 1(b) is the volume loss variation between base and ECAP after wear test under 10N and 30N. From the figure it can be undertood that wear resistace improved after ECAP, as the volume loss decreased from 10.28±1 mm^3 to 2.4±0.5 mm^3 and 14.14±1 mm^3 to 5.9±0.5mm^3 under 10N and 30N loading conditions respectively. The volume loss increased with increase in load from 10N to 30N which is in good agreement with the Archard's law, V=KLN/H where V= volume loss, K= wear coefficient, L=total sliding distance, N= applied load, H=hardness of the wear surface. It indicates that the volume loss depends on hardness and load applied. This decreased volume loss may be due to the combination of ageing and ECAP which produced higher hardness through grain refinement and precipitate fragmentation. It was reported that in Al 2014 alloy after ECAP the grain size decreased from ~10 μm to 0.23μm and the precipitate size reduced from 95±5nm to 45±5nm with uniform distribution its improved hardness [5].

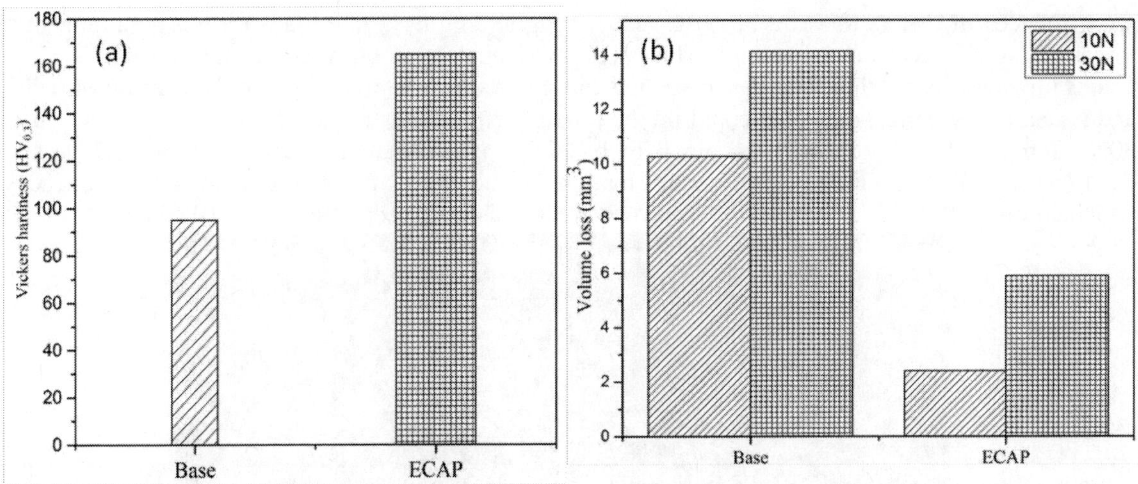

Figure1 (a) Hardness of Base and ECAPed sample (b) Volume loss of Base and ECAPed sample under 10N and 30N.

3.1.2. Coefficient of friction. Figure2 shows the variation in coefficient of friction (cof) of base and ECAPed sample. From the figure it can be observed that the coefficient of friction decreases after ECAP and it increases with increase in applied load. Under 10N the coefficient of friction vary between 0.35-0.45 for ECAP and 0.5-0.57 for base sample. The average value of cof decreases from 0.5±0.03 (base) to 0.43±0.04(ECAP). In case of higher loading conditions the coefficient of friction of base varied between 0.5-0.65 upto sliding distance of 1000m which may resulted from adhesion, thus higher wear loss. However as the sliding distance increases, the cof decreases. This decrease in cof may be attributed to the formation of oxide layer on the surface[6]. But in ECAPed sample the cof varied between 0.45-0.5. The decrease in cof after ECAP may be attributed to strain hardening during sliding [6]. As the friction between the surface induce plastic deformation, higher grain boundary area and fine precipitates hinder dislocation. This leads to strain hardening of the material that are in contact which results in reduction of volume loss.

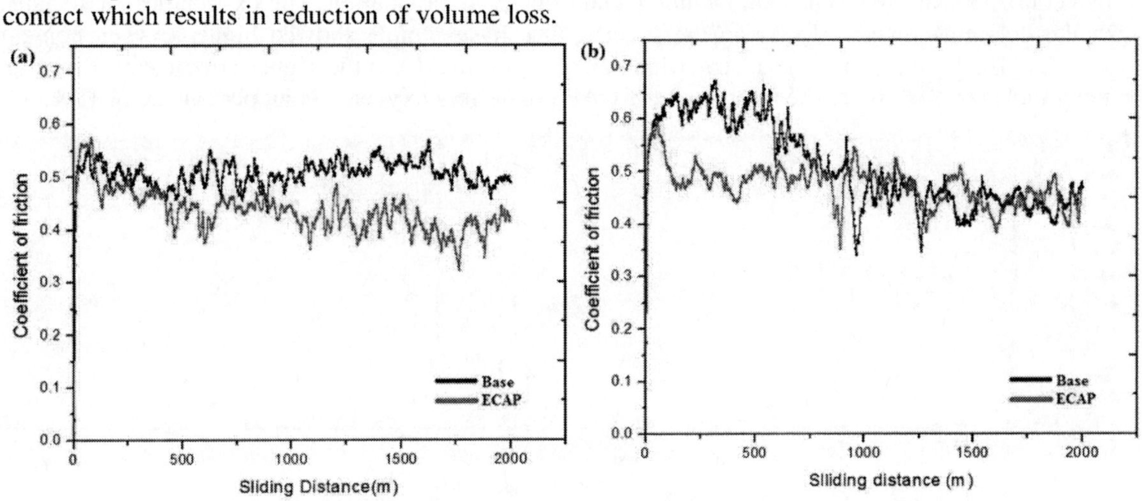

Figure2.Coefficient of friction of base and ECAP under (a) 10N (b) 30N.

3.1.3. Surface morphology. The figure3 shows the SEM images of wear tested surface of the base and ECAPed sample at applied load of 10N and 30N. The images demonstrate that the surface exhibit plastic deformation characterized by grooves formed by plowing in base under 10N and 30N. There are evidences of craters that can be observed. During sliding at higher loads (30N), high tangential forces acting between the contacting surfaces results in nucleation of cracks at the subsurface. These cracks propagate parallel to the pin surface and cause delamination that result in detachment of debris and formation of craters [7]. This could be the possible reason for the higher cof up to 1000m under 30N. Comparing the processed sample with base, it appears that the severity of wear is less in ECAPed sample. There are few plastic deformations visible with smooth morphology of wear tracks, which confirms that the ECAP process improves wear resistance properties in Al 2014 alloy.

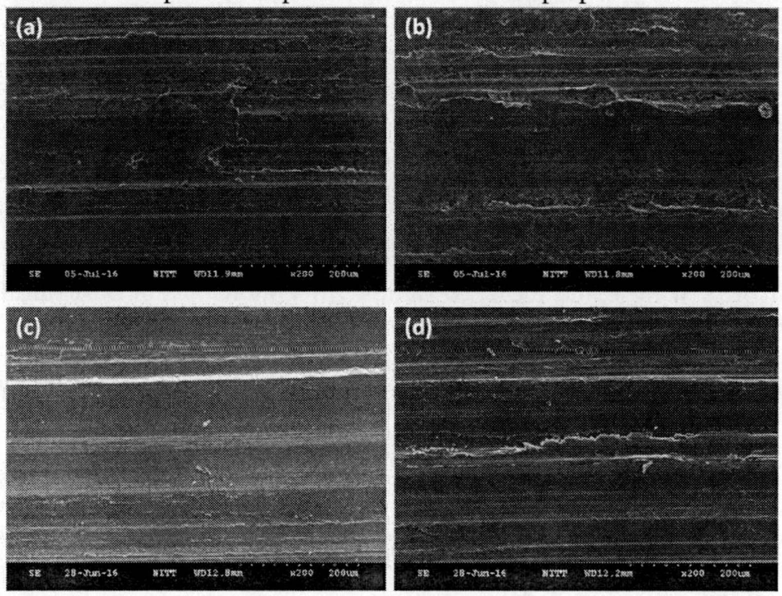

Figure3: SEM images of wear surface of (a)&(b) base under 10N and 30N (c)&(d) ECAPed under 10N and 30N.

3.1.4. EDS Analysis. The EDS spectra (Figure 4) showed presence of oxygen and iron on the surface which confirms that the mechanism include metal transfer and oxidation. The oxygen peak associates with the delamination and debris on the surface. The base sample showed higher oxygen content (24%) than ECAPed sample under 30N which may be resulted from the higher coefficient of friction in the initial stage. However the sample after ECAP posses less oxygen weight percentage of 15%.

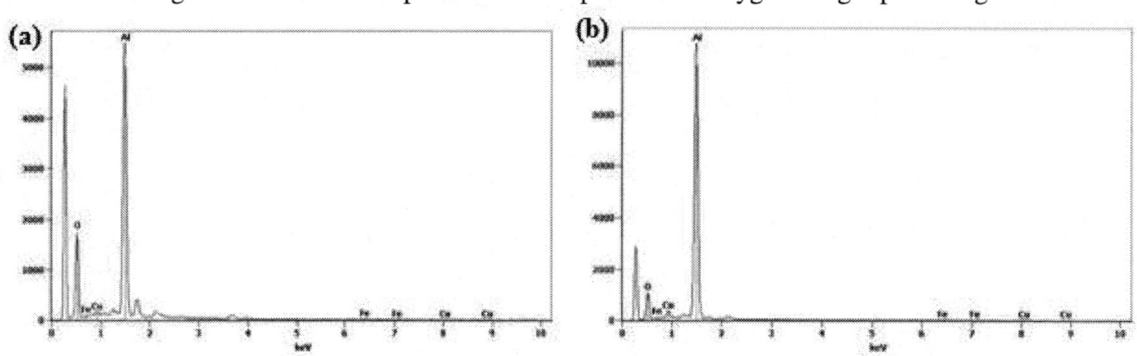

Figure 4: EDS analysis of surface after wear under 30N (a)base and (b) ECAP.

3.2. Corrosion behavior

3.2.1. EIS and PDP. Figure 5(a) show the Nyquist plot obtained from the EIS of base and ECAPed sample in 3.5% NaCl solution which is capacitive in nature. Sample processed by three passes of ECAP exhibit higher diameter arc which indicates its improved corrosion resistance after ECAP.

Impedance results showed two times constant behavior for both ECAP and base samples and the best fitted equivalent circuit R[R(Q)][R(Q)] is shown in figure 5(a) insert. This behavior may be due to the formation of defective oxide layer during the immersion in electrolyte. The corrosion parameters obtained from EIS is tabulated in table 1(a). From the table it is evident that the Rct and Rox values obtained for ECAP sample is twice and four times higher than base sample respectively. In addition, Qdl value reduced from 5900 to 2119 $\mu Ss^n cm^{-2}$ after ECAP from which it is understood that the passive layer formed on ECAPed sample is stable and thick than that of base. Figure 5(b) shows PDP curves of base and ECAP and the obtained values are listed in table 1(b). From the figure it is evident that the cathodic curves shifts to lower current for the ECAPed sample resulted in decreased corrosion current density to 3.83$\mu A/cm^2$ which is half the i_{corr} value of base. This is in good agreement with the results obtained from EIS. It has been reported that the grain boundaries and high dislocation density after ECAP act as active sites for the formation of oxide layer which accelerates passivation and reduces corrosion rates [8]. It also reduces the intensity of galvanic couple formed between grain boundary and grain interior, precipitates and grains due to their ultrafine grain size and precipitate fragmentation along with uniform distribution. This may be the possible reason for the improved corrosion resistance after ECAP.

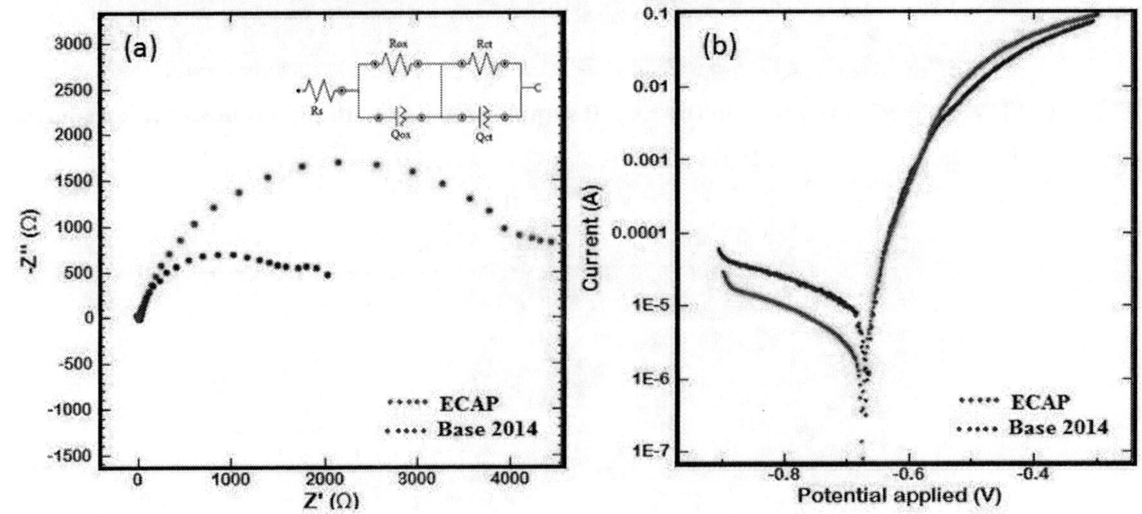

Figure 5: (a) PDP curve (b) Nyquist plot for Base and ECAP sample.

Table1: Corrosion parameters obtained from PDP and EIS.

	(a) Parameters obtained from EIS							(b) Parameters obtained from PDP	
Sample	**R_s Ωcm^2**	**R_{ox} Ωcm^2**	**Q_{ox} $\mu Ss^n cm^{-2}$**	**n_{ox}**	**R_{ct} Ωcm^2**	**Q_{dl} $\mu Ss^n cm^{-2}$**	**n_{dl}**	**E_{corr} (V)**	**I_{corr} ($\mu A/cm^2$)**
Base	3.3	1078	76	0.9	1000	5900	0.8	-0.67	8.6
ECAP	2.58	4162	30.89	0.85	2857	2119	0.72	-0.67	3.83

3.2.2. Surface morphology and EDS Analysis. Figure6 shows the SEM images of the surface obtained after PDP. The surface morphology of both base and ECAPed sample showed cracks and pits. But the pits observed in ECAPed sample was significantly small than that observed in base. The average size of the pits on the base was found to be 20±4μm and that of ECAPed sample was 3±2μm. The formation of pits may be due the galvanic couple formed between the cathodic precipitates and anodic matrix [9] which is evident from the EDS analysis as shown in figure 7. It shows 18% of copper and 49% of aluminum and 24% of oxygen in the spot analysis of pits. Also the EDS analysis on the surface of ECAP sample showed higher oxygen content of 24% than the base sample (8.26%). Hence it can be concluded that the improved corrosion resistance after ECAP is due to the oxide layer formed on the surface due to ultrafine grains which are consistent with the results obtained from EIS.

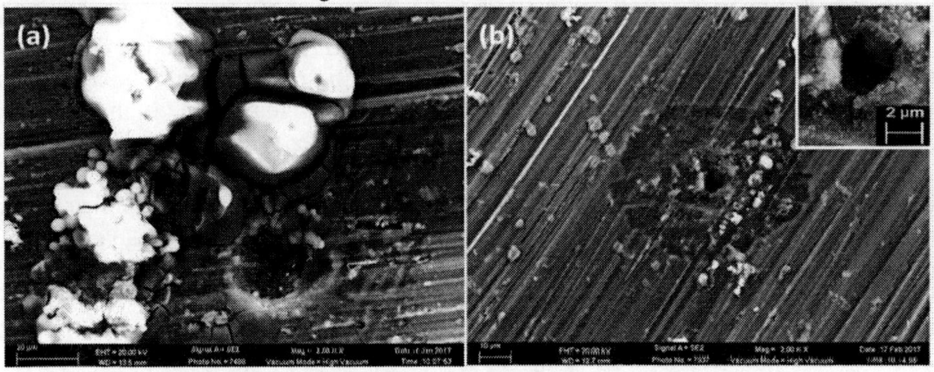

Figure6: SEM image of (a) base and (b) ECAP samples after potentiodynamic polarization test.

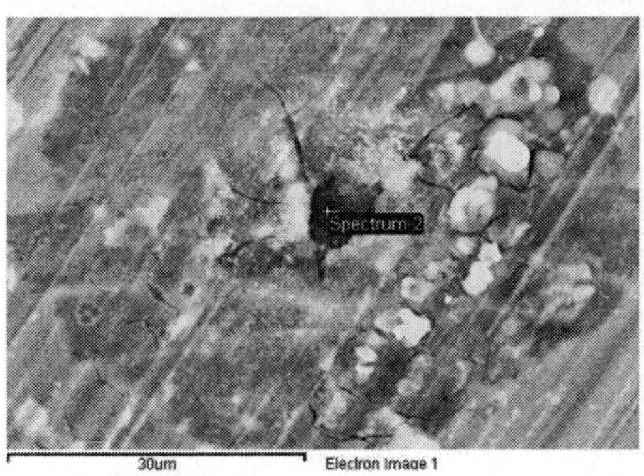

Elements	Weight %
O K	24.89
Al K	49.99
Si K	1.94
Cl K	1.03
Mn K	2.42
Fe K	0.52
Cu K	18.93

Figure 7: EDS analysis of pit on the ECAPed sample after PDP.

4. Conclusion

1. The pin on disc sliding wear test showed decreased volume loss and coefficient of friction after ECAP under both the loading conditions 10N and 30N due to improved hardness during ECAP. The SEM analysis of the worn surface revealed smooth morphology of wear tracks after ECAP and EDS analysis confirmed oxidation and metal transfer.
2. PDP results showed higher E_{corr} and lower i_{corr} values after ECAP. EIS results showed higher charge transfer resistance (R_{ct}) for ECAPed sample. SEM analysis revealed significantly smaller pits in ECAPed sample than that of base.
3. Higher oxygen content was observed after corrosion test in ECAPed sample than base sample in EDS analysis.
4. Hence it can be concluded from the above results that Al 2014 alloy showed improved wear and corrosion resistance after ECAP.

References

[1] Venkatachalam P, Ramesh Kumar S, Ravisankar B, Thomas Paul V and Vijayalakshmi M 2010 Effect of Processing Routes on Microstructure and Mechanical Properties of 2014 Al Alloy Processed by Equal Channel Angular Pressing, *Trans. Of Nonferrous Met. Scc China* **20** 1822-28.

[2] Gao N, Wang C T, Wood R J K and Langdon T G 2012 Tribological Properties of Ultrafine-grained Materials Processed by Severe Plastic Deformation, *J. Mater. Sci.* **47** 4779–97.

[3] Atef Korchef and Abdelkrim Kahoul 2013 Corrosion Behavior of Commercial Aluminum Alloy Processed by Equal Channel Angular Pressing, *International Journal Of Corrosion* Article Id 983261**2013** 1-11.

[4] Min-Kyong Chunga, Yoon-Seok Choi A, Jung-Gu Kima,, Young-Man Kimb and Jae-Chul Lee 2004 Effect of the Number of ECAP Pass Time on the Electrochemical Properties of 1050 Al Alloys *Materials Science And Engineering A* **366** 282–291.

[5] Venkatachalam P, Shibayan Roy, Ravisankar B, Thamos Paul V, Vijayalakshmi M and Suwas S 2012 Effect of Processing Routes on Evolution of Texture Heterogeneity in 2014 Aluminium Alloy Deformed by Equal Channel Angular Pressing (ECAP) *Materials Science and Technology* **28** 1446-58.

[6] Van Thuong N, Zuhailawati H, Seman A A, Huy T D and Dhindaw B K 2015 Microstructural Evolution and Wear Characteristics of Equal Channel Angular Pressing Processed semi-solid-cast Hypoeutectic Aluminum Alloys *Mater. Des.* **67** 448–456.

[7] Nam. P. Suh 1977 An Overview of the Delamination Theory of Wear, *Journal wear* **44** 1-16.

[8] Palumbo G, Thorpe S J and Aust K T 1990 On the Contribution of Triple Junctions to the Structure and Properties of Nanocrystalline Materials *Scripta Metallurgica et Materialia* **24** 1347-50.

[9] Jiang Jing-Hua, Ma Ai-Bin, Song Dan, N. Saito, Yuan Yu-Chun and Nishida Y 2010 Corrosion Behavior of Hypereutectic Al-23%Si Alloy (AC9A) Processed by Severe Plastic Deformation *Trans. of Nonferrous Metals Society of China* **20** 195-200.

IOP Conf. Series: Materials Science and Engineering **314** (2018) 012025 doi:10.1088/1757-899X/314/1/012025

Optimization of friction and wear behaviour of Al7075-Al_2O_3-B_4C metal matrix composites using Taguchi method

S Dhanalakshmi[1], N Mohanasundararaju[2], P G Venkatakrishnan[1] and V Karthik[1]

[1] Department of Metallurgical Engineering, Government College of Engineering, Salem – 636 011, Tamilnadu, India.
[2] Mahendra Institute of Technology, Mallasamudram, Namakkal – 637 503, Tamil Nadu, India.

Abstract: The present study deals with investigations relating to dry sliding wear behaviour of the Al 7075 alloy, reinforced with Al_2O_3 and B_4C. The hybrid composites are produced through Liquid Metallurgy route – Stir casting method. The amount of Al_2O_3 particles is varied as 3, 6, 9, 12 and 15 wt% and the amount of B_4C is kept constant as 3wt%. Experiments were conducted based on the plan of experiments generated through Taguchi's technique. A L27 Orthogonal array was selected for analysis of the data. The investigation is to find the effect of applied load, sliding speed and sliding distance on wear rate and Coefficient of Friction (COF) of the hybrid Al7075-Al_2O_3-B_4C composite and to determine the optimal parameters for obtaining minimum wear rate. The samples were examined using scanning electronic microscopy after wear testing and analyzed.

1. Introduction

Conventional materials have certain shortcomings in good combination of strength, stiffness, toughness and density. In order to conquer these shortcomings and to meet the demand of modern day technology, composites are extensively used. Out of which, Hybrid Metal Matrix Composites (HMMCs) possess high specific strength, toughness, impact strength and low sensitivity to temperature changes [1]. As a result, many of the current applications for HMMCs are widely in the field of aerospace and automobile components. Aluminium matrix is reinforced with ceramic particles like Al_2O_3, B_4C, SiC, TiB_2, etc, which increases the mechanical properties of the resulting composite materials by dispersion strengthening mechanism. Veeresh kumar et al. [2] have reported that addition of 2 – 6 % of SiC and Al_2O_3 reinforcements to the Al6061 and Al7075 alloys respectively resulted in enhanced density, hardness, tensile strength and wear resistance for the composites. Arunkumar and Swamy [3] reinforced the Al6061 matrix with varying percentage of fly ash and e-glass fibers, and concluded that the tensile properties, compressive strength and hardness of hybrid metal matrix composite increased with increasing fly ash content. Dhanalakshmi et al. [4] have found that the increasing the reinforcement percentage in the hybrid Al7075- Al_2O_3- B_4C composite increases the tensile strength and hardness of matrix.

The most accepted method for analyzing the wear properties is dry sliding wear test. Basavarajappa et al [5] studied the dry sliding wear behaviour of the Al2219 alloy, reinforced with SiC particles in 0–15 wt. % in three steps. The results showed that the wear rates of the composites are lower than that of the matrix alloy and further decrease with increasing SiC content. The dry sliding wear test is commonly carried out using pin-on-disc apparatus which includes various test parameters like normal load, rotational speed, time and sliding distance [6-8]. Reddy et al [9] studied the wear behaviour of hybrid metal matrix composites of Al 6061 alloy base material by incorporating 6% SiC and varying 2 to 4 % of graphite particulates. The wear resistance of the composites increases with addition of the SiC and graphite particle content. Radhika et al. [10] found Taguchi technique as a valuable technique to deal with responses influenced by multivariable. This method significantly reduces the number of trials that are required to model the response function compared with the full factorial design of experiments. The most important benefit of this technique is to find out the possible interaction between the factors. Basavarajappa et al [11] made an attempt to study the influence of wear parameters using L27 orthogonal array. Al2219 alloy was reinforced with SiC particle reinforcement varied from 0 to 15 wt% in steps of 5. Sliding distance is the wear factor that has the highest influence on the dry sliding wear of the composites followed by load, sliding speed and reinforcement on the dry sliding wear of the metal matrix composites. In the present work wear parameters have been optimized using Taguchi technique to obtain better tribological properties in the produced Al7075- Al_2O_3- B_4C hybrid metal matrix composite and to study the influence of the wear parameters.

2. Materials and methods

The Hybrid Aluminium Metal Matrix Composites (HAMMCs) are produced through Stir casting method. Aluminium alloy 7075 is used as the matrix material and its chemical composition is shown in Table 1. Boron carbide (B_4C) and Alumina (Al_2O_3) particles are used as reinforcement. The amount of B_4C addition has been kept constant (3 wt%) and the amount of Al_2O_3 addition was varied as 3, 6, 9, 12 and 15 wt% during the stir casting process. Table 2 shows the nomenclature for the prepared HAMMCs in the present study.

Table 1.Chemical composition of Al7075 alloy.

Elements	wt%
Si	0.4
Fe	0.5
Cu	1.6
Mn	0.3
Mg	2.5
Cr	0.15
Zn	5.5
Ti	0.2
Al	bal.

Table 2. Nomenclature of different composites and the corresponding compositions.

Composite Nomenclature	Al_2O_3 (%)	B_4C (%)	Al7075 (%)
C1	3	3	94
C2	6	3	91
C3	9	3	88
C4	12	3	85
C5	15	3	82

The melt temperature is maintained above 750°C to compensate the heat loss during the pouring operation. The reinforcement materials are preheated to more than 400°C and the mold is preheated to 300°C to reduce the temperature gradient. The composites are cast into cylindrical specimens.

3. Dry sliding wear test

Pin-on-disc wear test machine was used to study the wear behavior of the samples according to ASTM G99 standard. The EN31 steel with a hardness of 62 HRC was used as the counter disc. The wear test pins having dimension of 10 mm diameter and 30 mm long is prepared from the composites and held against the rotating steel counter disc during the test. Between each tests, the counter disc and the samples are thoroughly cleaned to remove any wear debris attached to the disc or sample. The surfaces of the pin samples were rubbed against emery paper prior to test in order to ensure effective contact of the flat surface with the steel disc. The weight of the sample before and after the test is measured using digital weighing balance with high accuracy of 0.0001g after thorough cleaning with the acetone solution. The wear rate is calculated from the weight loss and expressed in terms of wear volume loss per unit sliding distance. The surface morphologies of the pins are studied using scanning electron microscope after the wear test.

Design of experiments technique enables us to carry out modeling and analysis of the influence of process variables on the response variables. The wear parameters chosen for the experiment are applied load (L), sliding speed and (S) and sliding distance (D). The process parameters and their levels are shown in table 3. Based on Taguchi method, the experiments are carried out as per the standard L27 orthogonal array. The selection of the orthogonal array is based on the condition that the degrees of freedom for the orthogonal array should be greater than or at least equals sum of those of wear parameters.

Table 3. Process parameters and their levels.

Level	Load (N)	Speed (m/s)	Distance (m)
1	20	1.5	1000
2	40	3	2000
3	60	4.5	3000

4. Plan of experiments

In the present investigation L27 orthogonal array is chosen which has 27 rows and 13 columns. The selection of Orthogonal array depends on three items in order of priority, viz., the number of factors and their interactions, number of levels for the factors and the desired experimental resolution or cost limitations. A total of 27 experiments were performed based on the run order generated by the Taguchi model. The response for the model is wear rate and coefficient of friction. In Orthogonal array, first column is assigned to applied load, second column is assigned to Sliding speed and fifth column is assigned to sliding distance and the remaining columns are assigned to their interactions. The objective of model is to minimize wear rate and coefficient of friction. The responses were tabulated and results were subjected to Analysis of Variance (ANOVA). The Signal to Noise (S/N) ratio, which condenses the multiple data points within a trial, depends on the type of characteristic being evaluated. The S/N ratio for wear rate and coefficient of friction is analyzed using 'smaller the better' characteristic. The experimental observations are further transformed into Signal to Noise ratio. The response to be studied was the wear rate and coefficient of friction with the objective as smaller the better, which is calculated as logarithmic transformation of loss function as shown below,

$$\frac{S}{N} = -10\log\left(\frac{1}{n}\sum Y_i^2\right) \quad (1)$$

where 'n' is the number of observations, 'Yi' is the measured value of wear rate. It is suggested that quality characteristics are optimized when the S/N response is as smaller as possible.

5. Results and discussion

The effect of the tribological test parameters such as load, sliding speed and distance on wear rate and coefficient of friction have been optimized and evaluated using Taguchi technique, with appropriate Signal-to-Noise ratio (S/N) and Analysis of Variance (ANOVA). These design parameters are unique and inherent feature of the process that influence and decide the composite performance. The parameters that highly influence the wear rate and coefficient of friction have been determined by the rank value which has been obtained from the difference between the maximum and minimum value (delta) of the mean of S/N ratios.

5.1 Results of statistical analysis of experiments

By conducting the experiment as per the Orthogonal Array, the results for various combinations of parameters were obtained. The measured results were analyzed using the commercial software MINITAB 16 specifically used in DOE applications. Table 4 shows the L27 Orthogonal array for the composite C5 that exhibited better tribological properties than other composites (C1, C2, C3 and C4), which may be due to the fact that the surface of the matrix materials tend to get delaminated in the absence of harder reinforcement, thus increasing the wear. The experimental values are transformed into a signal to noise ratio to measure the quality characteristics. The effect of control parameters such as load, sliding speed and sliding distance on wear rate and COF has been analyzed using S/N ratio response table. The ranking of process parameters using signal to noise ratios that has been obtained for various parameter levels for wear rate and COF are given in the response tables 5 and 6 respectively for composite C5. The control factors are statistically significant in the Signal to Noise ratio and it has been observed that the load is a dominant parameter on the wear rate and COF, followed by sliding speed and distance. Figure (1-4) shows the influence of process parameters on wear rate and coefficient of friction graphically. The analysis of these experimental results using S/N gives the optimum conditions resulting in minimum wear rate and coefficient of friction. The optimum condition for wear rate and coefficient of friction are L1, S3 and D3.

5.2 Analysis of Variance Results for Wear Test

The experimental results were analyzed by ANOVA, which is used to examine the influence of the wear parameters, namely, applied load, sliding speed and distance that extensively affects the performance measures. By performing ANOVA, the independent factor which dominates over the other can be determined and the percentage contribution of that particular independent variable can be found out. Tables 7 and 8 show the ANOVA results for wear rate and coefficient of friction for three factors varied at three levels and interactions of those factors. This analysis is carried out for a significance level of $\alpha = 0.05$, i.e. for a confidence level of 95%. Sources with a P-value less than 0.05 were considered to have a statistically significant contribution to the performance measures. In the Analysis of Variance tables, the last column shows the percentage contribution (Pr) of each parameter on the total variation indicating their degree of influence on the result.

Table 4. Orthogonal Array of HAMMC - C5 (15% Al_2O_3 + 3% B_4C).

Exp.No.	Load (N)	Speed (m/s)	Distance (m)	Wear rate (mm^3/m)	S/N ratio (db)	Coefficient of friction	S/N ratio (db)
1	20	1.5	1000	0.002698	51.3792	0.319	9.9242
2	20	1.5	2000	0.002683	51.4276	0.301	10.4287
3	20	1.5	3000	0.002454	52.2025	0.293	10.6626
4	20	3.0	1000	0.002596	51.7139	0.297	10.5449
5	20	3.0	2000	0.002204	53.1358	0.290	10.7520
6	20	3.0	3000	0.001738	55.1990	0.284	10.9336

IOP Conf. Series: Materials Science and Engineering **314** (2018) 012025 doi:10.1088/1757-899X/314/1/012025

7	20	4.5	1000	0.001937	54.2574	0.288	10.8122
8	20	4.5	2000	0.001588	55.9830	0.276	11.1818
9	**20**	**4.5**	**3000**	**0.001305**	**57.6878**	**0.274**	**11.2450**
10	40	1.5	1000	0.003034	50.3597	0.327	9.7090
11	40	1.5	2000	0.002956	50.5859	0.319	9.9242
12	40	1.5	3000	0.002754	51.2007	0.311	10.1448
13	40	3.0	1000	0.003013	50.4200	0.319	9.9242
14	40	3.0	2000	0.002713	51.3310	0.313	10.0891
15	40	3.0	3000	0.002524	51.9582	0.295	10.6036
16	40	4.5	1000	0.002636	51.5811	0.305	10.3140
17	40	4.5	2000	0.002351	52.5749	0.294	10.6331
18	40	4.5	3000	0.002279	52.8451	0.283	10.9643
19	60	1.5	1000	0.003792	48.4226	0.346	9.2185
20	60	1.5	2000	0.003364	49.4629	0.335	9.4991
21	60	1.5	3000	0.002937	50.6419	0.318	9.9515
22	60	3.0	1000	0.003254	49.7516	0.340	9.3704
23	60	3.0	2000	0.002912	50.7162	0.327	9.7090
24	60	3.0	3000	0.002794	51.0755	0.309	10.2008
25	60	4.5	1000	0.003118	50.1225	0.323	9.8159
26	60	4.5	2000	0.002712	51.3342	0.318	9.9515
27	60	4.5	3000	0.002688	51.4114	0.286	10.8727

Table 5. Response Table for S/N Ratios – Smaller is better (Wear rate) – C5.

Level	Load (N)	Speed (m/s)	Distance (m)
1	53.67	50.63	50.89
2	51.43	51.70	51.84
3	50.33	53.09	52.69
Delta	3.34	2.46	1.80
Rank	1	2	3

Table 6. Response Table for S/N Ratios – Smaller is better (Coefficient of friction) – C5.

Level	Load (N)	Speed (m/s)	Distance (m)
1	10.721	9.940	9.959
2	10.256	10.236	10.241
3	9.843	10.643	10.620
Delta	0.877	0.703	0.661
Rank	1	2	3

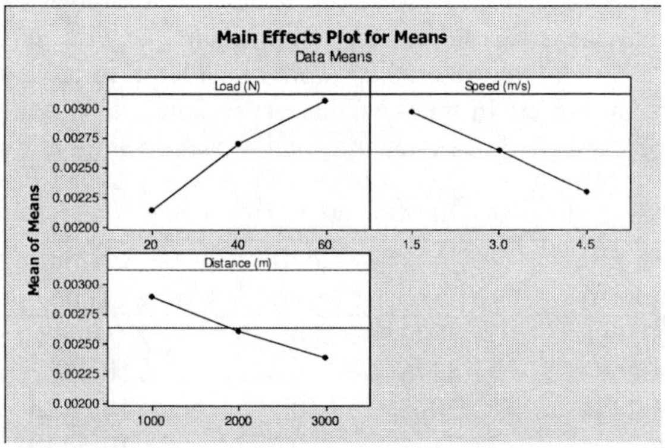

Figure 1. Main Effects plot for Means – Wear rate – C5.

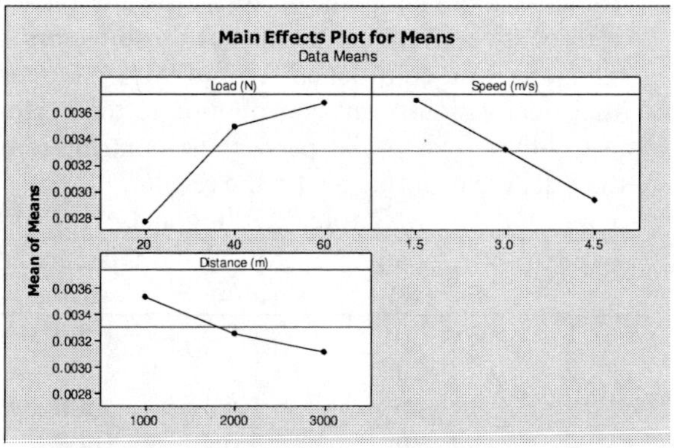

Figure 2. Main Effects plot for Means – Coefficient of friction – C5.

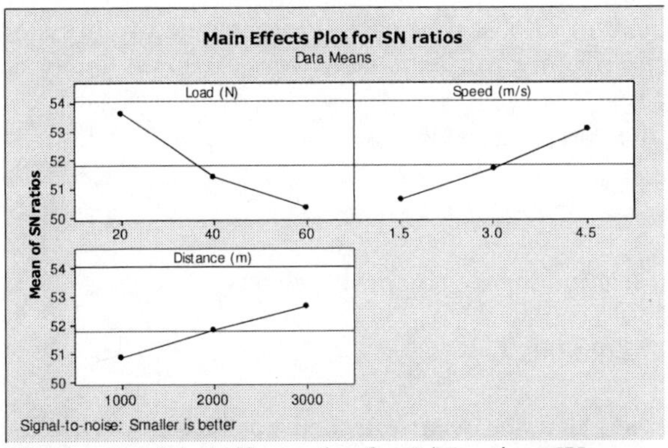

Figure 3. Main Effects plot for S/N ratio – Wear rate – C5.

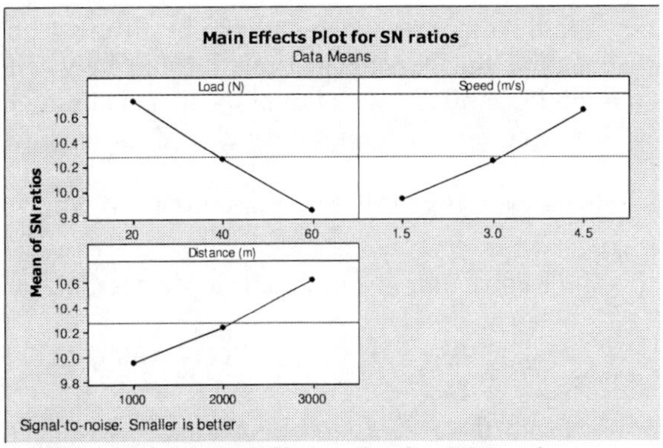

Figure 4. Main Effects plot for S/N ratio – Coefficient of friction – C5.

Table 7. Analysis of Variance for wear rate (mm^3/m) – C5.

Source	DF	Seq SS	Adj SS	Adj MS	F	P	Pr (%)
L (N)	2	0.0000039	0.0000039	0.0000020	108.19	0.000	50.64
S (m/s)	2	0.0000020	0.0000020	0.0000010	55.92	0.000	25.97
D (m)	2	0.0000012	0.0000012	0.0000006	32.47	0.000	15.58
L(N)*S(m/s)	4	0.0000003	0.0000003	0.0000001	4.00	0.045	3.89
L(N)*D(m)	4	0.0000001	0.0000001	0.0000000	0.84	0.539	1.29
S(m/s)*D(m)	4	0.0000001	0.0000001	0.0000000	0.64	0.649	1.29
Error	8	0.0000001	0.0000001	0.0000000			1.29
Total	26	0.0000077					100

Table 8. Analysis of Variance for Coefficient of friction – C5.

Source	DF	Seq SS	Adj SS	Adj MS	F	P	Pr (%)
L (N)	2	0.0043567	0.0043567	0.0021784	116.64	0.000	43.24
S (m/s)	2	0.0027570	0.0027570	0.0013785	73.81	0.000	27.36
D (m)	2	0.0024890	0.0024890	0.0012445	66.64	0.000	24.70
L(N)*S(m/s)	4	0.0000366	0.0000366	0.0000091	0.49	0.744	0.36
L(N)*D(m)	4	0.0002686	0.0002686	0.0000671	3.60	0.058	2.66
S(m/s)*D(m)	4	0.0000177	0.0000177	0.0000044	0.24	0.910	0.175
Error	8	0.0001494	0.0001494	0.0000187			1.48
Total	26	0.0100750					100

It can be observed from the Analysis of Variance results that load has the highest influence on wear rate and Coefficient of friction. Hence load is an important control factor to be taken into consideration during wear process followed by sliding speed and distance. However, the interactions between load and sliding speed, load and distance and sliding speed and distance have got negligible influences. The pooled errors associated in ANOVA table for wear rate and coefficient of friction has been approximately less than 3% for all the HAMMCs.

5.3 Multiple Linear Regression Models

Multiple Linear Regression models have been generated for wear rate and coefficient of friction using the statistical software MINITAB 16. This model gives the correlation between an independent / predictor variable and a response variable by fitting a linear equation to observed data. The generated Regression equation thus establishes relationship between the significant terms obtained from ANOVA analysis, namely, applied load, sliding speed and sliding distance.

The Regression Equation developed for wear rate for the composite - C5 is:

$$W_r = 0.00288589 + 2.32444e\text{-}005\ L - 0.00022437\ S - 2.55833e\text{-}007\ D$$

The Regression Equation developed for Coefficient of friction for the composite - C5 is:

$$C_f = 0.324037 + 0.000777778\ L - 0.00822222\ S - 1.17222e\text{-}005\ D$$

From the Regression equations, it has been observed that the wear rate and coefficient of friction associated with load (L) has been positive and it indicates that, as load increases, wear rate and coefficient of friction of the composite also increases. Moreover, sliding speed and distance has been negatively associated with the equation, which indicate that, as speed and distance increases, wear rate and friction decreases. Figure 5 (a-e) shows the wear track patterns (i.e. worn surfaces) of the HAMMCs that were studied using scanning electron microscope (SEM) for the optimal parameters L1S3D3 (20N, 4.5m/s, 3000m). The presence of pits or prows is clearly observed in the worn surfaces, which is an indication of adhesive wear. As the percentage of reinforcement increases, the volume of prows also decreases.

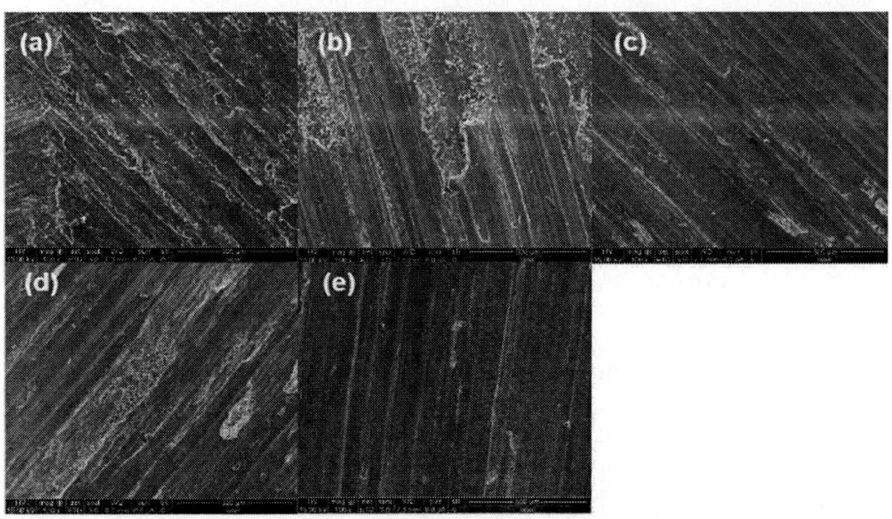

Figure 5 (a-e): SEM images of worn surface of HAMMCs C1–C5.

The effect of applied load is directly proportional to wear rate and friction, i.e., as applied load increases, wear rate and coefficient of friction also increases. This is because, the temperature at the interface between the disc and the pin increases with increase in the applied load. At low loads there is a probability of abrasive wear, where the reinforcing hard alumina particles remain undamaged without fracture during wear and thus act as load bearing elements. As the load increases, the induced stresses surpass the fracture strength of the particles causing their fracture. Thus material transfer from pin onto the disc can also occur due to the rubbing action of the fractured reinforcing particles against steel disc and the removal of material from the surface of the pin increases with increase in load.

IOP Conf. Series: Materials Science and Engineering **314** (2018) 012025 doi:10.1088/1757-899X/314/1/012025

As sliding speed increases, the wear rate and coefficient of friction decreases. This can be ascribed to the oxidation of aluminium alloy, which forms an oxide layer at higher interfacial temperature thus preventing the sliding, thereby decreasing the wear rate and coefficient of friction. The negative value of distance is indicative that increase in sliding distance decreases the wear rate as well as coefficient of friction and this can be attributed to the presence of hard reinforcing particles which provides abrasion resistance, resulting in improved dry sliding wear performance. The addition of reinforcement in the aluminium composites improves the friction and wear behaviour due to its self lubrication property. The reinforcement particles smear on the sliding pin surface and form a layer which reduces wear.

6. Conclusions

Taguchi's method is used to find the optimum conditions to achieve better wear resistance under dry sliding condition for the hybrid Al7075-Al_2O_3-B_4C composites produced through stir casting method. The following are the conclusions drawn from the present study.

- The wear resistance of the hybrid Al7075-Al_2O_3-B_4C composites increases with increasing Al_2O_3 weight percentage in the composite.
- ANOVA results show that load has the highest influence followed by sliding speed and distance, both on wear rate and coefficient of friction.
- From the Regression equations, it has been observed that load is positively associated; speed and distance are negatively associated with wear rate and coefficient of friction.

References

[1] Clyne T W 2000 An Introductory Overview of MMC Systems, Types, and Developments, *Comprehensive Composite Materials* Vol 3,ed-Clyne T W (Elsevier) pp.1-26.

[2] Veereshkumar G B,Rao C S P,Selvaraj N and Bhagyashekar M 2010 Studies on Al6061-SiC and Al7075-Al_2O_3 Metal Matrix Composites, *Journal of Minerals and Materials Characterization and Engineering* **9** 1 43 – 55.

[3] Arunkumar M B and Swamy R P 2011 Evaluation of mechanical properties of Al6061, flyash and e-glass fiber reinforced hybrid metal matrix composites, *ARPN Journal of Engineering and Applied Sciences* **6** 5 40-44.

[4] Dhanalakshmi S,Mohanasundararaju N and Venkatakrishnan P G 2014 Preparation and mechanical Characterization of stir cast hybrid Al7075-Al_2O_3-B_4C Metal Matrix Composites, *Applied Mechanics and Materials* **592-594**, 705-10.

[5] Basavarajappa S, Chandramohan G, Subramanian R and Chandrasekar A 2006 Dry sliding wear behaviour of Al 2219/SiC Metal Matrix Composites, *Materials Science* **24** 2/1 357-66.

[6] Baradeswaran A,Elayaperumal A and Issac R F 2013 A Statistical Analysis of Optimization of Wear Behaviour of Al-Al_2O_3 Composites Using Taguchi Technique, *Procedia Engineering* **64** 973–82.

[7] Ghosh S,Sahoo P and Sutradhar G 2012 Wear Behaviour of Al-SiC Metal Matrix Composites and Optimization Using Taguchi Method and Grey Relational Analysis, *Journal of Minerals and Materials Characterization and Engineering* **11** 1085-94.

[8] Uvaraja V C and Natarajan N 2012 Optimization on Friction and Wear Process Parameters Using Taguchi Technique, *International Journal of Engineering and Technology* **2** 4 694-99

[9] Reddy T P and Rao N A N 2014 Wear Optimization of Al/Sic/Gr Hybrid MMC Using Taguchi and ANOVA, *International Journal of Innovative Research in Science, Engineering and Technology* **3** 11 17205-10.

[10] Radhika N and Subramanian R 2011 Tribological Behaviour of Aluminium / Alumina / Graphite Hybrid Metal Matrix Composite Using Taguchi's Techniques, *Journal of Minerals and material Characterization and Engineering* **10** 5 427-43.
[11] Basavarajappa G and Chandramohan G 2005 Wear Studies on Metal Matrix Composites: a Taguchi Approach, *Journal of Materials Science and Technology* **21** 6 845-50.

Formability behavior studies on CP-Al sheets processed through the helical tool path of incremental forming process

H Markanday and D Nagarajan*

Department of Metallurgical and Materials Engineering,
National Institute of Technology Rourkela, Rourkela – 769008, Odisha, India.
*Corresponding author: nagarajand@nitrkl.ac.in

Abstract. Incremental sheet forming (ISF) is a novel die-less sheet metal forming process, which can produce components directly from the CAD geometry using a CNC milling machine at less production time and cost. The formability of the sheet material used is greatly affected by the process parameters involved and tool path adopted, and the present study is aimed to investigate the influence of different process parameter values using the helical tool path strategy on the formability of a commercial pure Al and to achieve maximum formability in the material. ISF experiments for producing an 80 mm diameter axisymmetric dome were carried out on 2 mm thickness commercially pure Al sheets for different tool speeds and feed rates in a CNC milling machine with a 10 mm hemispherical forming tool. The obtained parts were analyzed for springback, amount of thinning and maximum forming depth. The results showed that when the tool speed was increased by keeping the feed rate constant, the forming depth and thinning were also increased. On contrary, when the feed rate was increased by keeping the tool speed constant, the forming depth and thinning were decreased. Springback was found to be higher when the feed rate was increased rather than the tool speed was increased.
Keywords: Incremental sheet forming; CP-Al; Helical tool path; Thinning; Springback.

1. Introduction

The use of computers made huge revolution in the manufacturing sector and enabled several older sheet metal forming techniques to take lesser time and cost for manufacturing, besides maintaining better accuracy. However, the frequent necessity for the change in design of the components demanded by the end users is pushing backwards all the conventional forming processes due to significant raise in the cost of redesigning the toolings associated with the process and hence promoting novel forming processes, which incur less production cost and lead time. Incremental Sheet Forming (ISF) is a novel, die-less computer-aided manufacturing process, which uses computer numerically controlled (CNC) machines to produce components with very less lead time and production cost.

The idea of dieless forming technique with the use of single point forming tool using CNC machines was patented long before its commercial use [1]. The process uses a predefined tool path, created in CAM software using the CAD geometry of the product, to form the sheet metals in a CNC milling machine with the use of a hemispherical tool. The tool is in continuous contact with the sheet metal, thereby creating a narrow deformation region just beneath the tool area. A series of such narrow and incremental deformations form the final product. Due to these narrow deformation zones, the whole sheet is not under continuous state of stress and hence the formability of the sheet metal is increased largely, when compared with the conventional sheet metal forming process [2-4].

The formability of the material is largely affected by the process parameters used for experimentation [5-6]. Combinations of different process parameters like the tool speed, feed rate, tool diameter, and step size define the amount of thinning, forming depth, springback and surface finish obtained in the process [7-8].

Springback is a huge concern in metal forming process as it reduces the dimensional accuracy of the products. When the applied load is removed after plastic deformation or the product is taken out from the fixtures, the metal tends to deviate from the designed geometry, due to elastic recovery, which creates springback. It was found out that the springback is reduced in ISF process compared with the conventional sheet forming processes due to the localised plastic deformation and the ISF process using the fixture support aided in reducing the springback of the material further [2, 9-11].

Research on ISF has been initiated in the last decade and is still on the preliminary stage of investigation. ISF process parameters like tool speed, feed rate, tool diameter, wall angle, tool path plays an important role in determining the formability of the material and researchers are still trying to overcome issues like springback, dimensional inaccuracy, surface quality etc. Proper designing of tool path is a challenge in ISF process, which adversely affects the process parameters and the formability of the material [12-13]. The objectives of the present study are to analyse the formability behaviour of commercially pure aluminium (CP-Al) sheets formed incrementally using the helical tool path strategy for different process parameter combinations of tool speed (S) and feed rate (F), and study the influence of tool parameters on the springback and thinning of the formed components.

2. Experimental Procedure

The experimental set up employed for the ISF experiments is shown in figure 1(a) and (b). CP-Al sheet of dimension 250 × 240 × 2 mm^3 was mounted over the fixture and it was rigidly held using a supporting ring fastened with two clamps on both the sides. The whole setup was fixed to the bed of the CNC milling machine. An 80 mm axisymmetric dome geometry was created using the CAD software and the corresponding G-codes of the geometry to feed into the CNC milling machine was created in a CAM software using the helical tool path strategy. A hemispherical forming tool of 10 mm diameter was used for the experiments and coolant oil was used to reduce the frictional force between the tool and sheet.

Figure 1. (a) Experimental set up of the ISF process (b) Forming trial of the ISF experiment.

Different feed rate and tool speed combinations, as shown in table 1, were used in order to study the effect of these parameters on the formability of the material. The components were formed either up to the full forming depth of the designed geometry or until fracture. After the completion of forming process, the formed domes were scanned using a FAROARM 3D laser scanning profilometer, which rebuilt the whole profile as surface-by-surface. The scanned profile was then exported to Solid Works software and the thickness profile, springback and forming depth were analysed.

Table 1. Different combinations of tool speed (S) and feed rate (F) used for ISF experiments

Experiment No.	Tool Speed (rpm)	Feed Rate (mm/min.)
1	50	50
2	500	50
3	3000	50
4	500	1000
5	1000	1000

For thinning analysis, one-quarter of the dome through the centre was considered and the wall thickness values were measured at every 4 mm depth from the top surface, i.e. mouth, of the dome. Figure 2 shows the methodology of the thickness measurement for the S3000F50 profile. The maximum forming depth was estimated as the centre distance between the top surface, i.e. mouth, of the dome to the bottommost inner surface of the dome (Length AB in figure 2). Similarly, the springback on the formed profile was calculated by comparing the designed profile (obtained from CAD geometry) with the actual scanned 3D profile (obtained using the FAROARM laser scanner).

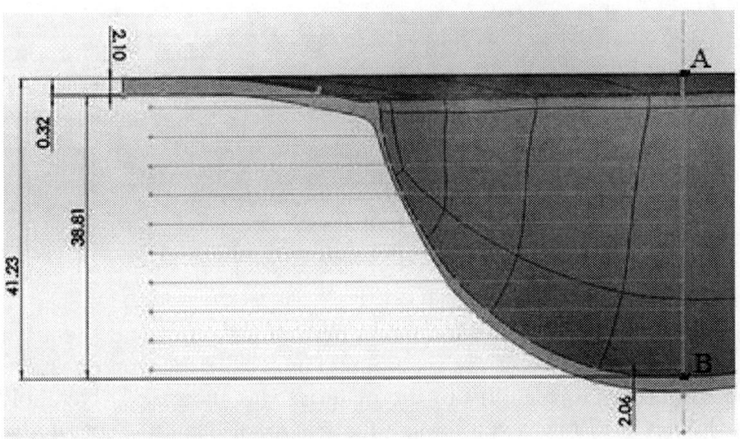

Figure 2. Methodology of thickness measurement of the formed profile.

3. Results and Discussion

3.1 Forming depth analysis.

Figure 3(a) shows the helical tool path followed for the ISF experiments. In helical tool path, the tool initially moves in a circular path, then steps down vertically at a constant value and proceeds again in the circular movement. For the present experiments, a constant vertical stepdown value of 0.5 mm was used, which constituted nearly 30,000 co-ordinate points for the whole tool path. Figure 3(b) shows the photograph of the incrementally formed component. The dimple line in the component denoted the tool movement during the stepdown. Table 2 shows the maximum forming depth achieved on different ISF parts formed using different process parameter combinations. All the samples were formed up to

IOP Conf. Series: Materials Science and Engineering **314** (2018) 012026 doi:10.1088/1757-899X/314/1/012026

the full forming depth and no cracking was observed in any of the components. At constant feed rate, when the tool speed was increased, the forming depth also increased. On contrary, when the feed rate was increased by keeping the tool speed constant, the forming depth was decreased. The increase in feed rate greatly reduced the forming time, whereas no change in forming time was observed for the increase in tool speed.

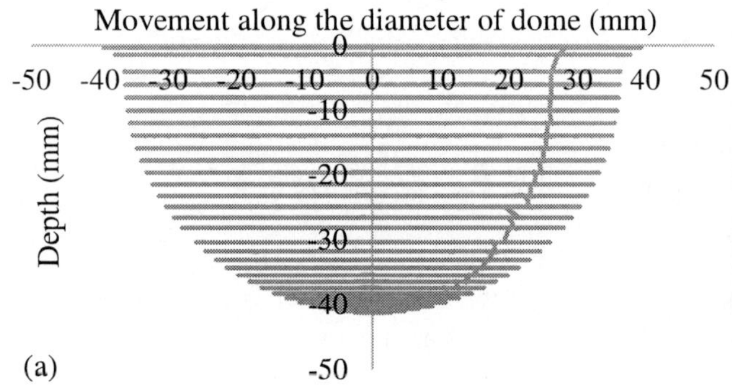

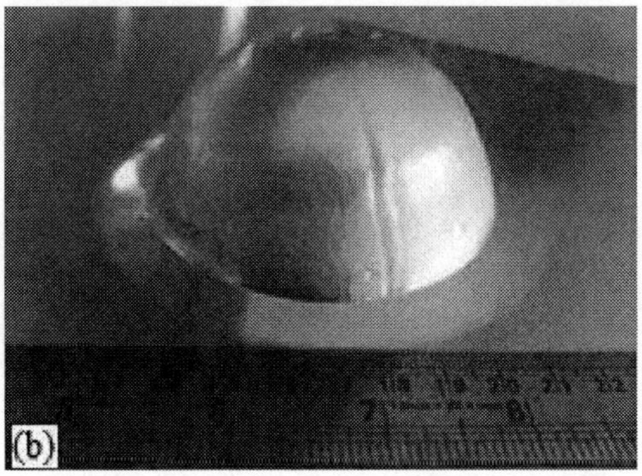

Figure 3. (a) Helical tool path followed for the ISF experiments **(b)** Formed dome after ISF.

Table 2. Maximum forming depth achieved on different ISF parts formed using different process parameter combinations.

Experiment no.	Parameters	Maximum forming depth (mm)
1	S50F50	40.67
2	S500F50	41.02
3	S3000F50	41.23
4	S500F1000	40.84
5	S1000F1000	40.33

When the tool speed was increased, the friction between the tool and sheet was increased, which eventually increased the temperature of the sheet metal under the tool region. As a result, the forming stresses were lowered and the formability of the material was increased. This holds true only for the lower feed rate. At higher feed rate, even though the frictional heating generated was high as a result of increased tool speed, the formability was decreased. It is speculated that the higher feed rate made the tool to reach the next deformation zone quicker than before the frictional heat dissipates to the subsequent deformation zone [14] and the frictional heat does not aid the forming process at higher feed rate, which eventually reduced the formability. The same argument applies to the increase in feed rate at constant tool speed also.

3.2 Thickness analysis.

Wall thinning is one of the important aspects to be considered in sheet metal forming, as plastic instability originates from this region. Figure 4 shows the thickness values at various points along the forming depth of the dome. At constant feed rate, when the tool speed was increased, the amount of thinning increased due to increased amount of deformation and subsequent increase in forming depth. Similarly, at constant tool speed, when the feed rate was increased, the amount of thinning decreased due to decreased amount of deformation and lower forming depth.

The thickness analysis showed that the thinning is always predominant on the sidewalls of the dome for all the process parameter combinations. The wall thickness of the formed part reduced drastically up to certain forming depth and increased afterwards for further forming depth. Maximum thinning was observed around 16 mm of the forming depth for all the process parameter combinations and it was found to be highest, i.e. ~80%, for the S1000F1000 combination. The reason behind the occurrence of maximum thinning at the same sidewall region for all the samples is that this is the region at which maximum wall angle of the formed part was achieved. This region is termed as "thinning band" in figure 5, and the wall angle of the part was reduced beyond this point. Similar observations were made by others during the ISF process [15]. Overall, the combined action of uniaxial stretching and shearing deformation modes, due to the vertical step down and circumferential movement of the tool, respectively, contributed to such a non-uniform thinning on the part.

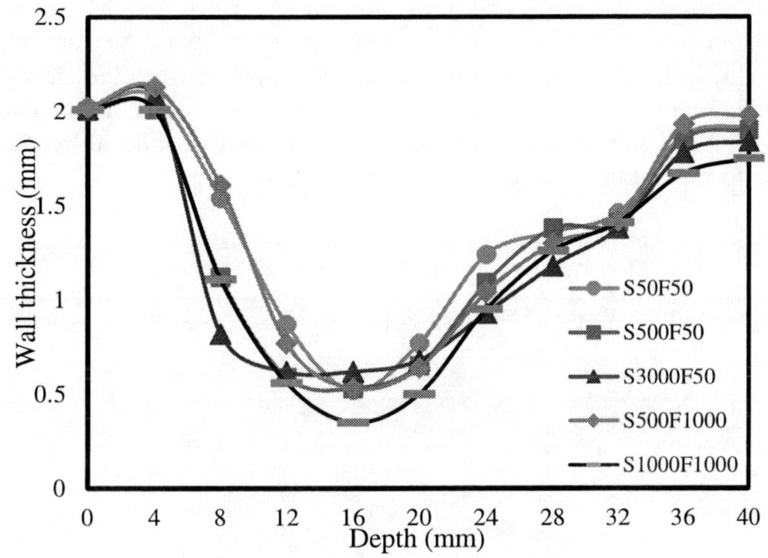

Figure 4. Thickness profile of ISF parts formed using different process parameter combinations.

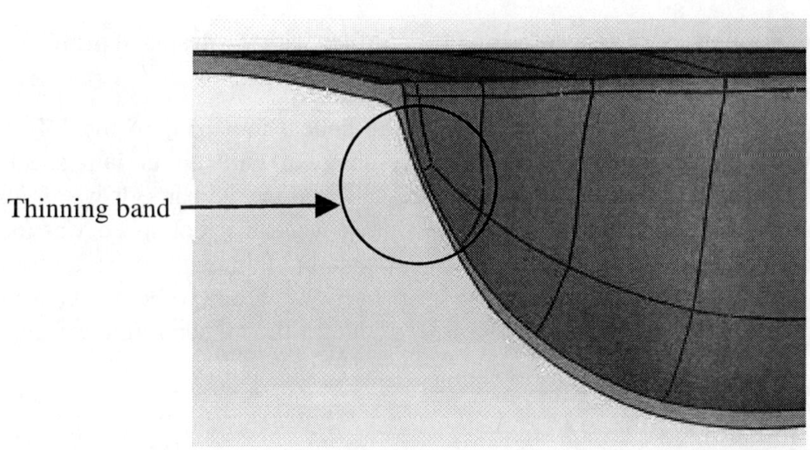

Figure 5. Thinning band depicting the region of maximum wall angle formed.

3.3 Springback analysis.

The comparison of actual 3D scanned profile of different incrementally formed parts with the designed profile is shown in figure 6. It was observed that the increase in tool speed, by keeping the feed rate constant, decreased the amount of springback in the part. On the other hand, the increase in feed rate, by keeping the tool speed constant, increased the amount of springback.

Overall, the springback was always greater near the sidewalls and mouth of the dome. Dimensional inaccuracy due to springback arises mainly because of the forming forces imparted by the tool while moving from one contact point to another and removal of clamps/ reduction in clamping forces after forming. During the ISF process, when the tool comes into contact with the sheet blank, the blank bends near the clamped edge and the huge elastic recovery at this bending region on removing the clamping forces created large springback near the mouth region of the dome. Similarly, the circumferential movement of the tool and the simultaneous increase in forming forces created springback on the sidewalls of the dome. However, the trade-off between the elastic recovery forces and frictional heating due to tool movement reduced the springback to considerable amount in this region. Microstructural investigations will be carried out in future in order to reduce the springback and thinning on the formed parts based on strain path optimisation.

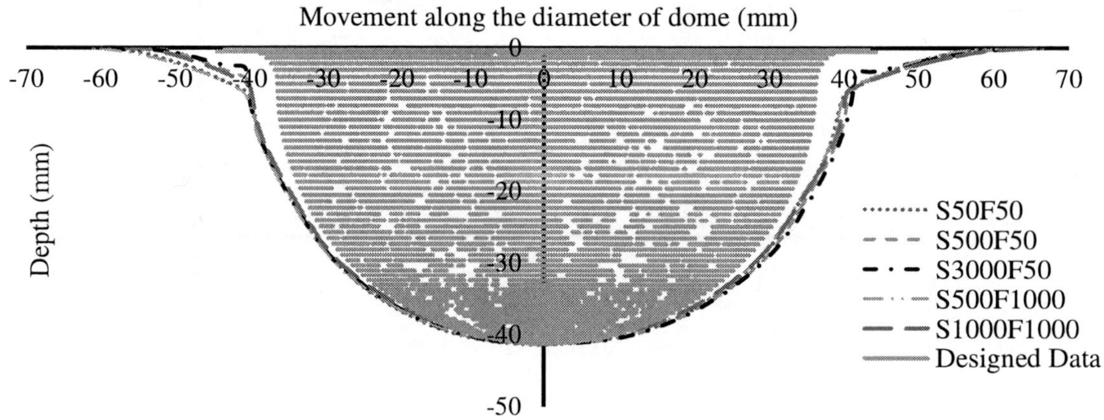

Figure 6. Springback analysis of actual ISF profiles with the designed profile.

4. Conclusions

An axisymmetric dome of 80 mm was formed using the helical tool path of the ISF process for different tool speed and feed rate combinations on commercial pure aluminium sheets, and the forming depth, amount of thinning and springback were analysed on the parts. It was found out that the forming depth and subsequently thinning were increased when the tool speed was increased by keeping the feed rate constant. On contrary, the forming depth and thinning were decreased when the feed rate was increased at constant tool speed. The springback was found to be higher when the feed rate was increased rather than the increase in tool speed, due to the higher forming forces imparted during the process.

5. References

[1] Leszak E. Apparatus and process for incremental dieless forming 1967 Patent US3342051A1.
[2] Jeswiet J, Micari F, Hirt G, Bramley A, Duflou J and Allwood J Asymmetric single point incremental forming of sheet metal 2005 *CIRP Ann. Manuf. Technol.* **54** 88-114.

[3] Nimbalkar D and Nandedkar V. Review of incremental forming of sheet setal components 2013 *Int. J. Eng. Res. Appl.* **5** 39-51.

[4] Tisza M. General overview of sheet incremental forming 2012 *J. Achiev. Mater. Manuf. Eng.* **55** 113-20.

[5] Park J and Kim Y Fundamental studies on the incremental sheet metal forming technique 2003 *J. Mater. Process. Technol.* **140** 447-53.

[6] Araghi B, Manco G, Bambach M and Hirt G. Investigation into a new hybrid forming process: Incremental sheet forming combined with stretch forming 2009 *CIRP Ann. Manuf. Technol.* 2009 **58** 225-8.

[7] Hamilton K and Jeswiet J Single point incremental forming at high feed rates and rotational speeds: Surface and structural consequences 2010 *CIRP Ann. Manuf. Technol* **59** 311-4.

[8] Ham M and Jeswiet J Single Point Incremental Forming and the Forming Criteria for AA3003 2006 *CIRP Ann. Manuf. Technol.* **55** 241-4.

[9] Allwood J, Braun D and Music O The effect of partially cut-out blanks on geometric accuracy in incremental sheet forming 2010 *J. Mater. Process. Technol.***11** 1501-10.

[10] Ambrogio G, Costantino I, De Napoli L, Filice L, Fratini L and Muzzupappa M. Influence of some relevant process parameters on the dimensional accuracy in incremental forming: a numerical and experimental investigation 2004 *J. Mater. Process. Technol.* **153-154** 1501-07.

[11] Moon Y, Kang S, Cho J and Kim T Effect of tool temperature on the reduction of the springback of aluminum sheets 2003 *J. Mater. Process. Technol.* **132** 365-8.

[12] Skjoedt M, Hancock M and Bay N Creating 3D Spiral Tool Paths for Single Point Incremental Forming 2007 *Key Eng. Mater.* **344** 583-90.

[13] Malhotra R, Reddy N and Cao J Automatic 3D Spiral Toolpath Generation for Single Point Incremental Forming 2010 *J. Manuf. Sci. Eng.*2010 **132** 061003.

[14] Ambrogio G and Gagliardi F Temperature variation during high speed incremental forming on different lightweight alloys 2014 *Int. J. Adv. Manuf. Technol.* **76** 1819-25

[15] Hussain G, Hayat N and Gao L An experimental study on the effect of thinning band on the sheet formability in negative incremental forming 2008 *Int. J M/c tool Manuf.* **48** 1170-78.

IOP Conf. Series: Materials Science and Engineering **314** (2018) 012027 doi:10.1088/1757-899X/314/1/012027

Microbiological corrosion of ASTM SA105 carbon steel pipe for industrial fire water usage

S Chidambaram[1], Ashok K[2], Karthik V[3] and P G Venkatakrishnan[4]

[1] Alumini Interaction Metallurgical Engineering Department, Government College of Engineering Salem, India
[2] Scientist, CSIR-National Metallurgical Laboratory Jamshedpur India
[3] Assistant Professor Metallurgical Engineering Department, Government College of Engineering Salem, India
[4] Professor and Head, Metallurgical Engineering Department, Government College of Engineering Salem, India

Abstract The large number of metallic systems developed for last few decades against both general uniform corrosion and localized corrosion. Among all microbiological induced corrosion (MIC) is attractive, multidisciplinary and complex in nature. Many chemical processing industries utilizes fresh water for fire service to nullify major/minor fire. One such fire water service line pipe attacked by micro-organisms leads to leakage which is industrially important from safety point of view. Also large numbers of leakage reported in similar fire water service of nearby food processing plant, paper & pulp plant, steel plant, electricity board etc...In present investigation one such industrial fire water service line failure analysis of carbon steel line pipe was analyzed to determine the cause of failure. The water sample subjected to various chemical and bacterial analyses. Turbidity, pH, calcium hardness, free chlorine, oxidation reduction potential, fungi, yeasts, sulphide reducing bacteria (SRB) and total bacteria (TB) were measured on water sample analysis. The corrosion rate was measured on steel samples and corrosion coupon measurements were installed in fire water for validating non flow assisted localized corrosion. The sulphide reducing bacteria (SRB) presents in fire water causes a localized micro biological corrosion attack of line pipe.

1. Introduction

The carbon steel has an optimum corrosion resistance for many fluid services including hydrocarbon. The microbiological influenced corrosion of carbon steel was complex and interdisciplinary in nature. In many number of industrial aqueous environments like crude oil and water, microbes metabolizes results in microbiological induced corrosion [1] [2]. There is no proactive inspection methodology for localized corrosion other than detailed corrosion prevention program of fire water pipelines [3] [4] [5]. The chemical processing industries utilizes carbon steel pipe for fresh or recycled water services. The

IOP Conf. Series: Materials Science and Engineering **314** (2018) 012027 doi:10.1088/1757-899X/314/1/012027

fire water service line pipe of chemical process industries carries fresh or recycled water in stagnant full load condition to nullify the severity of fire(s) during mishap, leakages or any other uncontrolled processing chemical reactions. In the present study one such frequent leakage of fire water network service of processing industry was investigated. Root cause and failure analysis of these case studies were performed in this investigation to improve the understanding related to process condition and failures [2] [10] [11] [12].

2. Materials and Methods

The material of construction for fire water service line pipe is ASTM SA105 carbon steel were produced by forging process and supplied in annealed condition. The operating pressure of fire water pipe was specified between 5kg/sq.cm to 7 kg/sq.cm depends on type of fire causes. The temperature of fire water was maintained between 25 deg C to 30 deg C with PH ranges from 6.5 to 9. The chemical composition of SA105 Grade B is shown in table1. The tensile strength, yield strength and elongation was specified as 480 MPa, 250 MPa and 30% respectively [1] [10] [11].

Table 1. Chemical composition of ASTM SA105 Grade B pipe.

Elements	Composition%
Carbon	0.35 max
Manganese	0.6-1.05
Phosphorous	0.035
Sulphur	0.040
Silicon	0.10-0.35
Copper	0.40 max
Nickel	0.40 max
Chromium	0.30 max
Molybdenum	0.12 max
Vanadium	0.08 max

3. Theory

The thermodynamic free energy equation of corrosion of line pipe was evaluated [1] and shown below

$$\log_e K_{eq} = \frac{nFE^0}{RT} \tag{1}$$

K_{eq} is equilibrium constant, n is number of electrons per atom of iron species
R is gas constant = 1.987 cal/K.mol,
T is absolute temperature 298 K and
F is faraday constant = 23060 cal/volt equivalent
The electromotive force series of ASTM SA105 Grade B line pipe corrosion reaction with standard potential at 25^{0}C is chosen as fire water operating at room temperature [1]

$$Fe^{2+} + 2e^- \rightarrow Fe \qquad E^0 = -0.440 \text{ volts} \tag{2}$$

The pourbaix diagram shown in figure1 is used to identify the type of scale film formed on steel surface.

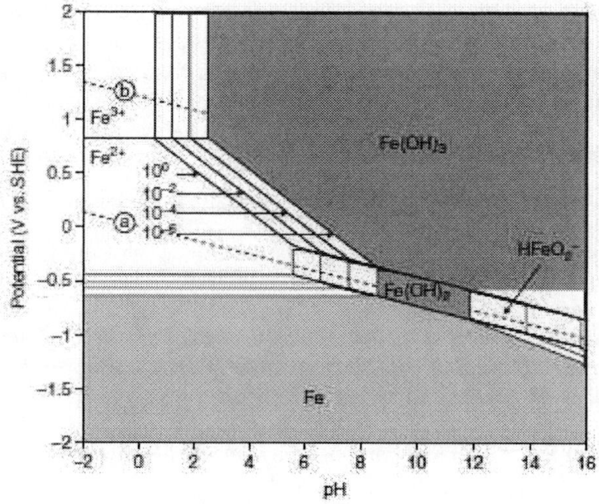

Figure1. Pourbaix (E vs. pH) diagram of carbon steel at 25 deg C [1].

The anaerobic sulphur reducing bacteria containing de-sulfovibrio desulfuricans can be written as,

$$SO_4^{2-} + 4H_2 \rightarrow S^{2-} + 4H_2O \tag{3}$$

The aerobic sulphur oxidizing bacteria containing thiobacillus thiosulfates and thiobaccillus thiooxidas can be written as,

$$2S + 3O_2 + 2H_2O \rightarrow 2H_2SO_4 \tag{4}$$

The sulphuric acid formation is localized in aerobic sulphur oxidizing bacterial reactions [3] [4] [5] [6] [7] [9].

4. Experimental procedure

The fire water samples were collected for chemical and bacterial constituent's analysis. The pH of fire water samples were measured using DBK digital pH meter with accuracy of 0.01pH. The turbidity and orthophosphate of fire water samples were measured using HACH calorimeter DR900 model. Upon addition of 0.25 ml molybdovanadate reagent (ammonium molybdate and sulphuric acid) to 10 ml fire water sample, the orthophosphate constituents in fire water were measured [12]. The oxidation reduction potential was measured using EZODO make of ORP5041model oxidation reduction potential meter. Free residual chlorine was measured by mixing 1:10 of chlorine testing reagent TESTCHLOR to fire water sample and interprets according to table2.The calcium hardness of fire water sample was measured by titration technique. Beaker solution contains addition of x ml fire water sample, 4ml sodium hydroxide (NaOH) and one drop Patton & reeders reagent ($C_{21}H_{14}N_2O_7S.3H_2O$). Pipette solution was prepared with 0.01M molar strength by addition of 3.7224gm EDTA (ethylenedinitrilotetraacetic acid disodium salt) to 1000 ml distilled water [12]. Titration was conducted using burette and pipette till solution colour changes from pink to blue and final calcium hardness was calculated using equation5 [8].

$$\text{Calcium hardness} = (\text{Initial reading} - \text{Final reading}) \times (1000\text{ml}/x \text{ ml of water sample}) \tag{5}$$

IOP Conf. Series: Materials Science and Engineering **314** (2018) 012027 doi:10.1088/1757-899X/314/1/012027

SA105 Grade B carbon steel specimens of size 2 x 4 cm were used for gravimetric measurements. The specimens were polished, degreased with trichloro ethylene, weighed and exposed in the fire water sample. After an exposure for seven days, the specimen was removed from fire water, washed with water and the corrosion products were cleaned with Clarke's solution. The specimen was dried, finally weighed and the weight loss was noted. The corrosion rates (milli metre per year) were calculated using the equation6 [1] [8] [9] [10] [11].

$$\text{Corrosion rate} = 87.6W/ATD \text{ (mmpy)} \tag{6}$$

where, W = Weight loss (in mg)
A = Area of the specimen (in cm2)
T = Time (in hrs)
D = Density of the carbon steel (7.86 g/cc)

One set of fire water samples were serially collected at six different locations, diluted and bacterial colonies like sulphate reducing bacteria (anaerobic bacteria) were enumerated by pour plate technique. The nutrient agar used for sulphur reducing bacterial growth nutrient medium [8]. Carbon steel equivalent C1020 grade carbon steel of 7.5 cm in length 1.5 cm in width and 0.25 cm in thick corrosion coupon was prepared and installed in line pipe under fire water service for measuring actual corrosion rate. Another set of fire water samples were collected from another location of fire water pipeline near to frequent failure reported area and mixed with neutralizing compounds (desulfotomaculum and desulfobacter) to nullify the effect of corrosion inhibitor in fire water sample. Thereafter bactaslyde were mixed with fire water sample to make solution and ensured air was completely evacuated in the tube. After mixed, the solution under incubate time starts and blackening of mixed solution begin from the bottom of tube. Thereafter sulphur reducing bacteria were estimated according to table5 [12]. The corrosion product scale film formed on inner side of steel pipe was scratched by using polymeric wire brush and its chemical composition was analysed by digestion technique.

5. Results and Discussion

From equation1equilibrium constant calculated as K_{eq}= 1.535. The pH of water was measured 7.30 and with available half negative electrode potential of 0.440 volts, thermodynamic parameters were mapped on pourbaix diagram shown in figure1. The iron Fe2+ scales were confirmed on the carbon steel surface. The turbidity of fire water sample was measured with respect to demineralised water assuming turbidity of demineralised water is 0 NTU (Nephelometric Turbidity Units).

Table 2. %Chlorine in fire water interpretation table.

Colour produced	Chlorine in ppm
White milky	0(no chlorine)
Fluorescent	0(no chlorine)
Orange	0.1(no chlorine)
Light brown	0.2(suit for use)
Red	0.5(suit for use)
Violet	0.6(not usable)
Black	0.8(not usable)
Dark Blue	1.0(not usable)

The oxidation reduction potential was measured as 225-235mv. Free residual chlorine of fire water sample was measured as zero ppm cl and white milky colour observed in solution. From the weight loss measurements made on carbon steel in fire water samples collected from entire fire service pipeline of different locations, it is observed that the corrosion rate is 5.0995 x 10^{-2} mmpy shown in table3. The results obtained from bacterial evaluation of the water samples indicate that bacteria such as SRB are responsible for the corrosion of the fire water pipeline from pour plate technique as shown in table4. The sulphur reducing bacteria per 100 ml measured and presented in table5 from bactaslyde mixed solution.

Table 3. Weight loss of ASTMSA105 Grade B.

Corrosion rate(10^{-2} mmpy)	Mean corrosion rate(10^{-2} mmpy)
5.7630	
4.8928	5.0995
4.6428	

Table 4. Sulphur reducing bacteria enumerated using pour plate technique.

Number of water samples	Sulfur reducing bacteria colony forming unit per ml
1	3.3 x 10^{-4}
2	4.2 x 10^{-4}
3	5.8 x 10^{-4}
4	6.3 x 10^{-4}
5	2.1 x 10^{-4}
6	3.9 x 10^{-4}

Table 5. Sulphur reducing bacteria interpretation table.

Blackening in hours	Sulphur reducing bacteria per 100 ml
24	100000
48	10000
72	1000
96	100
120	10
144	<10

The fungi and yeast will grow in bactaslyde either as pure yeast colonies or pure fungi colonies or fungi-yeast mixed colonies with yeast appears in white colour and fungal in either of white, green and brown colour of the solution. Neither fungi nor yeast were estimated from mixed solution, as any traces of white, green, brown colour was not observed in the bactaslyde mixed solution after 144 hours. The turbidity and orthophosphate of fire water sample was measured as 50 NTU and 0 ppm PO_4 respectively. The calcium hardness was measured using titration technique and using equation5, calcium hardness calculated as 74 ppm $CaCO_3$. The total bacteria count limit and sulphur reducing bacteria limit was measured 100000 counts/ml and 100 counts/100 ml, respectively for SA105 Grade B material. The corrosion coupon installed in line pipe under fire water service and the corresponding corrosion rate calculations were plotted in Figure 2. The corrosion coupon installed for period of one month and corrosion rate were calculated for one coupon, each month. The straight light line in figure 2 indicates corrosion rate limit of installed coupons, and corrosion rate is observed to be under control for the fire water service. After mixing of bactaslyde with water samples and solution was incubated minimum of 144 hours, the sulphide reducing bacteria count per 100 ml from sample water was interpreted and also plotted in figure 3. The straight light line in figure 3 indicates sulphide bacteria count limit. From figure 3 the sulphide reducing bacteria count per 100 ml is observed to be in the abnormal domain. The water analysis also reveals the presence of sulphide reducing bacteria in fire water network, which is considered to be one of the prime reasons for material degradation. The total bacteria count per ml from sample water is plotted in figure 4 and the straight light line in figure 4 indicates total bacteria count limit. From figure4 it is observed that the total bacteria counts per ml are under control.

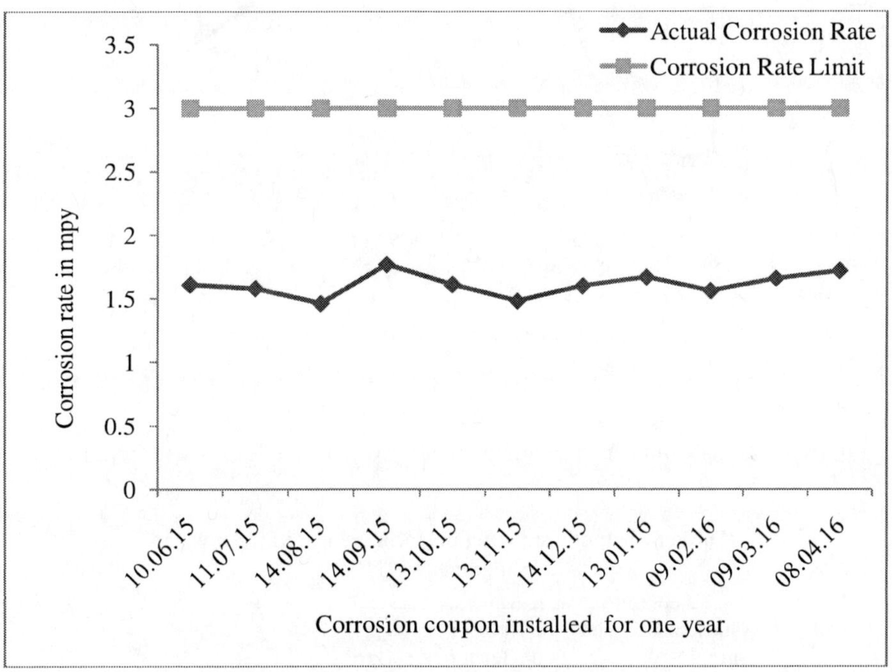

Figure2. Corrosion rate plotted for corrosion coupon in fire water.

The Algae observed vicinity to localized failure however barnacles or other aqueous organisms are not observed, as represented by arrow in figure 5a. The compositional analysis performed by digestion technique details are given below

Iron content (as elemental Fe) = 53.4 wt%

Sulphur content (as elemental sulphur) = 1.1 wt%
Silica content (as SiO2) = 1.3 wt%

Rest is unconvertible materials like mud, sand etc were also found. Anaerobic bacterium formed on internal side of pipe revealed from sulfur reducing bacteria count experimentation. The soil and vegetation contact of line pipe further increases the corrosion rate to localize near damage portion of the component. It can be attributed to the presence of organic products, cellulose and microorganisms present in soil and vegetation.

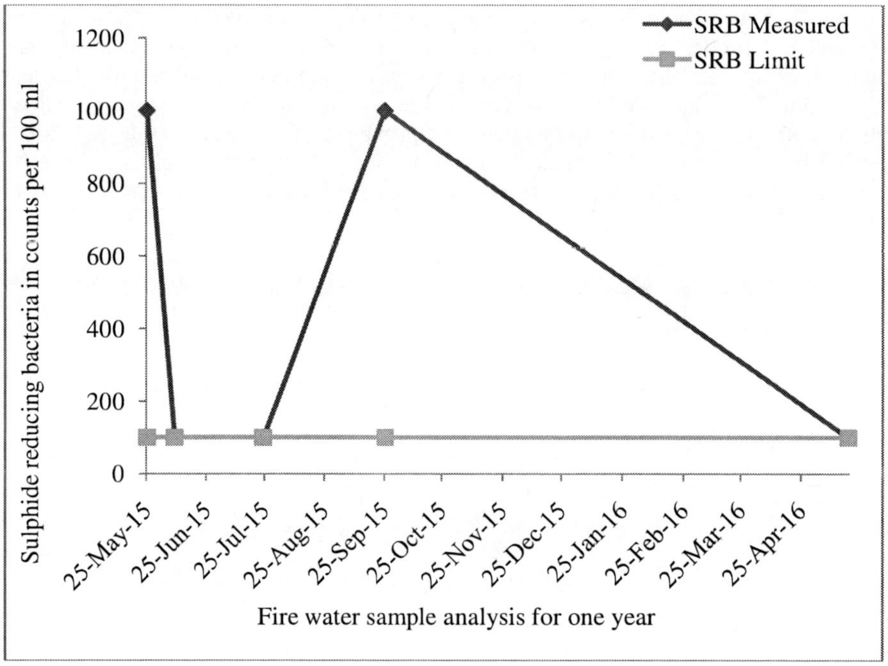

Figure3. Sulphide reducing bacteria plotted for water sample obtained from fire water.

Table 6. Sulphur reducing bacteria measured from water sample analysis.

Number of water samples	Sulphur reducing bacteria per 100 ml
1	100>SRB>1000
2	100
3	100
4	1000
5	100

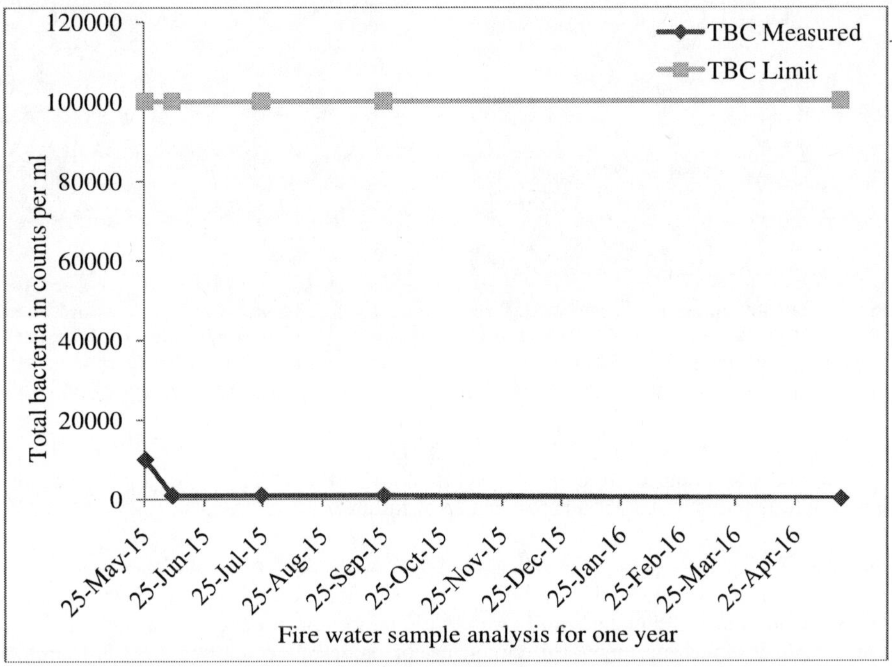

Figure4. Total bacteria plotted for water sample obtained from fire water.

Table 7. Total bacteria count measured from water sample analysis.

Number of water samples	Total bacteria per 100 ml
1	10000
2	1000
3	1000
4	1000
5	100

De-sulfovibrio desulfuricans forming sulphur reducing bacteria count above 100 counts per 100 ml in water sample is suspected to be the prime cause of failure. The desulfuricans anaerobic bacteria form on internal side of pipe, where sulphur tends to accelerates dis-solution of metal due to stagnant water deposits for long time and low fluid flow with or without mere oxygen under dark conditions leads to the growth of micro-organisms, which further promotes the micro-biological induced corrosion in line pipe steel under fire service.

Figure5a. Macroscopic view of fire water line pipe damaged by biological degradation.

Figure5b. Crude oil tank bottom carbon steel plate hemi spherically damaged by micro biological influenced corrosion of 250 mm diameter.

The pour plate technique and corrosion analysis data for last one year also clearly reveals that the presence of sulphide reducing bacteria is root cause of failure.

6. Conclusion

The process parameters of fire water sample like pH, turbidity, orthophosphate, free residual chlorine, oxidation reduction potential and calcium hardness are under control and not influencing corrosion of steel. From weight loss measurements (gravimetric) corrosion rate of carbon steel is found to be very low. The microbiological corrosion was also revealed by composition evaluation of internal corrosion product scale contains 1.1% sulphur which is prime source of anodic reaction. Based upon various chemical and microbiological experimental results it could be concluded that fire water service pipeline failed due to sulphide reducing bacteria. Sulphur reducing bacteria is responsible for microbiological influenced corrosion of fire water service pipeline. The fire water reacts with steel under dark conditions without or with little oxygen promotes sulphur reducing bacterial growth leads microbiological corrosion.

7. Recommendations

Fresh or recycled water could be treated with chlorine, bromine or iodine compounds. Frequent water flushing could avoid water stagnation greatly and reduces micro-organism growth to prevent the failure. Fire water could be treated with ClO2 poly silicate treatment to prevent sewage or foreign particles carry over and further reduce sulphur content in water. Corrosion protection by cathodic protection of carbon steel pipe shall be provided after appropriate external coating. Internal epoxy or polypropylene coatings are used to prevent carbon steel pipe from micro biological corrosion. Soil and vegetation contact with water piping shall be avoided to reduce the corrosion rate near damaged localized area and further retards aerobic bacteria's growth. The corrosion observed in the fire water pipelines can be controlled by the addition of biocides as well as suitable corrosion inhibitors.

8. Acknowledgments

The First Author wish to acknowledge the support of conducting various experiments and encouragement from Corrosion Laboratory, Department of Metallurgical Engineering, Government College of Engineering Salem, India

IOP Conf. Series: Materials Science and Engineering **314** (2018) 012027 doi:10.1088/1757-899X/314/1/012027

References

[1] Stephen C Dexter 1987 American Society of Metals **Volume 13** *Corrosion*, metals park (OH), ASM international chapter 2 p 114-117
[2] D H Pope and J G Stoecker 1986 *Microbiologically Influenced Corrosion in Process Industries*
[3] JDA Miller, Ed 1970 *Microbial Aspects of Metallurgy* American Elsevier
[4] P.F. Sanders and W.A. Hamilton, *Biological and Corrosion Activities of Sulfide Reducing Bacteria in Industrial Process Plant*
[5] API 571 2003 *Damage Mechanisms for Process Industries* chapter 4 p 86-91
[6] M G Fontana *Corrosion Engineering* Tata McGraw Hill chapter 9 p
[7] S.C. Dexter Ed 1986 *Biologically Induced Corrosion*, National association of corrosion engineers, p 47-68
[8] R. D. Port and H. M. Herro 1991*The NALCO Guide to Boiler Failure Analysis* McGraw Hill New York
[9] R. D. Port and H. M. Herro 1991*NALCO Guide to Cooling Water System Failure Analysis* McGraw Hill New York
[10] *1999 NACE Course Book on Corrosion Control in the Refining Industry* NACE International Houston
[11] L. Garverick *1994 Corrosion in Petrochemical Industry* ASM International metals park (OH)
[12] Betz *1980Handbook of Industrial Water Conditioning* Betz Laboratories PA

IOP Conf. Series: Materials Science and Engineering **314** (2018) 012028 doi:10.1088/1757-899X/314/1/012028

Preparation and Characterization of Mesoporous Nickel derived from Liquid crystalline Template and Evaluation of its Electro catalytic activity towards Methanol Oxidation

S Mohanapriya, R Renuka devi and V Raj
Advance Materials Research Lab, Department of Chemistry, Periyar University, Salem 636 011, India.

Abstract. Mesoporous Nickel has been prepared by electrodeposition using non-ionic surfactant based liquid crystalline template under optimized processing conditions. Physico-chemical properties of mesoporous nickel is systematically characterized through XRD, SEM and AFM analyses. Comparison of electrocatalytic activity of mesoporous nickel with smooth nickel was interrogated using cyclic voltammetry (CV), chronoamperometry (CA) and electrochemical impedance spectroscopy (EIS) analyses. Distinctly enhanced electrocatalytic activity with improved surface poisoning resistance related to mesoporous nickel electrode towards methanol oxidation stems from unique mesoporous morphology. This mesoporous morphology with high surface to volume ratio is highly beneficial to promote active catalytic centers to offer readily accessible Pt catalytic sites for MOR, through facilitating mass and electron transports.

1. Introduction

Owing to the simplified system design, high energy density and ease of transportation of fuel, Direct methanol fuel cells (DMFCs) are currently at forefront as renewable energy sources which involves electrochemical energy conversion and hence suitable for automotive as well as portable power applications [1-4]. Till now, Platinum (Pt) and its alloys represent most extensively used DMFC electrocatalysts on account of their superior catalytic activity and stability towards oxidation of methanol at anode side. [5-7]. But catalytic activity of Pt surface could easily deteriorated by the adsorption of CO generated during methanol oxidation reaction (MOR).

Promoting effect of Ni over Pt catalytic activity both in acidic and alkaline medium is well established in literature [8-10]. Earlier studies established the role of nickel in assisting the dissociative adsorption of water molecules to produce -OH_{ads} species, which facilitates the oxidation of CO adsorbed on Pt surface thereby favors regeneration of Pt catalytic centers. Many electrodes involving nickel as a component in their manufacture can be used as catalysts in fuel cells. It is commonly used as an electro- catalyst for both anodic and cathodic reactions in organic synthesis and water electrolysis [15–18]. One of the very important uses of nickel as a catalyst is for the oxidation of alcohols. Several studies of the electro- oxidation of alcohols on nickel have been reported. Alkaline direct methanol fuel cells are now receiving considerable attention because they allow the use of anion exchange membranes that reduce methanol cross-over – a serious problem with cation exchange membrane used in acidic methanol fuels [8]. Another reason for using alkaline medium is because the kinetics for both methanol oxidation and oxygen reduction reactions is found to be more facile in alkaline medium than in acidic one [9-12]. In recent years, the focus of research is towards finding cheaper alternative electrocatalysts and improving the overall cell performance. Non noble transition metal electrocatalysts such as copper [13-16], nickel [17-24] which oxidise methanol and other alcohols, offer good alternative electrocatalysts. The activity of nickel based electrocatalysts depends on the method of preparation of these electrodes. Rahim et al., [24] showed that only nickel dispersed in graphite is catalytically active for methanol oxidation while massive nickel is not. Also this nickel electrocatayst, after continuous cycling loses its performance due to possible loss of activity of nickel oxide [24]. This is believed to be due to increased thickness of NiO(OH) which acts as a barrier inhibiting the charge transfer process for methanol oxidation. Alternative approaches to the preparation of nickel electrocatalysts that can minimise the buildup of such a barrier layer, would be of great interest in the development of non-noble metal electrodes for methanol oxidation.

The purpose of the present work is to establish the electro-catalytic oxidation of methanol on mesoporous nickel in a solution of 1.0 M NaOH. Surfactant mesophases come under the category of soft template systems and electrodeposition through them is a very useful and versatile method for the synthesis of nanostructured materials. In addition, lyotropic liquid crystalline phases possess a long-ranged spatially periodic architecture with lattice parameters in the range of a few nanometers. This makes them ideal candidate systems for synthesis of nanomaterials. In this study, we report a simple electrochemical method of preparing a high surface area mesoporous nickel deposit, using a new hexagonal liquid crystalline phase as template. This room temperature deposited nickel has been characterized by electrochemical techniques such as cyclic voltammetry (CV) and electrochemical impedance spectroscopy. Besides, scanning electron microscopy (SEM), Atomic force microscopy (AFM) and X-ray diffraction (XRD) are also used for surface characterization. Catalytic efficiency of synthesized mesoporous nickel is evaluated by methanol oxidation using the techniques such as CV, Impedance etc., Morphology and composition of mesoporous nickel cloud easily be tuned through controlling parameters involved in electrodeposition.

2. Experimental

2.1.Chemicals

Triton X-100 (Spectrochem), Poly(acrylic acid) (PAA) (Aldrich), sodium hydroxide pellets (Emerck), nickel(II) chloride (EMerck), nickel sulphamate (Grauer and Wheel), boric acid (Sarabhai chemicals) were used in this study. All chemical reagents used were AnalaR (AR) grade. Millipore water having a resistivity of 18 M cm was used in all the experiments.

2.2. Preparation and characterization of hexagonal liquid crystalline phase

Hexagonal liquid crystalline phase was prepared from a ternary mixture of Triton X-100, poly(acrylic acid) and water. We have carried out our studies at two different weight compositions viz. (1) Triton X-100 42% + water 58%. The mixture was stirred in a magnetic stirrer at a temperature of 33–35° C for 1 h and was allowed to cool down to room temperature. When viewed under a polarized light microscope, the mixture shows the characteristic birefringence of a hexagonal liquid crystalline phase stable up to 29° C. All the electrodeposition using this phase as template has been done at the room temperature of 25° C.

2.3. Nickel electroplating through hexagonal liquid crystalline

The electrodeposition of mesoporous nickel was performed in a standard three-electrode glass cell using Al as a working electrode, a saturated calomel electrode (SCE) and Pt foil served as a reference and counter electrodes respectively. Mesoporous nickel was synthesized through electrodeposition at a constant of -1 V for different time period. For electrodeposition, we have prepared the aqueous phase of the above described hexagonal liquid crystalline system from the standard nickel sulphamate bath [21] of the composition: 300 g l^{-1} nickel sulphamate, 6 g l^{-1} nickel chloride and 30 g l^{-1} boric acid.

3. Results and discussion

3.1 Micro-structural analyses

Interesting electrocatlytic properties is exhibited by mesooporous nickel owing to their high surface area, unique morphology and high conductivity. Schematics of synthesis of steps involved in the process of preparation of mesoporous nickel is illustrated in **Figure 1**. Triton X 100 forms micelles aggregated in the aqueous phase. Nickel ions encapsulated in the micelles of Triton X100 undergoes electrochemical reduction in such way that meso domains are created over the surface. Textural properties of nickel deposit are modified due to entrapment of nickel ions inside the micelles.

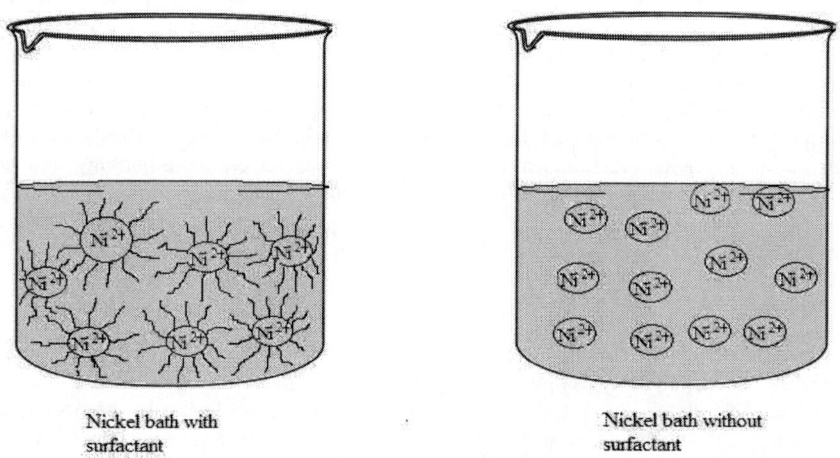

Figure-1a. Pictorial representation of nickel bath with and without surfactant.

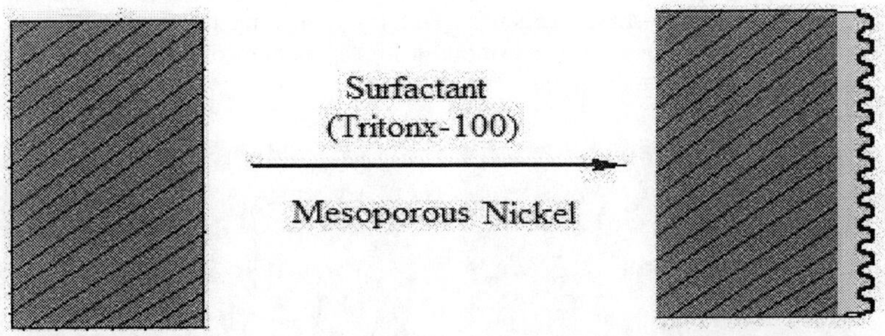

Figure-1b Schematic representation for preparation of Mesoporous nickel.

Mesoporous nickel was characterized by X-ray powder diffraction (XRD). It can be noticed from **Figure 2** that XRD pattern of nickel deposited electrodes clearly reveal the characteristic peaks expected for nickel with face centered cubic (fcc) structure in addition to the reflections observed for bare Al electrode. Mesoporous nickel deposited show the peaks at 44.3° and 51.8° corresponding to (111) and (200) planes respectively and are indicated by the asterisk mark in the figure.

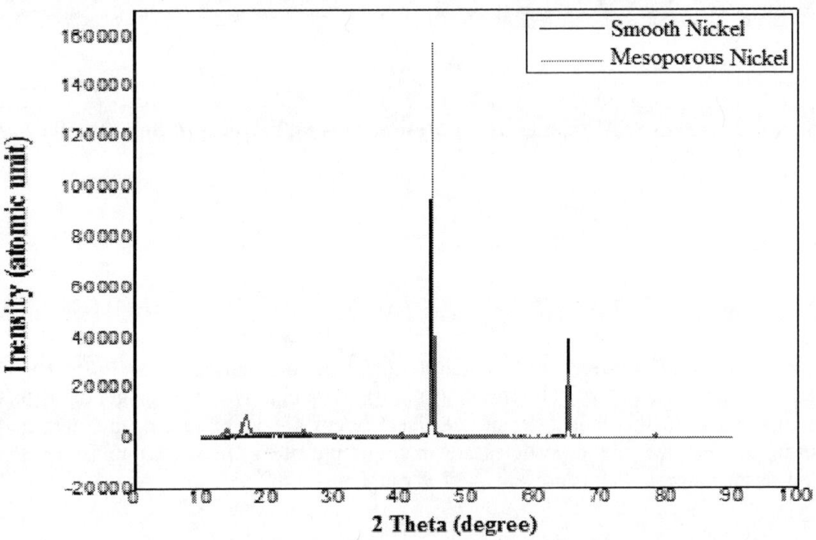

Fig 2 X-ray diffraction of Smooth Nickel and Rough Nickel.

The ratio of intensities of (111) and (200) planes associated with these nickel electrodes with and without surfactant suggests that there is a preferential growth and orientation of nickel film along (200) direction in the case of surfactant electro deposition. It can be noticed from the figure that XRD pattern of nickel deposited electrodes clearly reveal the characteristic peaks expected for nickel with face centered cubic structure.

IOP Conf. Series: Materials Science and Engineering **314** (2018) 012028 doi:10.1088/1757-899X/314/1/012028

SEM pictures of nickel deposited on with and without addition of Triton-X 100 are provided in **Figure 3.** The morphology of an individual deposit depends upon the electrolyte used for deposition. These images clearly display the distinguishable structural features indicating the homogeneous growth of nickel deposits. In effect, template deposited Nickel exhibits a higher amount of uniform growth compared to the nickel electrode produced without the template.

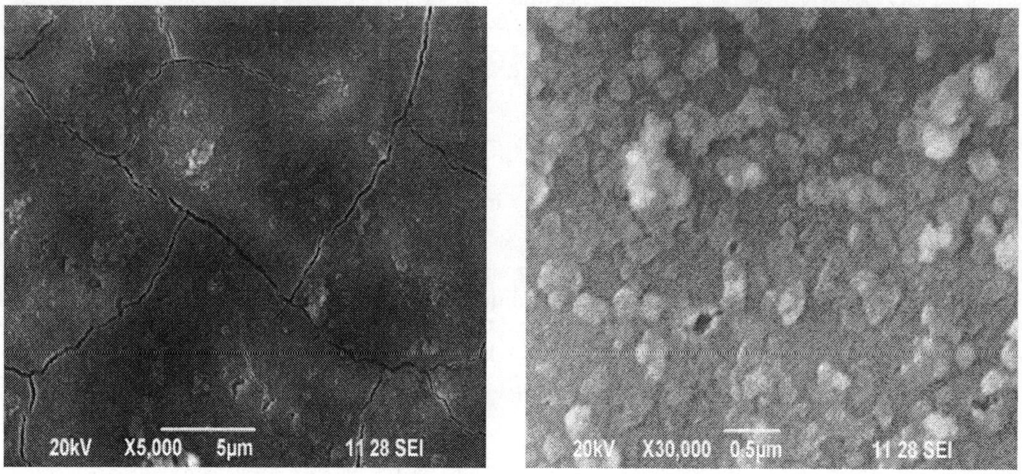

(a) **(b)**

Figure-3 SEM images of (a) Mesoporous nickel and (b) Smooth nice

The particles are same-sized spheres with smooth surfaces, the surface area can be related to the average equivalent particle size by d =6000/(ρ.s) (in nm), where d is the average diameter of spherical particle in manometer; ρ is the theoretical density of nickel (g/cm^3) and s represents the measured surface area of the powder in m^2/g. The size of the particles are in the range of 5- 10 μm. Compare to the both images mesoporous nickel surface morphologies presents higher roughness.

AFM studies also validate the above observation as presented by **Figure 4.** AFM studies are used to demonstrate the difference in topography of mesoporous nickel and smooth nickel. Difference in the surface topographies related to mesoporous nickel and smooth nickel are evaluated through AFM analysis. Mesoporous structure helps to improve electronic conductivity which is highly advantageous for efficient electrocatalysis. The modulations in the height profile indicate that synthesized mesoporous nickel is associated with high degree of roughness. Surface modulations on the mesoporous nickel drastically increase the roughness factor associated with it, however nickel deposited using electrolyte bath without addition of surfactant exhibits two times lower roughness.

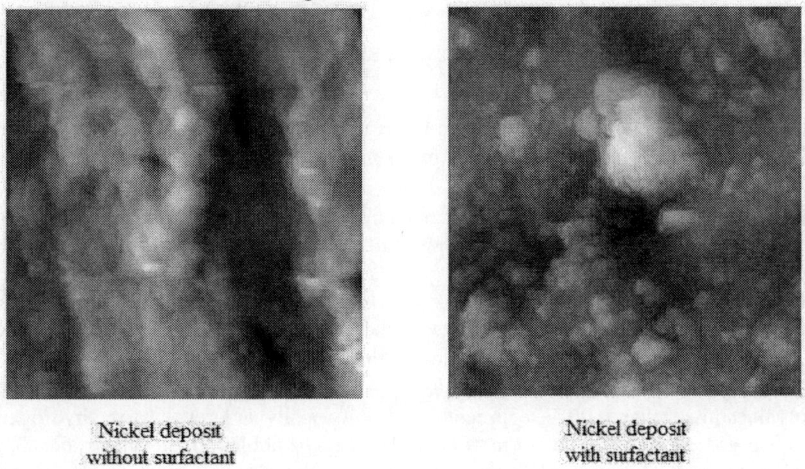

Figure-4 AFM images of smooth nickel and mesoporous nickel.

It is because of this unique porous structure eletrocatalytic activity of mesoporous nickel is higher. Distinct porous structure of mesoporous deposits, which substantially increase the surface area. SEM and XRD studies are in good agreement with AFM results.

3.2 Electrochemical studies

The electrochemical activity of mseoporous nickel electrode was studied by Cyclic Voltammetry, Amperometry and Electrochemical Impedance techniques. The cyclic voltammograms (CV) of nickel deposited with and without the addition of Triton-X100 were recorded in alkaline medium. The CV measurements were conducted in 0.1 M NaOH at scanning rate of 100 mV/s. The applied potential was limited between 1 and -1V.

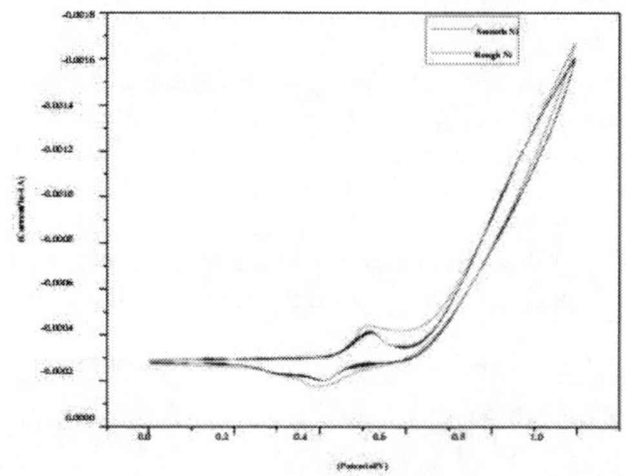

Figure-5 Cyclic Voltametry curves of mesoporous nickel and smooth nickel scanned from -1V to 1V at a scan rate of 50 mV/s performed in 0.1M NaOH.

When discussing the electrochemical behavior of nickel, it is convenient to consider the three oxide phases formed when a positive potential is applied to the metallic nickel electrode. A cyclic voltametry (CV) profile of nickel in aqueous alkaline solution reveals features corresponding to the formation of three oxide species as the potential is increased from −1 to 1 V vs hydrogen evolution reaction. The anodic peak corresponding to the formation of the first surface oxide, α-$Ni(OH)_2$, is present in the V range of −1to 1 V. If the CV scan is reversed at a potential lower than 0.50 V, the full reduction of α-$Ni(OH)_2$ to metallic nickel will occur, giving rise to a cathodic peak in the CV profile at $-1\leq E \leq 1$ V . When the potential is brought to values lower than 0.00 V, appreciable current density as a result of the hydrogen evolution reaction (HER) is observed along with the formation of H_2 (gas) bubbles at the electrode. The HER overlaps in potential with the reduction of α-$Ni(OH)_2$. If the forward CV scan is continued to potentials higher than 0.50 V, the α- $Ni(OH)_2$ already present on the electrode is converted to β- $Ni(OH)_2$, and additional β-$Ni(OH)_2$ is formed. This species is formed in the E region between 0.50 V and 1.30 V, which is known as the passive potential region for nickel. Of these two $Ni(OH)_2$ structures, the β phase is more stable. Formation of this phase is irreversible, meaning that β-$Ni(OH)_2$ cannot be removed from the surface of the electrode by simply reversing the potential and reducing the oxide electrochemically. Once formed, β-$Ni(OH)_2$ can be removed through chemical etching, mechanical polishing, and thermal reduction in a hydrogen environment. The α and β $Ni(OH)_2$ species have different crystallographic structures; the α phase incorporates either water or alkali cations from the electrolyte and has a lattice constant that is larger than that of β-$Ni(OH)_2$. The final oxidation species to be formed is NiOOH. at $E > 1.30$ V; the β-$Ni(OH)_2$ species is oxidized reversibly to β- NiOOH which has a similar crystallographic structure and lattice constant. If an increasing positive potential is continuously applied, a γ-NiOOH phase is formed. This species adopts a crystallographic structure similar to that of α-$Ni(OH)_2$ and can be electrochemically reduced back to β-$Ni(OH)_2$.

From **Figure 5** it is evident that mesoporous nickel deposits presnts higher current density compared to smooth nickel prepared without asddition of surfactant. It has been reported that periodic mesoporous structures with high surface area can have charge storage capacity by one order over that of the bulk materials. Increased electrochemical surface area of mesoporous deposits are due to presence of Triton-X100 molecules during deposition porcess. These surfactant molecules acts as a template during deposition, creates mesopores in the nickel surface and favors the formation of highly acctive mesoporous nickel deposits. Owing to the larger surface area elelctrochemically actvie surface area is alos higher for the mesoporous nickel deposits.

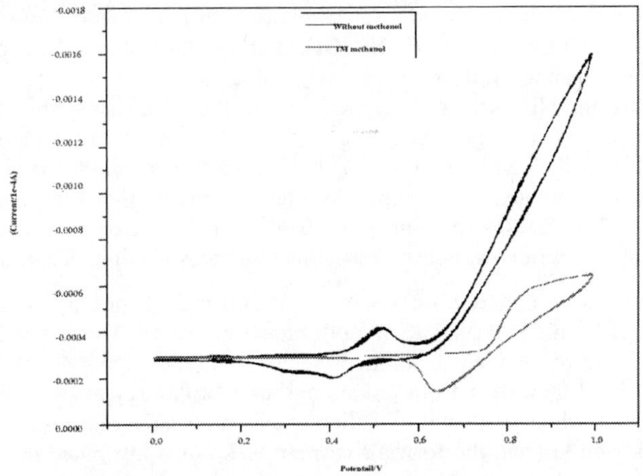

Figure 6. Cyclic voltametric curves of smooth nickel without and with 1M methanol scanned from -1V to 1V at a scan rate of 50 mV/s performed in 0.1M NaOH.

Higher roughness associated with mesopores nickel is due to the lagere number of electrolyte accessing channels formed over mesoporous nickel due to more avalibility of electrolyte acess-electrochemical surface area. In other words through the mesoporous deposits, electrolyte could easily reach the nickel surface compared to smooth nickel.

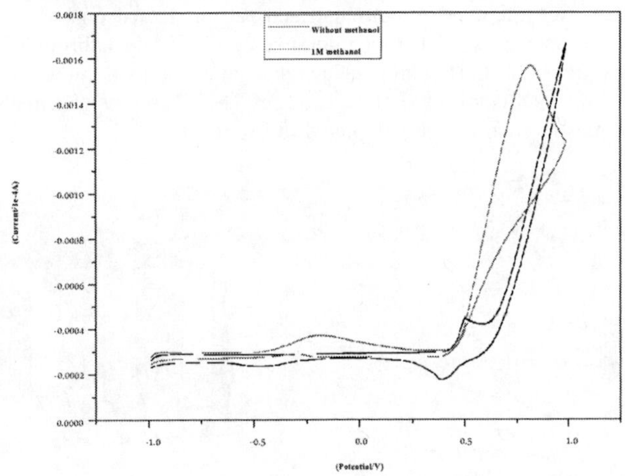

Figure 7. Cyclic Voltametric curves of mesoporous nickel without and with 1M methanol scanned from -1V to 1V at a scan rate of 50 mV/s performed in 0.1M NaOH.

Electro catalytic response of prepared nickel deposits are evaluated through the addition of 1 M methanol. CV curves of nickel deposits before and after addition of 1M methanol in 0.1 M aqueous NaOH solution are given in **Figure 6** and **Figure 7** respectively. Since NiOOH was constantly being reduced back to $Ni(OH)_2$ once methanol oxidation started, some anodic current could always be attributed to the continuous re-oxidation of $Ni(OH)_2$ back to NiOOH. Furthermore, during the cathodic scans, lingering methanol oxidation caused some portion of the electrode surface to be prematurely reduced to the Ni^{2+} oxidation state before the Ni^{3+}/Ni^{2+} reduction occurred, thus decreasing the amount of NiOOH on the surface and the magnitude of the corresponding peak. On comparing electro catalytic response of methanol on mesoporous nickel and smooth nickel, it is clear that anodic peak current associated with methanol electro oxidation is higher for mesoporous nickel. Increase in the anodic peak current on the addition of 1 M methanol is 0.0012 mA/cm^2 and 0.0002 mA/cm^2 for mesoporous nickel and smooth nickel respectively. Improved current response clearly demonstrates significance of mesoporous structure.

The onset of methanol oxidation occurs at about 500 mV at mesoporous nickel electrode whereas methanol oxidation begins from 550 mV with smooth nickel electrode. The lower onset potential associated with mesoporous nickel clearly evidences higher electro catalytic activity for MOR. Superior catalytic activity of mesoporous nickel may stem from unique porous morphology associated with higher ESCA. It is evident that peak potential at which methanol oxidation occurs on mesoporous nickel is 0.75 V but it is 0.85 V for smooth nickel. As well known, the forward current peak (I_f) is attributed to the oxidation of methanol molecule and the backward current peak (I_b) to the oxidation of adsorbed intermediates such as CO, CHxOH ($0 < x < 2$), CHxO, $HCOO^-$ and so on. Hence, amount of oxidizable intermediates adsorbed over the catalytic surface could be predicted by the relative magnitude of backward peak. $(I_f)/(I_b)$ ratio is a measure of efficiency of a catalyst to tolerate towards poisoning of surface due to adsorption of incompletely oxidized intermediates. $(I_f)/(I_b)$ value for mesoporous nickel and smooth nickel is 1.3 and 1.17 respectively, which are higher than that of pure Pt as reported earlier. Higher $(I_f)/(I_b)$ value indicates that most of the intermediate carbonaceous species were oxidized to carbon dioxide in the forward scan on mesoporous nickel electrode. Increased CO tolerance of mesoporous nickel electrode may be attributed due to basic difference in methanol oxidation between mesoporous and nonporous structure, which could be explained on the basis of morphology dependant CO tolerance as demonstrated earlier. Based on the results, it could be inferred that mesoporous architecture increases poisoning tolerance of the catalyst through influencing the availability of continuous binding sites for C-H bond cleavage and enhances more number of readily accessible active catalytic sites for methanol oxidation reaction (MOR). The schematic illustration of mechanism of CO removal over mesoporous nickel is depicted in **Figure 8**.

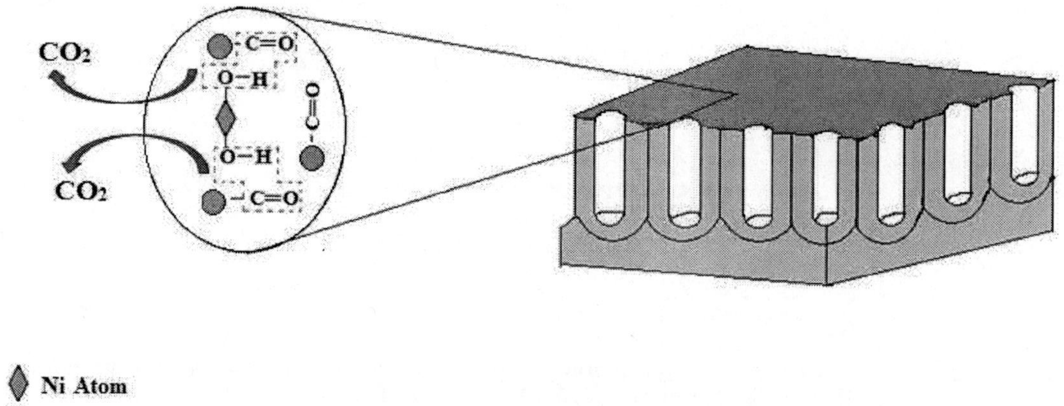

Figure-8 Mechanism of electro-oxidation of methanol on mesoporous nickel catalyst.

Various concentration of methanol 2M, 3M, 4M, and 5M were added to smooth nickel and mesoporous nickel. Smooth nickel has good electro catalytic activity, but surfactant Tritonx-100 nickel has high electro catalytic activity **Figure 9** and **Figure 10** respectively.

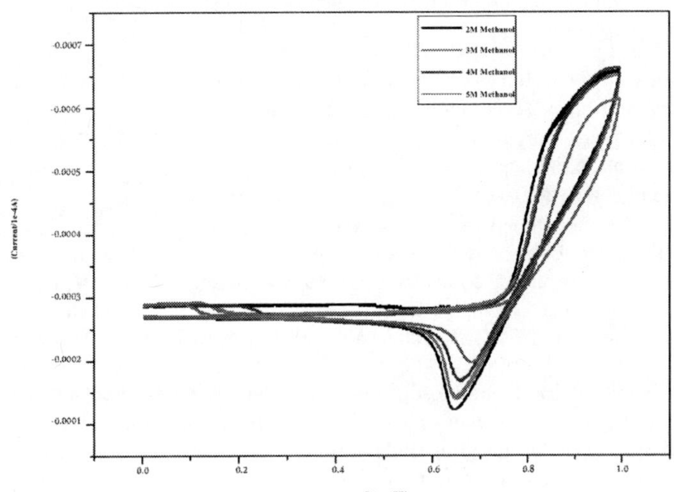

Figure-9 Cyclic Voltametric curves of smooth nickel in various concentration of methanol scanned from -1V to 1V at a scan rate of 50 mV/s performed in 0.1M NaOH.

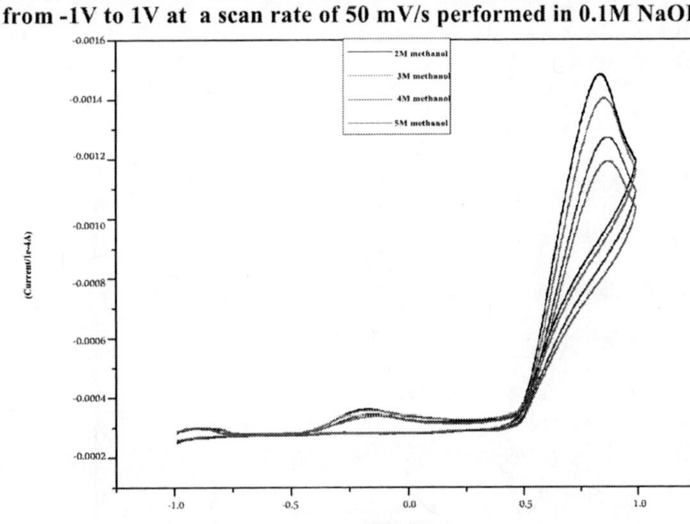

Figure-10 CV curves of mesoporous nickel in various concentration of methanol scanned from -1V to 1V at a scan arte of 50 mV/s performed in 0.1M NaOH.

CV was used to examine the electrochemical properties of mesoporous nickel in aqueous alkaline electrolyte and to relate such acquired insight to their chemical composition and surface characteristics. A

IOP Conf. Series: Materials Science and Engineering **314** (2018) 012028 doi:10.1088/1757-899X/314/1/012028

typical CV curves for bulk nickel in aqueous NaOH solution in the −1 to +1 V potential (E) range shows the following anodic features: (i) oxidation of metallic nickel to α-$Ni(OH)_2$ at $0.20 < E < 0.40$ V (ii) concurrent conversion of α-$Ni(OH)_2$ to β-$Ni(OH)_2$ and oxidation of metallic nickel to β-$Ni(OH)_2$ at $0.50 < E < 1.30$ V; (iii) increase of the oxidation state of Ni from +2 to +3 through the oxidation of β-$Ni(OH)_2$ to β- NiOOH at $1.30 < E < 1.55$ V and (iv) oxygen evolution reaction (OER) at $E \geq 1.55$ V. The β phase of $Ni(OH)_2$ is the most stable and thermodynamically favored oxide of nickel; it is the passive layer that develops on the surface of metallic nickel upon contact with the ambient environment. The conversion of α-$Ni(OH)_2$ to β-$Ni(OH)_2$ is irreversible and once it has taken place, the cathodic peak corresponding to the reduction of α -$Ni(OH)_2$ is not observed anymore. The reduction of β-$Ni(OH)_2$ cannot be accomplished electrochemically and can be achieved at elevated temperatures in the presence of H_2(g). The anodic and cathodic peaks characteristic of α-$Ni(OH)_2$ formation and reduction are not observed because a layer of β-$Ni(OH)_2$ has developed as the result of several prior CV scans. The CV curves for nickel mesoporous displays the same features as those observed for smooth nickel in alkaline media. The CV curves for mesoporous nickel reveals pronounced differences as compared to the CV curves for smooth nickel, especially in the anodic scan. Compared to the mesporous nickel, the peak is shifted towards higher potentials and overlaps the region of OER. The value of Is for OER is greater than in the case of nickel mesoporous; the specific current for the cathodic feature is greater than in the case of nickel mesoporous but the difference is small. Nickel and mesoporous nickel have very different surface morphologies and the rough, bubbly surface of mesoporous nickel gives rise to a larger surface area than smooth nickel. The charges under the two cathodic peaks are similar, thus indicating that both samples have similar amounts of β-NiOOH on the surface. Bearing in mind that the difference in Qs values for smooth nickel is less than 40% and that mesoporous nickel has a significantly larger real surface area, we conclude that β-NiOOH does not make up the entire surface of mesoporous nickel.

3.3. Chronoamperometry

In order to evaluate the long term activity of the catalyst, the steady state current responses of mesoporous nickel and smooth nickel were recorded.

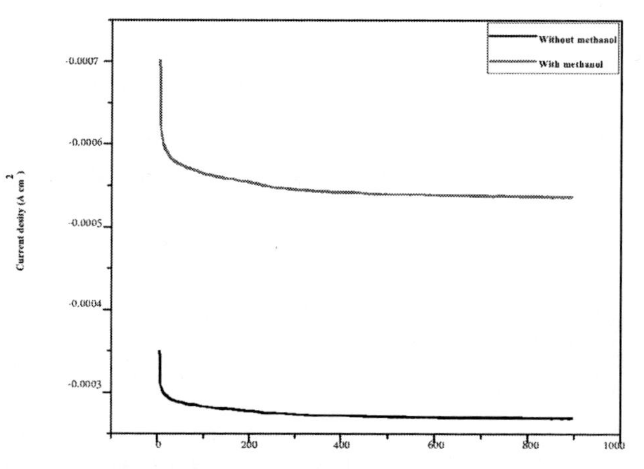

Figure-11. Amperometric curves of using mesoporous nickel without and with 1M methanol.

Initial rapid current decay for the both catalysts is due to double layer capacitance. Transient current due to methanol oxidation attains a steady state after 820 s and 1600 s with mesoporous nickel electrode and smooth nickel electrode respectively, indicating better catalytic performance of the former. It could be noticed from chronoamperometric results that addition of 1M methanol shows an increased electrocatalytic response on mesoporous nickel surface but increase current response is not very high with smooth nickel. As evident from the figures addition of 1M methanol causes increase in current 0.07 A/cm^2 with mesoporous nickel and it is 0.0001 mA/cm^2 for smooth nickel.

Methanol oxidation current tends to decrease due to the accumulation of adsorbed species on catalyst surfaces due to the decomposition of methanol molecules. On comparing the current responses, it is obvious that mesoporous nickel electrode possesses enhanced catalytic durability. Unique mesoporous structure comprising continuous Meso channels is highly favorable to provide easy transport paths of electrons and reaction intermediates therefore enhances MOR activity of mesoporous nickel electrode.

3.4. Impedance studies

To further extract information about the electro catalytic process, EIS studies are performed. **Figure-12** and **13** represent Nyquist plots recorded at 500 mV dc-offset potential in 0.1 M NaOH with methanol for mesoporous nickel and smooth nickel electrodes respectively.

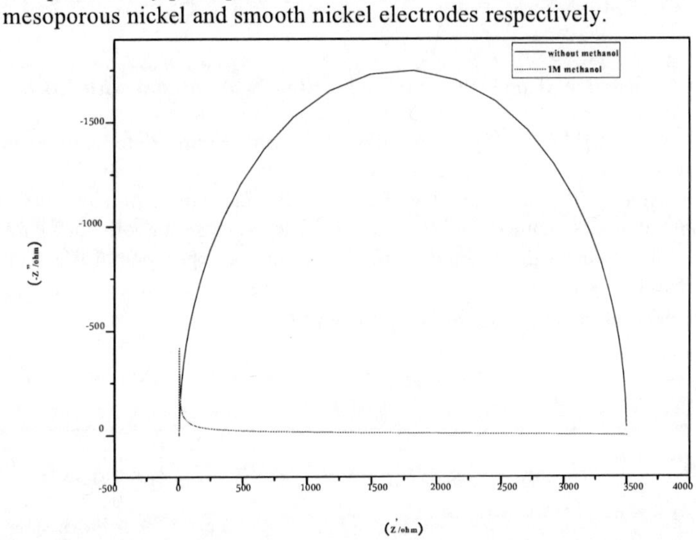

Figure-12 Impedance of Mesoporous nickel deposition, without and with 1M methanol.

General pattern of Nyquist plot obtained with electrodes under study is unaltered on changing the concentration of methanol from 0.1 to 4 M, pointing that mechanism of methanol oxidation is not affected by methanol concentration. Conversely, diameter of semicircle has been dramatically decreased on increasing the concentration of methanol denoting that charge transfer resistance (R_{ct}) regularly drops.

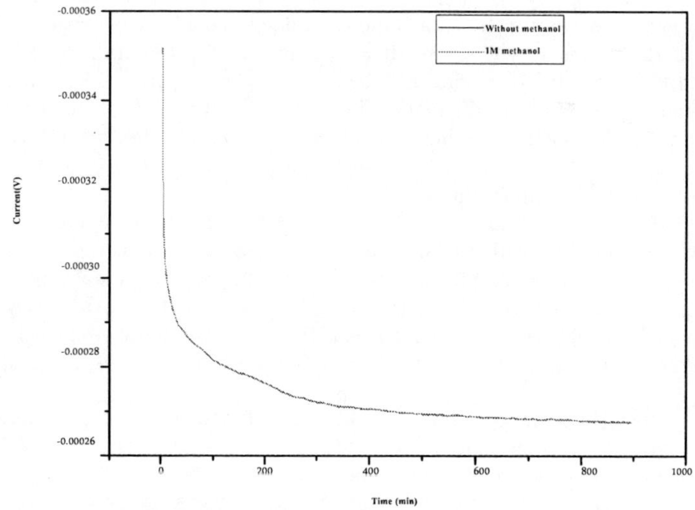

Figure-13 Impedance of smooth nickel deposition, without and with 1M methanol.

EIS analysis was carried out by fitting the data with appropriate equivalent circuit as shown in Table-1. Equivalent circuit is composed of solution resistance (R_s), double layer capacitance (CPE_1) and charge transfer resistance (R_{ct}) associated with MOR. Porous nature of electrode is attributed by the presence of additional elements namely CPE_2 and R_f. The values for all the parameters R_{ct}, CPE_1, CPE_2, and their associated % error determined by the fitting of the experimental EIS data are summarized in Table 1 and 2. The parallel combination of the charge-transfer resistance (R_{ct}) and CPE take into account for methanol adsorption and oxidation on porous thin film.

Table 1 Numerical values of elements in the equivalent circuit fitted with Nyquist plots of mesoporous nickel.

R_s/Ω^{-1}	$CPE_1(\times10^6)/\Omega^{-1}s^n$	n_1	R_{ct}/Ω^{-1}	$CPE_2(\times10^5)/\Omega^{-1}s^n$	n_2	R_f/Ω^{-1}	Concentration of Methanol (M)
14.8	4.186	0.91	1866.7	9.93	0.92	32380	0
12.78	5.636	0.87	875.8	7.95	0.94	632.7	1
11.65	6.10	0.88	153.3	4.28	0.91	449.7	2
11.5	2.68	0.92	102.2	2.998	0.93	427.2	3
11.58	4.22	0.96	82.2	1.746	0.89	499.1	4

Table 2 Numerical values of elements in the equivalent circuit fitted with Nyquist plots of smooth nickel.

R_s/Ω^{-1}	$CPE_1(\times 10^6)/\Omega^{-1}s^n$	n_1	R_{ct}/Ω^{-1}	$CPE_2(\times 10^5)/\Omega^{-1}s^n$	n_2	R_f/Ω^{-1}	Concentration of Methanol (M)
15.1	5.767	0.88	3381.9	2.5	0.89	89342	0
15.02	3.761	0.89	1505	4.536	0.9	6861.4	1
12.5	2.504	0.85	771.4	3.089	0.91	739.5	2
12.3	5.766	0.91	537.8	5.292	0.89	324.8	3
12.32	4.34	0.89	331.7	4.7	0.95	241.9	4

The parallel combination Rct with CPE leads to a depressed semicircle in the corresponding Nyquist impedance plot. It is noteworthy that R_{ct} is an order lower for mesoporous nickel than that of smooth nickel, which reflects the enhanced electro catalytic activity of the former. The Electrochemical Impedance spectroscopic (EIS) has a power tool for studying the electrochemical behavior of electrode.

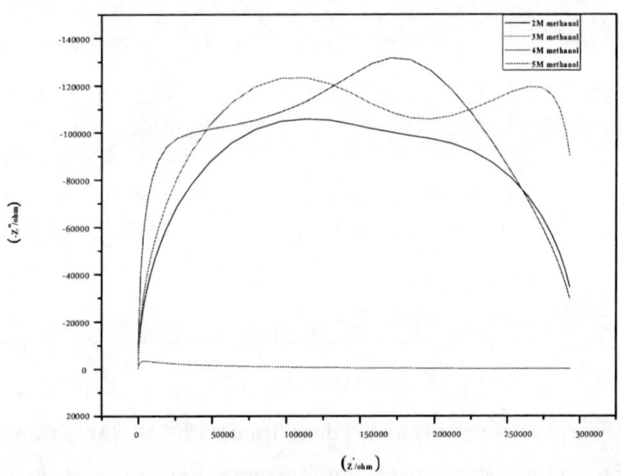

Figure-14 Impedance of smooth nickel deposition, 2M, 3M, 4M, and 5M methanol.

It can be seen that the R_{ct} values are found to decrease in the order 1M methanol > without methanol. Without methanol is low resistance compared to 1M methanol. The diameter of the semicircle is proportional to the value of the impedance. The smaller the value of impedance, the better will be the conducting property of the coating. This can be further confirmed by its low resistance and high capacitance. The equivalent circuit shown **Figure-15** is a simplified electrochemical model which has been used to fit the impedance data obtained for the composite coatings present on substrate. As seen from the equivalent circuit, R_s refers to the resistance of the solution, C_{dl} is the electric double layer capacitance and R_{ct} is the charge transfer resistance that represents the electrochemical activity of the electrode. Based on equivalent circuit model proposed, these EIS curves were best fitted.

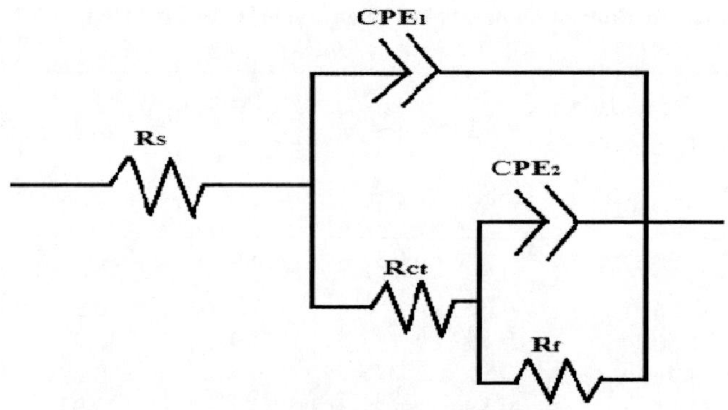

Figure-15 Electrochemical equivalent circuit.

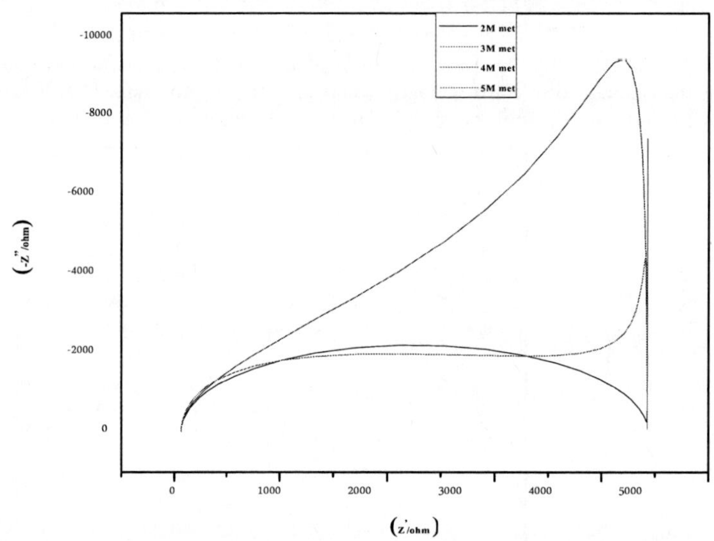

Figure-16 Impedance of mesoporous nickel deposition with 2M, 3M, 4M, and 5M methanol.

Various concentration of methanol solution in Tritonx-100 surfactant the deposited nickel has low resistance and high capacitance.

4. Conclusions

In a nutshell, a facile and simple approach is adopted the produce mesoporous nickel through electrodeposition using surfactant Triton-X100. The developed methods is scalable and reproducible Triton-X100 template could be easily removed by washing the electrode with water. Mesoporous nickel has been prepared by surfactant assistant assisted electrochemical deposition of nickel under optimized process conditions. Surface characteristics of the mesoporous nickel and smooth nickel were systematically characterized through XRD, SEM, and AFM analyses. Mesoporous morphology is highly beneficial to offer readily accessible catalytic sites for methanol oxidation reaction (MOR), through facilitating mass and electron transport. The prepared nickel was used as electrocatalyst for DMFC. Comparison of electrocalytic activity of mesoporous nickel with smooth nickel was interrogated using cyclic voltammetry (CV), chronoamperometry (CA) and electrochemical impedance spectroscopy (EIS) analyses. Distinctly enhanced electrocatalytic activity with improved CO tolerance associate with mesoporous nickel electrode towards methanol oxidation stems from readily accessible high surface area associated with mesoporous structured, which facilitate mass transport of both the reactants and products.

Acknowledgements
One of the authors S. Mohanapriya is grateful to University Grants Commission (UGC), Government of India, for providing fund under the scheme of 'UGC-Dr. D. S. Kothari Post Doctoral Fellowship'. (Ref: No. Award Letter-No.F.4-2/2006 (BSR)/CH/14-15/0102 dated 5-5-2015).

References

[1] Liu H, Song C, Zhang L, Zhang J, Wang H and Wilkinson D P 2006 *J. Power Sources* **155** 95.
[2] Neburchilov V, Martin J, Wang H and Zhang J 2007 *J. Power Sources* **169** 221.
[3] Mohanapriya S, Sahu A K, Bhat S D, Pitchumani S, Sridhar P and Shukla A K 2009 *Energy Environ.Sci.* **2** 1210.
[4] Mohanapriya S, Bhat S D, Sahu A K, Manokaran A, Vijayakumar R,
Pitchumani S, Sridhar P and Shukla A K 2010 *Energy Environ. Sci.* **3** 1746.
[5] Mohanapriya S, Sahu A K, Bhat S D, Pitchumani S, Sridhar P, George C, Chandrakumar N and Shukla A K 2011 *J. Electrochem. Soc.* **158** A1.
[6] Mohanapriya S, Bhat S D, Sahu A K, Manokaran A, Pitchumani S, Sridhar P and Shukla A K 2009 *J. Bionanoscience* **3** 131.
[7] Suganthi S, Mohanapriya S and Raj V 2016 *J. Appl. Polym. Sci.* **133** 43514.
[8] Iwasita T 2007 *Electrochim. Acta* **47** 3663.
[9] Radmilovic V, Gasteiger H A and Ross P N 1995 *J. catal.* **154** 98.
[10] E. Antolini E, Jose, Salgado R C and Gonzalez R 2006 *Applied Catalysis B: L Environmental* **63** 137.
[11] Wasmus S and Kuver A 1999 *J electroanal. Chem.* **461** 14.
[12] Iwasita T, Hoster H, John-Anacker A, Lin W F and Vielstich W 2000 *Langmuir* **16** 522.
[13] Wang K, Gasteiger H A, Markovic N M and Ross P N 1996 *Electochim. Acta* **41** 2587.
[14] Hu Y, Zhang H, Wu P, Zhou H and Cai C 2011 *Phys. Chem. Chem. Phys.* **13** 4083.
[15] Liu X J, Cui C H, Gong M, Li H H, Xue Y, FanF J and Yu S H 2013 *Chem. Commun.* **49** 8704.
[16] Xu D, Liu Z, Yang H, Liu Q, Zhang J, Fang J, Zou S and Kai Sun 2009 *Angew. Chem. Int. Ed.* **48** 4217.
[17] Mohanapriya S, Suganthi S and Raj V 2017 *J. Porous Mater.* **24** 355.
[18] Abdel Rahim M A, Hassan H B and Abdel Ham R M 2007 *Fuel cells* **4** 298.
[19] Niu Z, Wang D, Yu R, Penga Q and Li Y 2012 *J. Chem. Sci.* **3** 1925.
[20] Ganesh V and Lakshminarayanan V 2004 *Electrochimi. Acta* **49** 3561.
[21] Xia Y, Xiong Y, Lim B and SkrabalakS E 2009 *Angew. Chem. Int. Ed. Engl.* **48** 60.
[22] Mohanapriya S, Tintula K K, Bhat S D, Pitchumani S and Sridhar P 2012 *Bull. Mater. Sci.* **35** 297.
[23] Julia van D, Brandy Kinkead P, Yoseif M and Gregory J 2014 *ACS Appl. Mater. Interfaces* **6** 15.
[24] Mohanapriya S, Suganthi S and Raj V 2017 *J. Porous Mater.* **24** 355.
[25] Xing W, Li F, Yan Z F and Lu G Q 2004 *J. Power Sources* **134** 324.
[26] Skowronski J M and Wazny A 2006 *J. New Mater. Electrochem. Syst.* **9** 345.
[27] Abdel rahim M A, Abdel Hameed R M and Khalil M W 2004 *J. Power sources* **134** 160.

Coating of Bio-mimetic Minerals-Substituted Hydroxyapatite on Surgical Grade Stainless Steel 316L by Electrophoretic Deposition for Hard tissue Applications

Dharman Govindaraj, Mariappan Rajan,*

*Biomaterials in Medicinal Chemistry Lab, Department of Natural Products Chemistry, School of Chemistry, Madurai Kamaraj University, Madurai, 625021, India.

*Email id- rajanm153@gmail.com

Abstract. Third-era bio-implant materials intend to empower particular live cell reactions at the atomic level, these materials represented with a resorbable and biocompatibility that bodies recuperate once they have been embedded. Necessitate to decrease expenses in public health services has required the utilization of surgical grade stainless steel (SS 316L) as the most inexpensive choice for orthodontic and orthopaedic implants. 316L SS is one of the broadly used implant biomaterials in orthodontic and orthopaedic surgeries. Yet, frequently those discharge for toxic metal ions is confirm from the implants and hence a second surgery is required will remove those implant material. One approach to managing the discharge of toxic metal ions is to coat the implant substance with bio-mimetic minerals in hydroxyapatite (HA). Bio-mimetic minerals such as magnesium (Mg), strontium (Sr), also zinc (Zn) were revealed with animate bone growth furthermore restrain bone resorption both *in vitro* and *in vivo*. The present work deals with the electrophoretic deposition (EPD) for multi minerals substituted hydroxyapatite (M-HA) on the surface treated 316L SS under distinctive temperatures (27°C, (room temperature), 60 and 80°C). The resultant coatings were characterized by FT-IR, XRD, SEM-EDX, adhesion strength and leach out analysis.

1. Introduction

Around Different bio-implant materials, 316L SS is generally utilized within orthodontic and orthopedic implants applications due to its cheap, beneficial biocompatibility, phenomenal mechanical properties, and facile fabrication [1]. To this reason, it is significant in creating strategies on enhancing that bioactivity and corrosion safety for these implants. Yet the formation of chromium oxide layer on the 316L SS surface might have been unabated will endure for that physiological condition and consequently those discharges of metallic ions, for example, chromium, iron also nickel are indicated in the human body. So, the surface treatment for the 316L SS implant is essential to stop those discharges about metallic ions in physiological medium [2]. Additionally, the 316L SS are bio-inert, keeping in mind the end goal to guarantee the bioactivity of these sorts of materials, the surface of the 316L SS bioimplants are typically coated with biomimetic materials, for example, apatite. Apatite [HA] is the generally utilized bioceramic for coating metallic implants and is most researched amid the earlier decade [3-4].

IOP Conf. Series: Materials Science and Engineering **314** (2018) 012029 doi:10.1088/1757-899X/314/1/012029

The biological properties of the apatite-based biomaterials might be improved by the inclusion of divalent ions that have invigorated various studies on ion-substituted coatings to encourage positive

biological effects. In particular, the beneficial impact of Mg^{2+}, Sr^{2+}, Zn^{2+} is an attractive and relevant divalent minerals ions for substitution into HA and is directly linked with the mineralization of calcified tissues, specifically amid the early phases of osteogenesis where its lessening causes bone loss and bone weakness, and encourage osteoblast propagation. With respect to the above focuses, Mg, Sr, Zn substituted HA is relied upon to have preferable bioactivity over pristine HA [4, 5].

The strong chemical adhesion of pristine/substituted HA makes its bio-implant coating to support the fresh osteoblast cell development [6]. Though owing to the constant interface with the harsh environment, the apatite coating degrades throughout long-term implantation also brings about the toxic metal ions leach out of underlying 316L SS bio-implants. Therefore, surface treatment of implant previous to the progress of bio-ceramic coating is vital in the hindrance of implant toxic metal ions leach out and to survive through long-term implant situation [2]. In the current study, the inclination of sort 316L SS in H_2SO_4 to frame passive layer is investigated and the M-HA electrophoretic deposition of the surface treated 316L SS is achieved.

2. Experimental Details

Type 316LSS electrodes of 10 mm x 10 mm x 2 mm size (brought from Sigma-Aldrich) were used as the metal substrate for the electrophoritic deposition (EPD). The 316L SS electrodes were mechanically polished with 120-1000 silica carbide sheets until all apparent marks were detached. Following smoothing, the electrode was washed with acetone and sonicated before exterior treatment. The whole electrode surface was subjected to acid etched of entire immersion in an acid with H_2SO_4 60 min at 27°C (room temperrature. The EPD process was carried out at room temperature, 60 and 80oC from 2.0 g of M-HA nanoparticles (M-HA synthesis procedure adapted from our earlier reports [4,5]) suspension in 100 ml isopropanol. A flag shape platinum sheet was used as the anode and the working electrode (316L SS) was used as the cathode. The gap between the anode and cathode was set aside to be 1 cm. The voltage used was between 90 for a fixed time interval of 5 min. The as-coated samples were subjected to different characterization methods such as FT-IR (Nicolet 380, Perkin Elmer,), XRD (Bruker D8), SEM (SEM-JEOL JSM-6400), adhesion (Model 5569, Instron) and metal ions leach out tests (ICP-AES analysis).

3. Results and Discussion

3.1. Functional group analysis

Figure1 represents the FT-IR spectra of M-HA coated on 316LSS implants under various temperatures. From the FT-IR, all M-HA coated samples were observed similar patterns of the spectrum. The bands for phosphate at 935 (v_2), 532 (v_4), 606 (v_1), and 1091(v_3) cm^{-1} correspond to bending stretching vibration modes respectively [7]. The bands at 3459 and 630 cm^{-1} suggests the existence of stretching modes of the OH- group of M-HA [4-5, 7]. All these acquired bands confirmed the coating of M-HA on 316LSS implant.

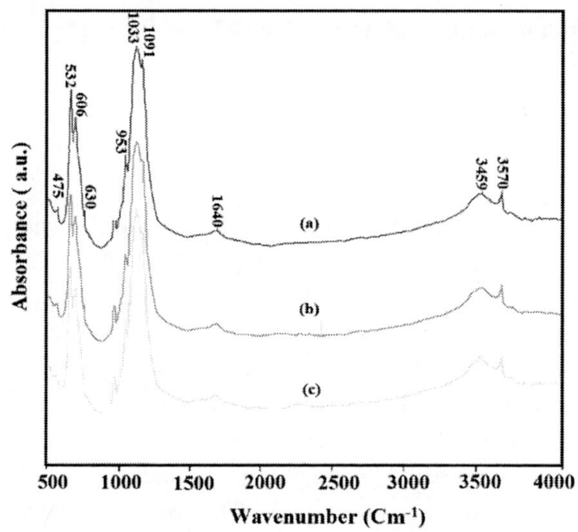

Figure 1. FT-IR spectra acquire for the M-HA coating on 316L SS at different coating temperature for (a) RT (b) 60 min and (c) 80°C.

3.2. Phase analysis

Figure 2 reveals the usual XRD patterns of the M-HA coating on the acid etched 316L SS at RT, 60 and 80°C. The major diffraction peaks of M-HA are apparent between 25° and 55° interval. Moreover the EPD temperature raises the intensity of M-HAP coating rise (Figure 2.

a-c) [8]. XRD patterns clearly affirm that the coating temperature plays an important part in the control of crystallinity of the films.

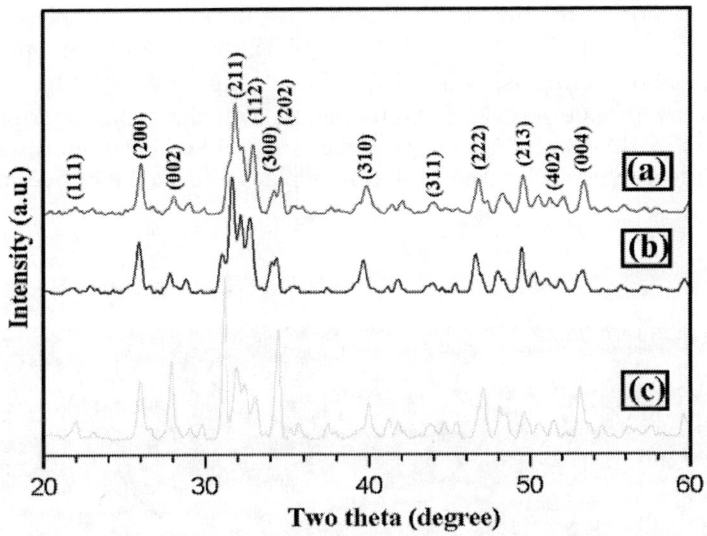

Figure 2. XRD spectra acquire for the M-HA coating on 316L SS at different coating temperature for (a) RT (b) 60 min and (c) 80°C.

3.3. Surface Morphology

The surface morphology of acid etched 316L SS, M-HA-RT, M-HA-60, and M-HA-80 coatings on acid etched 316L SS are represented as Figure 3a–h. Figure 3a demonstrate acid etched implant and the morphology shows the existence of a great amount of excellent pores on the surface, which would ensure strong adherent coating on implant substrate. SEM images obviously confirm that the coating temperature shows a vital role in the influence of morphology of the films. It is furthermore held that excessive temperature can support the coating of M-HA on the exterior owing to significant dispersion movement [9]. From on this, the EPD deposition of M-HA on 316L SS was carried out at RT and raising the temperature from 60 and 80 ^{0}C. The FESEM morphology as illustrated in Figure 3b for the M-HA coating on the acid etched implant at RT consists of nano-flaks of size varying between 55 and 82 nm. When, the temperature was increased to 60 & 80°C for the M-HA coating on the acid etched implant, the coating morphology became very much agglomerated as revealed in Figure

3(c-d). Figure 3(e-g) indicates the cross-sectional appearance of M-HA-RT, M-HA-60, M-HA-80 coating, and EDX analysis of M-HA-80. Additionally, there are no gap between the coating and implant substrate. This recommends that the adhesion between the coating and implant is great enough to maintain the forces during *in-vivo* surgery.

Figure 3h demonstrate the EDX analysis of the coated samples which authenticate the existence of Mg, Sr, Zn, Ca, P and O in the M-HA-80 coating on specimens. The Ca: P proportion of M-HA-80 coating is 1. 68 [10]. The strong peaks of Mg, Sr, Zn, Ca, P and O represent the existence of a definite M-HA-80 coating over the metal exterior. Adhesion strengths of the M-HA-RT and M-HA-60 coatings are predictable as 7.9 and 8.3 MPa correspondingly; whereas that of the M-HA-80 coating on acid etched 316L SS sample is found to be 11.5 MPa.

Figure 3. SEM images of the (a) acid etched 316L SS (b) M-HA-RT (c) M-HA-60 (d) M-HA-80 (e) Cross sectional view of M-HA-RT coating (f) Cross sectional view of M-HA-60 (g) M-HA-80 coating (h) EDX Analysis of M-HA-80.

3.4. Leach out analysis of ICP-AES analysis

The stimulated leach out the investigation of HA-RT, HA-60 and HA-80 nanoparticle films on acid etched 316L SS samples in simulated body fluid solution after incubated for 1 hr at a marked potential of 455 mV vs. SCE was carried out employing inductively coupled plasma atomic emission spectrometry (ICP-AES) and is demonstrated in Figure 4. A considerable quantity of metal ions such as Cr, Fe, Ni, and Mo were released out from the pristine 316L SS. There is no restriction layer on the 316L SS implant exterior to anticipate the attack of chloride ions existing in the SBF solution. While the 316L SS implant substrate is coated with M-HA nanoparticles, a number of toxic ions that release out are found to be lesser than that of the pristine 316L SS implant and also comprise the release out of the significant amount of Ca and P ions from the M-HA-RT, M-HA-60 and M-HA-80 coatings. Though, the release of Ca ions could not be anticipated from the particular coating of the M-HA-RT, M-HA-60, and M-HA-80. In the M-HA-80 coated 316L SS substrate, the leach out of toxic metal ions is considerably reduced contrasted to that of the acid etched 316L SS and individual coatings. This consideration serves to the affirmation of the construction of chromium-enriched inert oxide layer by the M-HA-80 coating on acid etched 316L SS surface which reduces a number of toxic metal ions release.

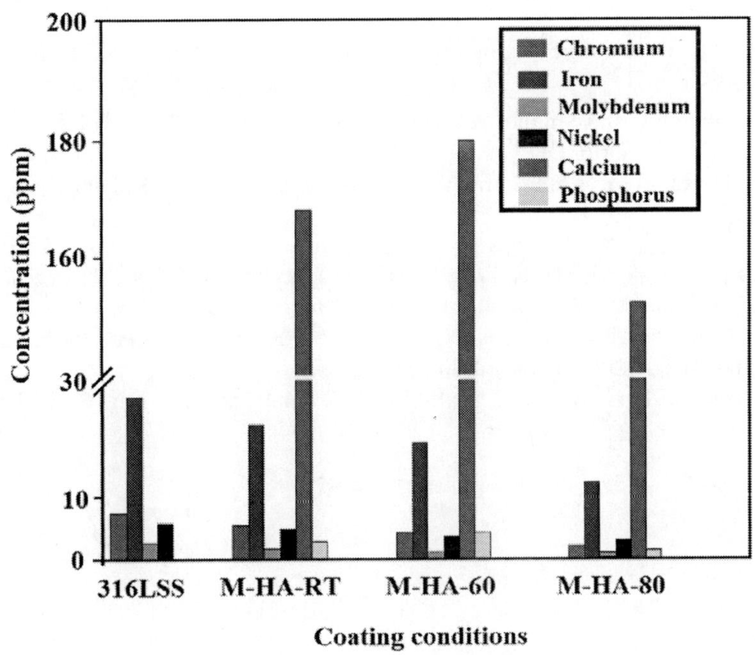

Figure 4. Metal leach out study of acid etched 316L SS, M-HA-RT, M-HA-60 and M-HA-80coating on acid etched 316L SS implants.

4. Conclusion

Minerals substituted hydroxyapatite coating was effectively produced on acid etched 316LSS surface by the EPD process. The FTIR, XRD and SEM results validated the development of M-HA-coated on 316LSS implant. Phase and morphologies for the M-HA films were influenced by the coating temperature. The leach out investigation revealed that the M-HA-80 coating on 316L SS implant surface was controlled the rate of dissolution of poisonous metal ions. The as-coated M-HA on the acid etched surface of implants showed enhanced adhesion strength of 11.5 MPa. In conclusion, the M-HA-80 coating on acid etched 316L SS improves its performance in hard tissue applications.

5. Acknowledgement

One of the authors, M. Rajan, is grateful to the University Grant Commission (UGC), Government of India, for providing financial assistance under the scheme of “DST-SERB” (Ref: YSS/2015/001532).

6. References

[1] Balamurugan A, Rajeswari S, Balossier G, Rebelo A H S and Ferreira J M F 2008 Mater. Corros. **59** 855
[2] Zhou S, Schöneich C and Singh S K 2011 AAPS *PharmSciTech*, **12** 411
[3] Tobin E J 2017 Adv. Drug Delivery Rev 2017 DOI: 10.1016/j.addr.2017.01.007
[4] Govindaraj D, Rajan M, Munusamy M A, Balakumaran M D, Kalaichelvan P T 2015 RSC Adv **5** 44705
[5] Govindaraj D, Rajan M, Munusamy M A, Higuch A 2015 RSC Adv. **5** 58980
[6] Balamurugan A, Balossier G, Kannan S, Michel J, Rajeswari S 2007 *Mater. Sci. Eng.*, C **27** 162
[7] Govindaraj D, Rajan M, 2016 Materials Today: Proceedings, **3** 2394
[8] Shum H C, Bandyopadhyay A, Bose S, Weitz D A 2009 *Chem. Mater.* **21** 5548
[9] Yuan Q, Golden T D 2009 Thin Solid Films **518** 55
[10] Cox S C, Jamshidi P, Grover L M, Mallick K K 2014 *Mater. Sci. Eng.* C, **35** 106

IOP Conf. Series: Materials Science and Engineering **314** (2018) 012030 doi:10.1088/1757-899X/314/1/012030

Effect of mechanical vibrations on the wear behavior of AZ91 Mg alloy

V Chaturvedi[1], U Pandel and A Sharma
Department of Metallurgical and Materials Engineering,
Malaviya National Institute of Technology Jaipur, Rajasthan-302017, India

Abstract. AZ91 Mg alloy is the most promising alloy used for structural applications. The vibration induced methods are effective and economic viable in term of mechanical properties. Sliding wear tests were performed on AZ91 Mg alloy using a pin-on- disc configuration. Wear rates were measured at 5 N and 10N at a sliding velocity of 1m/s for varied frequency within the range of 5- 25Hz and a constant amplitude of 2mm. Microstructures of worn surfaces and wear debris were characterized by field emission scanning electron microscopy (FESEM). It is observed that wear resistance of vibrated AZ91 alloy at 15Hz frequency ad 2mm amplitude was superior than cast AZ91 Mg alloy. Finer grain size and equiaxed grain shape both are important parameters for better wear resistance in vibrated AZ91 Mg alloys. FESEM analysis revealed that wear is considerably affected due to frictional heat generated by the relative motion between AZ91 Mg alloy and EN31 steel surface. No single mechanism was responsible for material loss.

Keywords: AZ91Mg alloy, mechanical vibration, wear resistance, wear mechanism

[1] Corresponding author, E-mail address: vatsala26340@gmail.com (V. Chaturvedi)

1. Introduction Magnesium and its alloys are getting attention in the field of automotive and aerospace industries due to lightest weight material among all the structural materials [1]. Grain size affect the mechanical properties like hardness, tensile strength, wear. The vibration induced methods are effective and economic viable in term of mechanical properties. There are different techniques like mechanical vibrations, electromagnetic vibrations and ultrasonic vibrations [2]. In automobile applications such as automotive brakes, piston and cylinder bores of engine, material are subjected to sliding motion. There are several factors, which influence the tribological behavior of sliding system material. These include properties of sample material along with counterface material and interaction with environment as well as experimental conditions like applied load, sliding velocity, sliding distance. Extensive researches [3], have been carried out on wear and its mechanisms. Chen and Alpas [1], investigated dry sliding wear of AZ91 alloy against AISI52100 steel counterface using a block on- ring configuration. Tests were done within a load range of 1- 350N and sliding velocity within a range of 0.1- 0.2 m/s. They observed two wear regimes; mild wear regime and severe wear regime. The mild wear regime includes oxidational and delamination whereas severe wear regime include surface damage and debris formation. They also investigated the role of contact surface temperature on the mild to severe wear transition. J.An et.al [4] studied dry sliding tests on as cast magnesium alloys $Mg_{97}Zn_1Y_2$ and AZ91 using a pin-on- disc apparatus. Tests were performed at a sliding velocity of 0.785 m/s and within a load range of 20-380N. Abrasion, oxidation, delamination, thermal softening and melting wear mechanisms were observed. Under the given conditions, for the $Mg_{97}Zn_1Y_2$ alloy the dominant wear mechanisms in the load range of 20–200N is abrasion and delamination. In the load range of 240–280 N, thermal softening is an important wear mechanism. Surface melting is the wear mechanism as the load is over 280 N. Morsy [3] performed tests within a sliding velocity range of 0.2- 1.8m/s and loads in the range of 50- 350 N against stainless steel as a counterface and found linear increment in the wear rate on increasing the sliding velocity and applied load in the mild wear region. On the other hand, in the severe wear regime, there was almost proportional rise in the wear rate with sliding velocity and load. Aung et.al [5] studied dry sliding wear behavior of AZ91 alloy at low sliding speed in the range of 0.01- 1.0m/s. Tests were done on pin-on-disc apparatus at 10N load for 1 and 10km sliding distances against JISG4303SUS440C stainless steel. Abrasive wear was dominant at low sliding speeds (<0.1m/s) whereas at high sliding speeds (>0.1m/s) frictional heat causes oxidational wear. Oxidational and delamination wear conditions were found due to protective layers of agglomerated oxidized debris and due to load bearing components along with transmitted shear strains during sliding respectively. The debris containing large flakes of delamination wear were least oxidized. Das et.al [6] studied the sliding wear of wrought AZ31 Mg alloy to delineate the micromechanisms of wear. Dry pin-on- disc tests were performed against tool steel as conterface at 673K. It was found that subsurface grains beneath the contact surface were subjected to large plastic strains and experienced dynamic recrystallization and growth. Zafari et.al [7] investigated the tribological behavior of AZ91 and AZ91 + 3RE Mg alloys. Tests were conducted at a load of 20N, sliding speeds of 0.4m/s and 1m/s and at wear temperature of 25- 250°C. There was around 8% and 60% reduction in wear rates for AZ91 and AZ91 + 3RE alloys at 25°C whereas at 100°C and 0.4m/s, AZ91 showed 58% reduction.

In the light of above, study of tribological behavior has become a prime concern in the field of automotive and aerospace industries. In the present investigation, experimental work was designed to study the effect of the applied load and sliding velocity on the wear transitions. The present study emphasizes the sliding wear behavior of as-cast and vibration imposed AZ91 magnesium alloy. A wear mapping approach has been undertaken to represent the wear regimes and the main mechanism of wear in each regime.

2. Experimental methods AZ91 magnesium alloy was selected for present investigation. The alloy was prepared in the bottom pouring type casting furnace under a shielding atmosphere of argon gas as shown in Figure 1. The vibrations were imposed at constant amplitude of 2mm and varied frequency of 5, 10, 15,

20 ad 25Hz to investigate the effect on wear behavior. The chemical composition of the as-cast AZ91 magnesium alloy is given in Table 1.

Table 1. Chemical composition of AZ91 Magnesium alloy (wt %)

Cu	Si	Fe	Ni	Mn	Zn	Al	Mg
0.003	0.060	0.102	0.026	0.011	1.53	8.27	Bal.

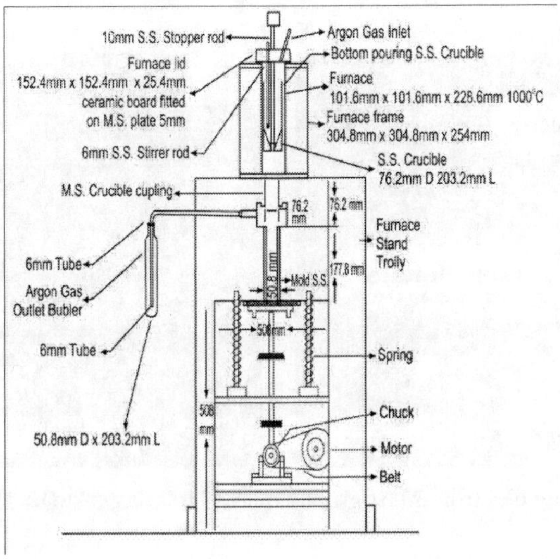

Figure1. Experimental set up for mechanical vibration treatment of magnesium alloy

Wear tests were conducted on pin-on- disc apparatus. Wear test samples in the form of cylindrical pin (7mm dia and 20mm length) were machined from the cast plates. The wear test samples were wet grounded using SiC paper from 320 to 1500 grit step-by-step followed by polishing with 1.0μm and 0.5μm alumina suspension using a slow speed polishing machine. The polished surfaces were ultrasonically cleaned in acetone and dried prior to test. The narrow circular faces of the specimens were kept in contact with the slider disc. The counterface disc was made of EN31 steel (0.9-1.20% C, 0.10-0.35% Si, 0.30- 0.75% M, 0.05%S, 0.05%P, 1.0- 1.60%Cr). The diameter of the disc was 50 mm and was cleaned with acetone before each wear test. The tests were carried out at a sliding velocity of 1 m/s and applied loads at 5N and 10N for a total sliding distance of 1500m. Mass losses from the surfaces of specimens were determined as a function of sliding distance and sliding velocity at each load. The mass losses were calculated from the differences in weight of specimens measured before and after the sliding tests (after removing any loose debris) using an analytical balance having an accuracy of 0.01 mg. Volumetric wear loss was estimated by dividing the mass loss to density of the alloy. A separate specimen was used to measure the mass loss for each load. Microstructural analysis on the worn surfaces and the loose debris particles were undertaken using a field emission scanning electron microscope (FESEM).

3. Results and discussion

IOP Conf. Series: Materials Science and Engineering **314** (2018) 012030 doi:10.1088/1757-899X/314/1/012030

3.1. Cast and vibration imposed microstructure The microstructures of the as cast AZ91 Mg alloy and vibration imposed AZ91 Mg alloy at 15Hz frequency and 2mm amplitude are shown in Figure 2(a) and Figure 2(b). The as cast AZ91 alloy consists of coarse dendritic and non- uniform distribution of α-Mg primary phase, divorced eutectic β-$Mg_{17}Al_{12}$ phase and secondary precipitated β-$Mg_{17}Al_{12}$ phases. The eutectic phase precipitates in the form of network at grain boundaries. On imposing vibrations, coarse eutectic phase is refined and became discontinuous and new granular phases formed. Vibrations promote the fine dendritic structure and faster cooling of the melt [8]. Also a high heat transfer produced by vibrations inside the molten metal leads to a high solidification rate of the molten metal. The cooling rate of the molten metal without vibration is slower in comparison to vibration imposed molten metal. At 15 Hz frequency, vibration produced fine grain sizes by imparting numerous nucleation sites in the melt [9].

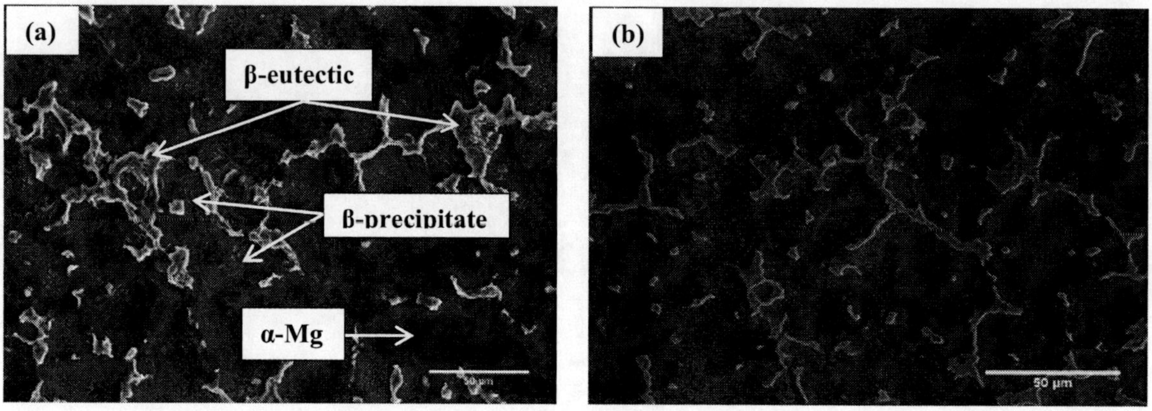

Figure 2. Scanning electron micrographs of AZ91 (a) as cast (b) 15 Hz-2mm amplitude

3.2. Mechanical properties and density Table 2 summaries the Brinell hardness, ultimate tensile strength (UTS), ductility and density of as-cast and vibration imposed at constant amplitude of 2mm and varied frequency of 5, 10, 15, 20 ad 25Hz AZ91 magnesium alloy. It was found that on imposing vibrations the mechanical properties increase up to certain frequency and then decreases on further increasing the frequency. Dislocation motion can be impeded by large grain boundary area. Therefore, fine grained material is harder and stronger than coarse grained material. The microstructure of AZ91 magnesium alloy contains α- Mg ad intermetallic compound $Mg_{17}Al_{12}$ at the grain boundaries. The improvement in the hardness of the alloy is due to the uniform distribution and features of these brittle intermetallic compounds. Hence, both cooling rate and uniformly distributed β- $Mg_{17}Al_{12}$ are responsible for the increase in hardness and tensile properties [10].

Table 2. Results of various mechanical properties at different frequencies of mechanical vibrations

Frequency of vibration (Hz)	Hardness (BHN)	Density (g/cm³)	Ductility (% elon.)	UTS (MPa)
0	58	1.795	1.8	140
5	60	1.797	1.78	145
10	61	1.8	1.75	150

15	78	1.812	2.5	168
20	59	1.798	2.0	151
25	62	1.796	1.9	145

3.3. Friction Coefficient and Wear behavior

3.3.1. Effect of applied load on the coefficient of friction and wear rates. At lower load of 5N, the value of coefficient of friction (COF) was maximum for as cast alloy (0.5912 ± 0.0081) during sliding at $1ms^{-1}$ and it dropped to 0.407 ± 0.007 when the alloy was given vibration at a frequency of 5Hz. The friction gradually decreased with the increase in frequency, and reaches minimum to 0.2559 ± 0.004 at a frequency of 15Hz. This is due to the refinement and uniform distribution of the β- phase. This value of coefficient of friction once again increases on increasing the frequency of vibration up to 25Hz. This increase in the value of COF is because of the coarse and little porous structure of the alloy at higher frequencies. Similar pattern was observed for 10N load (Figure 3 a).

The specific wear rates of the AZ91 alloy are plotted against the applied loads at 5N and 10N for the tests conducted at constant sliding velocity of 1m/s in Figure 3b. At 5N load, there was a change in the slope of the wear rate curve at low frequency of 5Hz. The specific wear rate for the as cast alloy and 5 Hz frequency of vibration was 1.07 (± 0.007) x10-6mm3/Nm and 0.79 (± 0.008) x10-6mm3/Nm respectively. On further increasing the frequency, there was a decrement in wear rate and it reaches to minimum to a value of 0.51 (± 0.005) x10-6mm3/Nm when the vibration frequency was 15Hz. Again, this is happening due to the fine and uniform distribution of the β- phase and sound casting. The wear rates increased as frequency of vibration increases further to 20Hz and 25Hz, similar trend was also observed at 10N of load. The frictional force decreased between the two sliding surfaces as the actual area of contact rose towards the nominal area, on augmenting the load on the pin. This decrement in the frictional force and real surface area in contact resulted in lower wear. When sample comes in contact with rotating disc, frictional heat generates. Due to which softening of the pin surface may take place. During continuous sliding and frictional heat the sub-surface cracks may be formed preferably at the voids. These sub-surface cracks when run parallel to the sub-surface, some particles get detached from the sample and transferred to the rotating disc. These particles are present as third party between the sample and the rotating disc and become responsible for abrasion of samples. After plastic deformation, such abrasion grooves are formed as shown in Fig. 4.27 and Fig. 4.28. The results commended that wear rate of AZ91 magnesium alloy considerably get affected by frictional heat that is been generated by the relative motion between AZ91 magnesium alloy and EN31 steel surface. At both the loads specific wear rate is increasing upto 15 Hz and then decreasing on further increasing the frequency to 20 Hz and 25 Hz. According to E. Iianaganar and et.al [11] the actual area of contact to nominal area would increase on increasing the load leading increased frictional force between two sliding surfaces. The increased frictional force and real surface area in contact cause higher wear. Therefore, it could be said that on increasing the applied load, the wear resistance of the alloy with vibration at constant amplitude and varied frequency decreases. The best results were obtained at 15Hz- 2mm amplitude of vibrations at both the loads.

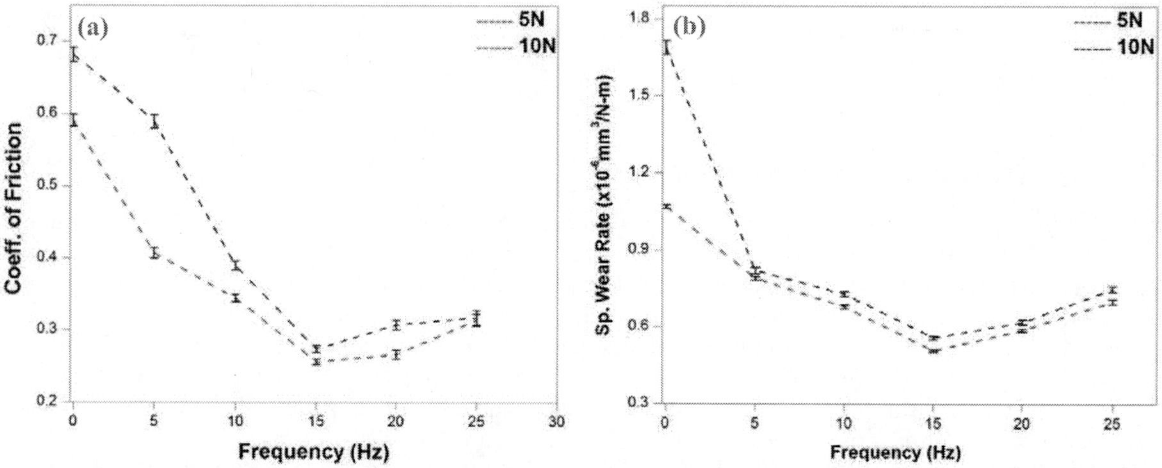

Figure 3 Effect of frequency of vibration at constant amplitude of 2mm on (a) coefficient of friction (b) Sp. wear rates of AZ91 at 5N and 10N loads

3.3.2. Examination of worn surfaces. The worn pin surfaces under scanning electron microscope examination recognized four wear mechanisms operating under various processing conditions. They are abrasion, delamination, oxidation and plastic deformation. In the following sections, the observed wear mechanisms in relation to the processing conditions are identified for better understanding of the tribological behavior of AZ91 magnesium alloy. Figure 4.1 (a-d) and Figure 4.2 (a-d) show the FESEM micrographs of worn surfaces of as-cast and vibrated (5Hz, 15Hz, 25Hz) AZ91 magnesium alloy at 5 N and 10 N loads respectively. The wear track generated on as-cast AZ91 magnesium alloy is uneven/ irregular with deep and wide abrasive grooves as shown in Figure 4.1 (a). When the as cast AZ91 alloy pin slides on EN31 steel disc, it is slightly plastically deformed and wear debris is formed [12]. Since, as-cast AZ91 pin is softer than tool steel disc, the loose worn debris embedded at the contact region of as-cast AZ91 pin. The entrapped debris particles produce further damage on both surfaces as a third-body abrasive and the debris itself undergoes delamination during the sliding. It is due to high friction caused by coarse grain size as well as inhomogeneous hardness of matrix and secondary phases. Similarly, at 10 N load worn surfaces of as-cast AZ91 magnesium alloy showed the same features and trends as observed at 5 N load (Figure 4.2 (a)). The features of abrasive grooves, embedded debris and fracture fragments are more prominent than observed at 5 N loads. It is obvious due to applied higher load, which helps to initiate high friction during wear process. Abrasive grooves, fracture fragments (delamination), oxides and embedded debris are also observed in vibration induced alloy at 5Hz (Figure 4.1 (b)). However, these features are slightly lesser than that observed in as-cast AZ91 magnesium alloy. From the Figure 4.1 (c), it can be seen that as the frequency of the vibration is increased to 15 Hz, wear track generated are smooth with mild abrasive grooves. The reason for the low friction is due to optimum surface hardness and fine grain structure which eliminate adhesion. The fine distributed second phase particles in vibrated (15Hz) AZ91 magnesium alloy could also be beneficial for high wear resistance, as it has the tendency to provide some hardness in the matrix. As the frequency of vibration is further increased to 25Hz in Figure 4.1 (d), severe plastic deformation can be seen in large number. It is in accordance to the increase the volume fraction of secondary phases after imposition of vibration at 25Hz, which provides inhomogeneous surface hardness. Similarly, at 10 N load worn surfaces of vibrated (5Hz, 15Hz, 25Hz) AZ91 magnesium alloy showed the same features and trends as observed at 5 N load (Figure 4.2 (b-d)). The features of abrasive grooves, embedded debris and fracture fragments are more prominent than observed at 5 N loads. These features are indicating the occurrence of severe metallic wear. The transition to severe wear accompanied by a significant increase in the roughness of worn surface of the samples. The severely deformed metallic layers extruded along the sliding direction [11]. It is observed that AZ91 magnesium alloy vibrated at 15Hz at both the loads (5N and 10N) could resist adverse conditions of wear better in comparison to the other vibrated alloys under similar conditions. In case of AZ91 Mg alloy vibrated at 15Hz, the presence of fine eutectic phase strengthens the magnesium matrix. In such cases, improvement in wear resistance was due to fine grains. Therefore, it is clear that the decrease in grain size results increase in wear resistance of grain refined AZ91 alloy. So, grain size is a vital parameter for better wear resistance [13].

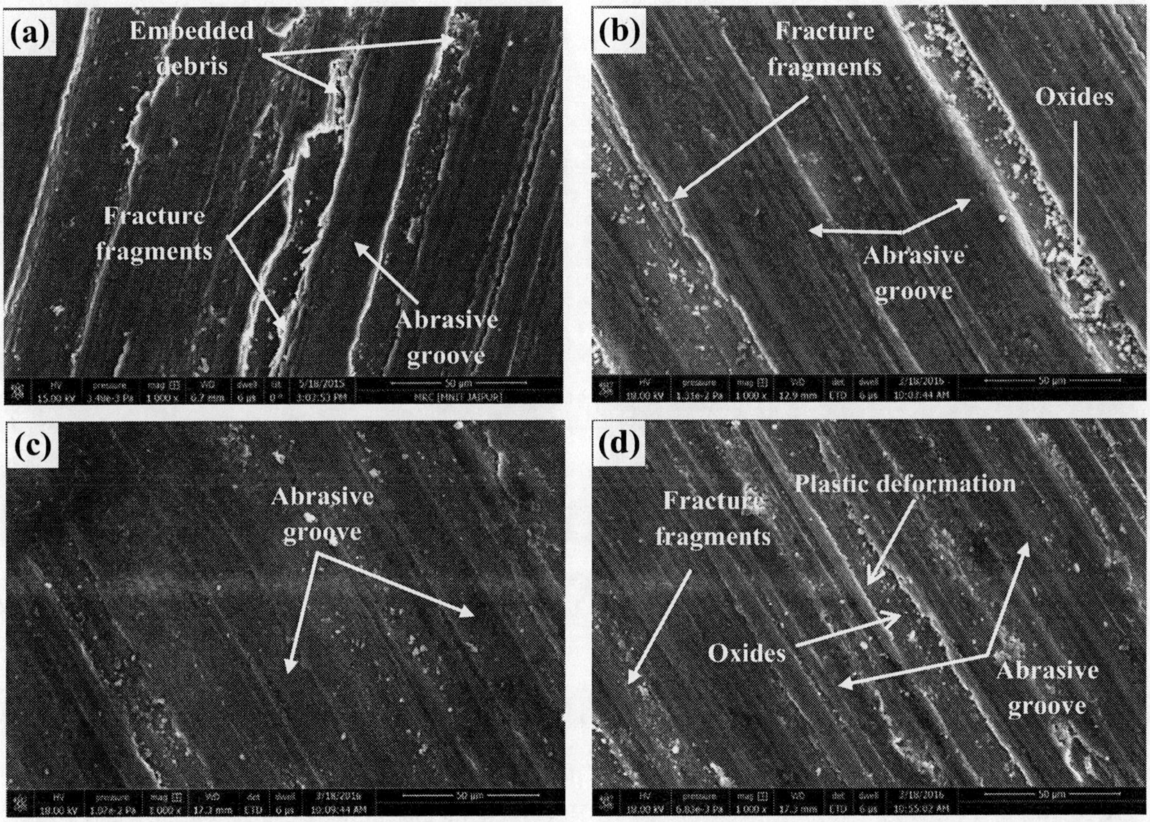

Figure 4.1 FESEM micrographs of worn surfaces of AZ91 magnesium alloy at 5 N load: (a) as-cast (b) 5Hz (c) 15Hz, (d) 25Hz.

IOP Conf. Series: Materials Science and Engineering **314** (2018) 012030 doi:10.1088/1757-899X/314/1/012030

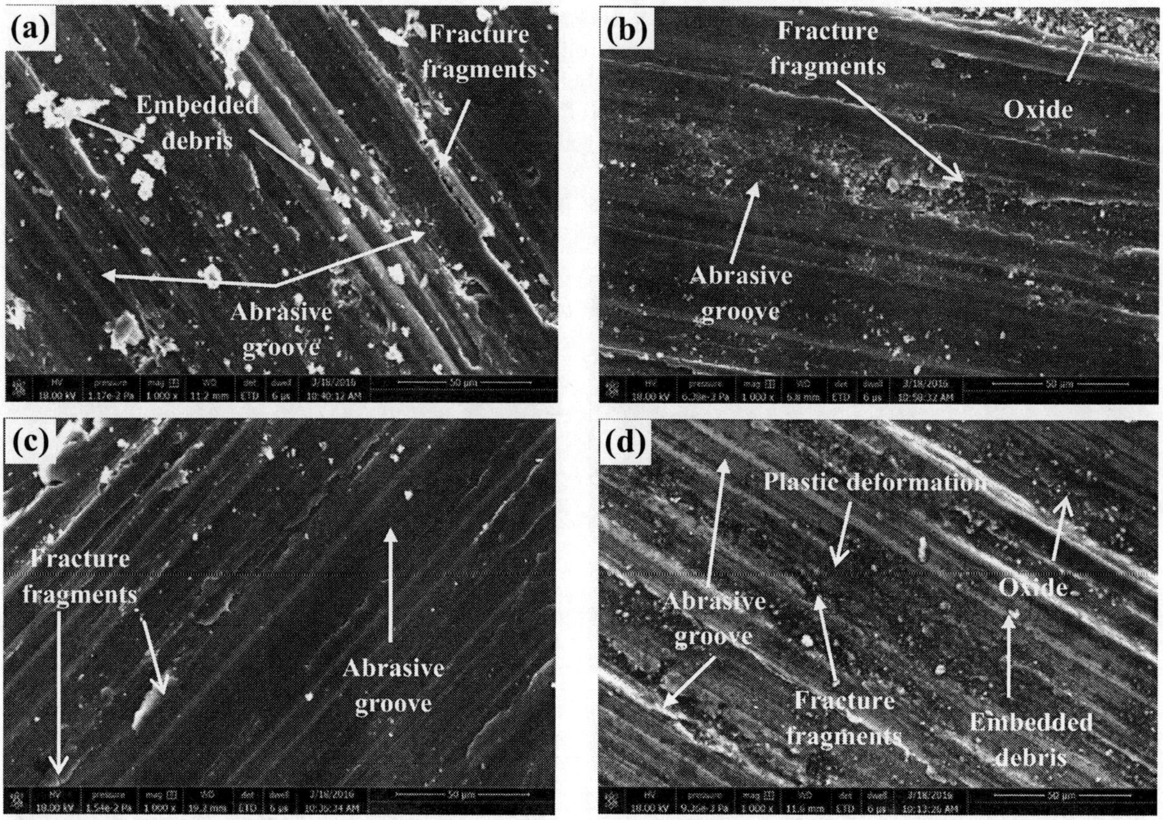

Figure 4.2. FESEM micrographs of worn surfaces of AZ91 magnesium alloy at 10 N loads: (a) as-cast (b) 5Hz (c) 15Hz (d) 25Hz.

3.4 Examination of wear debris FESEM micrographs of wear debris of as-cast and vibrated (15Hz) AZ91 magnesium alloy collected after sliding wear test at 5 N and 10 N are shown in Figure 5.1 and Figur 5.2 respectively. The as-cast AZ91 magnesium alloy at both 5 and 10 N load shows coarse debris as shown in Figure 5.1 (a) and Figure 5.2 (a). While, vibrated (15Hz) AZ91 magnesium alloy tested at 5 N and 10 N show the fine debris as shown in Figure 5.1 (b) and Figure 5.2 (b). Generally, fine debris is beneficial in term of wear resistance because fine particles in the debris reattached to form a transfer film on the surface of disc coming under sliding contact.

IOP Conf. Series: Materials Science and Engineering **314** (2018) 012030 doi:10.1088/1757-899X/314/1/012030

Figure 5.1. FESEM micrographs of worn debris of AZ91 magnesium alloy collected after wear test at 5 N load: (a) as-cast, (b) 15Hz.

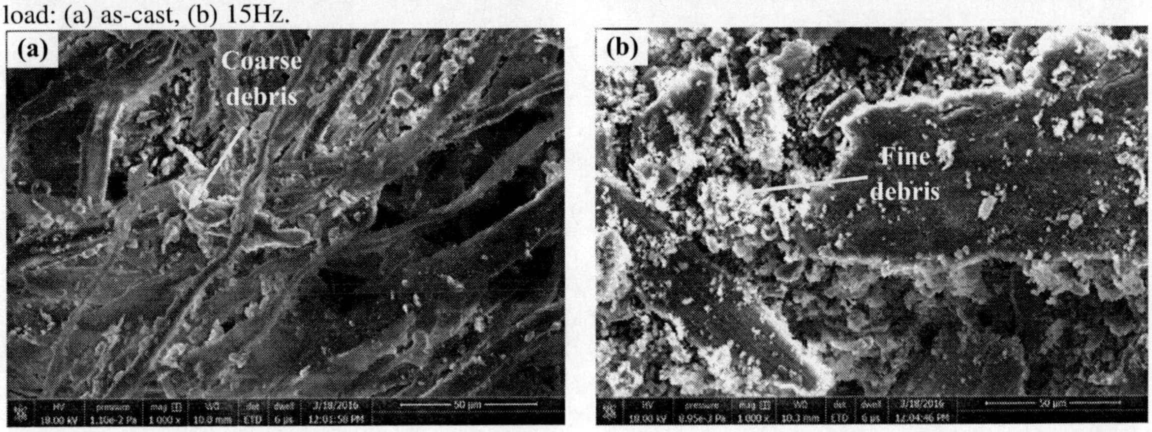

Figure 5.2. FESEM micrographs of worn debris of AZ91 magnesium alloy collected after wear test at 10N load: (a) as-cast, (b) 15Hz.

4. Conclusions

- The wear resistance of vibrated AZ91 Mg alloys increased with decrease in grain size. This was attributed to the fact that due to finer grains, grain boundary strengthening took place.
- The results commended that wear rate of AZ91 magnesium alloy considerably get affected by frictional heat that is been generated by the relative motion between AZ91 magnesium alloy and EN31 steel surface.
- It is observed that no single mechanism of wear is responsible for material loss.
- Debris generated from pin surfaces without imposing vibrations was larger in comparison to respective vibration imposed samples.
- Finer grain size and equiaxed grain shape both are important parameters for better wear resistance in AZ91 Mg alloys.

References

[1] Chen H, Alpas A T *Wear,* **246** (2000)106.
[2] Tamura T, Matsuki T and Miwa K *The Min., Met. & Mater. Soci.,* (2011) 827.
[3] Morsy A W E, *Mater. Sci. Eng. A,* **473** (2008) 330.

[4] An J, Li R G, Lua Y, Chena C M, Xua Y, Chena X and Wang L M *Wear,* **265** (2008) 97.
[5] Aung N N, Zhou W and Lim L EN *Wear,* **265** (2008) 780.
[6] Das S, Morales A T and Alpas A T *Wear*, **268** (2010) 94.
[7] Zafari A, Ghasemi H M and Mahmudi R *Materials and Design,* **54** (2014) 544- 552.
[8] Mishra S S, Sah S S and Ray V *IJTRE,* **3** (2015) 131.
[9] Chen Y, Hsu W N and Shih R J *Mater. Trans.,* **50** (2009) 401.
[10] Chaturvedi V, Sharma A and Pandel U *Mater. Res.Express,* **4** (2017).
[11] Ilanaganar E and Anbuselvan *Materials today Proceedings,* (2016).
[12] Bhaskar G V, Sharmila B and Chetty P N *IJERT,* **3** (2014) 2778- 2782.
[13] Meena P C, Sharma A, Singh S, *La Metallur. Ital.- n*, **1** (2015).

Development of hydroxyapatite/polyvinyl alcohol bionanocomposite for prosthesis implants

V Karthik[1*], S K Pabi[2], S K Roy Chowdhury[3]

[1]Department of Metallurgical Engineering, Government College of Engineering, Salem – 636011, Tamil Nadu, India.
[2]Department of Metallurgical and Materials Engineering, Indian Institute of Technology Kharagpur, Kharagpur – 721302, West Bengal, India
[3]Department of Mechanical Engineering, Indian Institute of Technology Kharagpur, Kharagpur – 721302, West Bengal, India

*Corresponding Author: Email: karthik@gcesalem.edu.in, Contact: +919788444987.

Abstract. Hydroxyapatite ($Ca_{10}(PO_4)_6(OH)_2$) has similar structural and chemical properties of natural bone mineral and hence widely used as a bone replacement substitute. Natural bone consists of hydroxyapatite and collagen. For mimicking the natural, in the present work, a sintered porous hydroxyapatite component has been vacuum impregnated with Polyvinyl alcohol (PVA), which has better properties like biocompatibility, biodegradability and water-solubility. Hydroxyapatite powders have been made into nanosize to reduce the melting point and hence the sintering temperature. In the present investigation high energy ball mill is used to produce nano-hydroxyapatite powders in bulk quantity by optimizing the milling parameters using stainless steel grinding media. Pellets of 10 mm diameter have been produced from nano-hydroxyapatite powders under different uniaxial compaction pressures. The pellets have been sintered to form porous compacts. The vacuum impregnation of sintered pallets with PVA solution of different strength has been done to find the optimum impregnation condition. Microhardness, compressive strength, wear loss and haemocompatibility of hydroxyapatite ceramics have been studied before and after impregnation of PVA. The nano-hydroxyapatite/PVA composites have superior mechanical properties and reduced wear loss than the non-impregnated porous nano-hydroxyapatite ceramics.

Keywords: Bionanocomposite, nano-hydroxyapatite, prosthesis implants, haemocompatibility, poly vinyl alcohol.

1. Introduction

Prosthesis implants are used to replace the injured part of the human body. Bone implants play the role of natural bone's functions in our body, so the properties such as composition, structure, strength, fracture toughness and other properties should match the properties of natural bone. To mimic natural bone with prosthesis implants we should clearly know the bone's hierarchical structure and composition. Bone mainly consists of collagen fibres and an inorganic bone mineral in the form of

IOP Conf. Series: Materials Science and Engineering **314** (2018) 012031 doi:10.1088/1757-899X/314/1/012031

small crystals. Dry bone consists of approximately 60-70 % of bone mineral, 30-40 % of collagen, and the remaining 2-5 % of other substances such as proteins and inorganic salts. The composition of the mineral component can be approximated to hydroxyapatite (HA), which has the chemical formula $Ca_{10}(PO_4)_6(OH)_2$. However, it is not exactly HA but it is an approximation since hydroxyapatite has a Ca:P ratio of 5:3 (1.66), whereas bone mineral actually has a Ca:P ratio ranging from 1.37-1.87 and has internal crystal disorder. The other main constituent in the bone is the collagen. It has a triple helical structure, and specific points along the collagen fibres serve as nucleation sites for the bone mineral crystals [1].

The human body is not an easy environment for a material to function in for prolonged periods. It must be able to operate for many years at a temperature of 37°C in a very moist environment. Also, bone growth can only occur in acidic conditions as these are necessary for mineral deposition, so the implant must be capable of withstanding this for many years without corroding. Implant materials must be designed to minimize the adverse reactions associated with introducing a foreign material to the body. The materials used in each component of the prosthesis implant must have suitable properties to allow them to replace the natural tissue and continue to perform the same functions. HA is the main mineral constituent of teeth and bones. HA ceramics does not exhibit any cytoxic effects. It shows excellent biocompatibility with hard tissues and also with skin and muscle tissues [2]. HA is not only bioactive but also osteoconductive, nontoxic, nonimmunogenic, and its structure is crystallographically similar to that of bone mineral [3]. The flexural strength, strain-to-failure, and fracture toughness values of HA-ceramics are significantly less than those of bone, whereas the elastic modulus is much higher. To improve the mechanical compatibility of HA ceramics, the composite approach can be successfully adopted. The polymer-HA composites can be successfully used in most of the clinical applications. Polyvinyl Alcohol (PVA) is widely used in pharmaceutical, biomedical and biochemical applications due to its many desirable characteristics, such as biocompatibility, biodegradability, and water-solubility [4][5]. The objective of the present work is to develop a hydroxyapatite (HA)/polyvinyl alcohol (PVA) bio-nano-composite for prosthesis implant) by impregnation of polyvinyl alcohol solution into the open pores of the sintered implant produced from nano-hydroxyapatite powders.

2. Materials and Methods

Nano-HA powders for bio-implants have been produced successfully in bulk quantity using high energy planetary ball mill froom the micron-sized HA powders. The ball to powder weight ratio (BPR) of 10:1 and 15 h milling were optimum for producing nanostructured HA (crystallite size ~120 nm) with low contamination level using stainless steel grinding media. Unmilled (micron-sized) and milled (nano-sized) HA powders have been compacted under uniaxial compaction pressures of 50-250 MPa and sintered to form porous 10 mm dia pellets. Sintering at 950°C for 2 h in normal air atmosphere was optimum for producing 30-40 % of open porosity which was found to be good for PVA impregnation. The 4 % PVA solution is found to impregnate the porous ceramics effectively during the vacuum impregnation process to form nano-HA/PVA composite. Microhardness of the pellets has been obtained using Vickers microhardness tester (UHL VMHT). 300 g load has been applied for 10 s to make the indentation on the samples. The pellets have been tested in UTM (Tinius Olsen model H50K-S) for their compressive strength. The standard pin-on-disc apparatus (Ducom, India TR-20) with a surgical grade stainless steel disc of 230 mm diameter has been used for wear tests. The carboxymethyl cellulose solution (1 gram in 100 ml of distilled water) has been used as lubricant while testing, because of its similarity of rheological properties with synovial fluid. The wear tests have been carried out for 20 minutes at sliding velocity of 0.24 m/s under the load of 10 N [6]. The haemolysis count is one of the important parameter in testing the biocompatibility of materials. Haemolysis indicates premature destruction of red blood cells when they come in contact with water or other foreign elements. The haemolysis counts of the developed composites have been obtained with the help of goat's blood. The haemolysis count test procedure can be found elsewhere [7].

3. Results and Discussion

The XRD patterns of HA powders milled at different time intervals with 10:1 BPR has been shown in Figure 1. XRD pattern of unmilled sample (0 h) is compared with XRD patterns of milled samples. It shows continuous peak broadening with increasing milling time, which indicates the refinement of hydroxyapatite powder. The extra peaks are from some new phases such as $Fe_3(PO_4)_2$ and $FePO_4$, which are formed during milling due to wear of grinding balls and vials. It indicates the contamination in the starting material due to milling. Figure 2 shows the photograph of sintered pellets prepared from nano-hydroxyapatite powders under different uniaxial compaction pressure (50MPa, 100MPa, 150MPa, 200MPa and 250MPa). Higher compaction pressure increases the bulk density and mechanical properties, whereas the amount of porosity in the pellets decreases.

Nano-hydroxyapatite pellets have been sintered at 950°C for 2 h to achieve 30-40 % of open porosity. The change in the open porosity percentage with compaction pressure has been shown in Figure 3. Total amount of open porosity in the nano-hydroxyapatite pellets decrease with increasing compaction pressure. Microhardness of the pellets increases with increasing compaction pressure (Figure 4). The variation of the compressive strength of the nano-HA pellets and nano-HA/PVA composite with compaction pressure is shown in Figure 5.

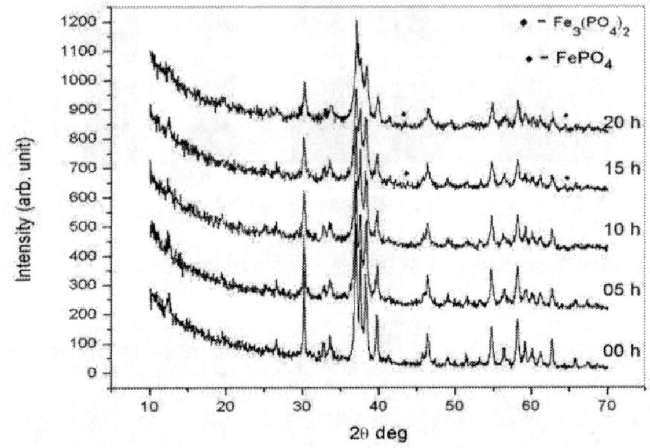

Figure 1. XRD patterns of as-milled hydroxyapatite powder (BPR 10:1)

Figure 2. Photograph of nano-HA pellets after sintering (Sintering temp: 950°C, sintering time: 2h, compaction pressure: 50MPa to 250MPa from left to right)

Compressive strength increases with increasing compaction pressure. The compressive strength of the nano-HA pellets ranges from 105 – 119 MPa. Composite of nano-HA and PVA has compressive strength of 140 – 163 MPa, which depends on the compaction pressure.

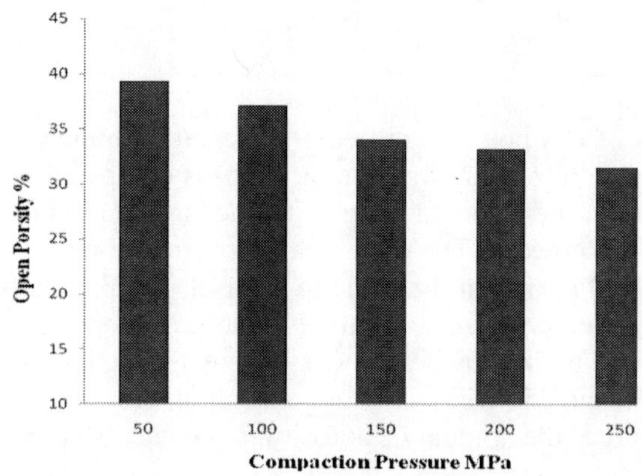

Figure 3. Open porosity Vs. Compaction pressure plot for nano-hydroxyapatite pellets

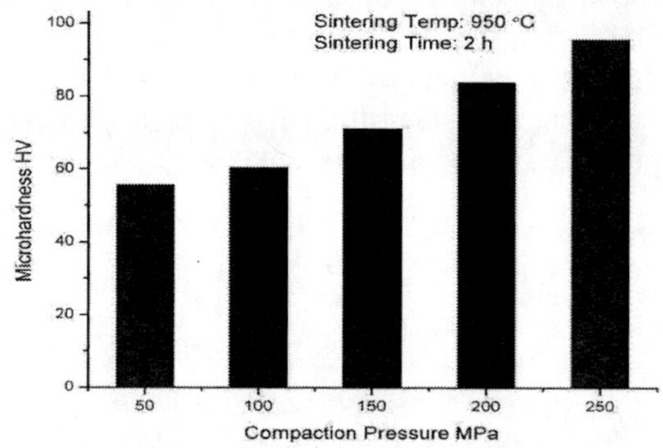

Figure 4. Effect of compaction pressure on microhardness values of nano-HA pellets

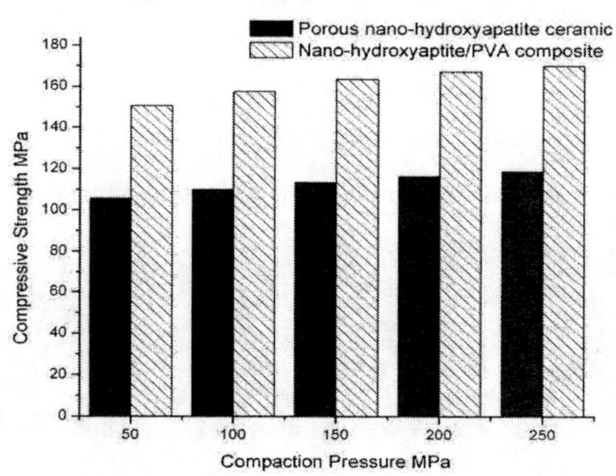

Figure 5. Variation in compressive strength of porous nano-HA ceramics and nano-HA/PVA composites with compaction pressure

Higher compaction pressure produces denser pellet and thereby increases the hardness and strength. Higher compaction pressure gives more densification in the pellets, leads to reduced porosity. The high porosity in the pellet produced at lower compaction pressure cause earlier fracture and lead to low compressive strength. Polyvinyl alcohol impregnation in the nano-HA pellets has been found to improve its compressive strength (Figure 5).

The wear loss decreases with increasing compaction pressures (Figure 6). The polyvinyl alcohol impregnation into the ceramic matrix improves the wear resistance of the pellet. The wear volume has been found comparatively low for nano-hydroxyapatite/PVA composite when compared to pure nano-hydroxyapatite pellets. The pellet produced with higher compaction pressure gives low porosity and high strength, therefore the wear loss is very low. The impregnated PVA reduces the wear of hydroxyapatite.

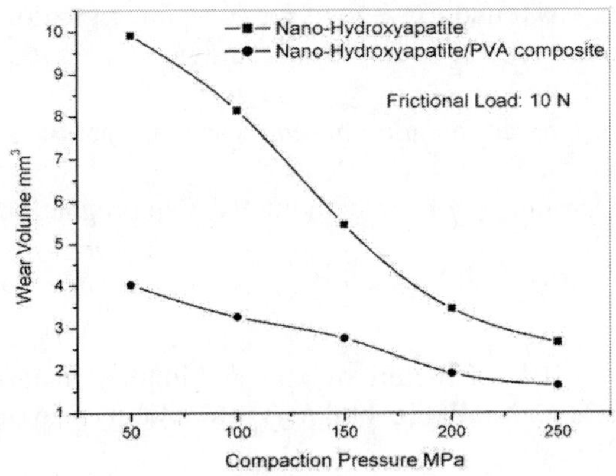

Figure 6. Wear volume Vs Compaction pressure plots for nano-HA pellets and nano-HA/PVA composite

Haemocompatibility is the compatibility of the material with the blood. The haemolysis percentage has been calculated for starting materials and final product using goat's blood. The haemolysis percentages for starting materials (PVA and nano- HA powder) and products (porous nano-HA pellets and nano-HA/PVA composite) have been shown in Figure 7.

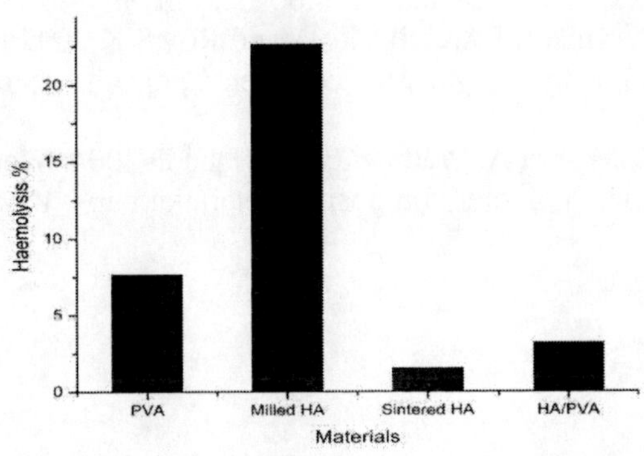

Figure 7. Haemolysis percentage (%) of starting materials and products

The accepted norm is that if the haemolysis percentage is less than 10 the test material is taken as haemocompatible and if it is less than 5 the material is highly haemocompatible (Roy Chowdhury et al. 2007). The composite has haemolysis percentage of ~3 % and is well below 5 so it is a highly haemocompatible material.

IOP Conf. Series: Materials Science and Engineering **314** (2018) 012031 doi:10.1088/1757-899X/314/1/012031

4. Conclusions

- Nano-hydroxyapatite powders for bio-implants have been produced successfully in bulk quantity using high energy planetary ball mill.
- Nano-HA powders have been compacted under uniaxial compaction pressures of 50-250 MPa and sintered to form porous compacts. The mechanical properties and wear resistance of the porous compacts increased with compaction pressure.
- Sintering at 950°C for 2 h in normal air atmosphere was optimum for producing 30-40 % of open porosity which was found to be good for PVA impregnation
- The 4% PVA solution is found to impregnate the porous ceramics effectively to form nano-HA/PVA composite.
- PVA impregnation improved the microhardness and compressive strength of porous nano-HA ceramics and reduced the wear loss.
- Sintered nano-HA ceramics with or without PVA impregnation have manifested excellent haemocompatibility.

5. References

[1] Cambridge University 2011. Structure of bone and implant materials [Online]. England: University of Cambridge. Available: http://www.doitpoms.ac.uk/tlplib/bones/index.php [Accessed 01.10.2015 2015].

[2] Suchanek W and Yoshimura M 1998. Processing and properties of hydroxyapatite-based biomaterials for use as hard tissue replacement implants. *J. Mat. Res.* **13** 94-117.

[3] Murugan R and Ramakrishna S 2007. Nanoengineered Biomimetic Bone-Building Blocks. In: Mansoori, G. A., George, T., Assoufid, L. & Zhang, G. (eds.) Molecular Building Blocks for Nanotechnology. *Springer New York.*

[4] Pan Y and Xiong D 2009. Friction properties of nano-hydroxyapatite reinforced (polyvinyl alcohol) gel composites as an articular cartilage. *Wear* **266** 699-703.

[5] Peng Z, Kong L X and Li S D 2005. Non-isothermal crystallisation kinetics of self-assembled polyvinylalcohol/silica nano-composite. *Polymer* **46** 1949-55.

[6] Roy Chowdhury S K, Kulkarni A C, Basak A and Roy S K 2007. Wear characteristic and biocompatibility of some hydroxyapatite–collagen composite acetabular cups. *Wear* **262** 1387-98.

[7] Roy Chowdhury S K, Mishra A, Pradhan B and Saha D 2004. Wear characteristic and biocompatibility of some polymer composite acetabular cups. *Wear* **256** 1026-36.